MW00637425

PRINCIPLES OF
ELECTRICAL SAFETY

PRINCIPLES OF ELECTRICAL SAFETY

PETER E. SUTHERLAND

IEEE PRESS

Library of Congress Cataloging-in-Publication Data:

Sutherland, Peter E.
  Principles of electrical safety / Peter E. Sutherland.
     pages cm
  ISBN 978-1-118-02194-1 (cloth)
  1. Electrical engineering–Safety measures. 2. Electricity–Safety measures. 3. Electric apparatus
and appliances–Safety measures. I. Title.
  TK152.S8174 2015
  621.3028′9–dc23
                                        2015012677

Printed in the United States of America

10 9 8 7 6 5 4 3 2 1

To all the victims

CONTENTS

LIST OF FIGURES

LIST OF TABLES

PREFACE

From the beginning of my career in electric power engineering, safety has been an important topic. The first training provided by my new employer, after we had completed the paperwork and received an introduction to the company and its many products and services, was a series of safety training talks and films. The graphic nature of this material can cause some discomfort, but it was looked on as the only way to communicate the severity of the problem.

When arc flash protection became the law in the United States, instituted in a peculiar fashion by OSHA, which meant that industry had to follow the private industry consensus standard NFPA 70E, all of this changed. Soon I found myself toting two large duffel bags of safety gear to the arc flash hazard and electrical safety classes I was teaching to electricians, engineers, and managers in industry. The most telling moments of the course were in the showing of a video, "The Mark Standifer Story" (Standifer, 2004), which was a somber moment in the proceedings, after which I had to pause and let people reflect on what they had just witnessed. This was the story of a man who was going to work on energized high voltage electrical equipment, when he was injured by a severe arc flash, receiving second- and third-degree burns over 40% of his body. After months of excruciating treatment and rehabilitation in a burn center, he recovered fully, and was able to tell his story. Mark Standifer is now an electrical safety speaker and trainer, spreading the message of electrical safety. After that video, the course covered many aspects of electrical safety, shock hazards, and arc flash protection. Teaching the course was the beginning of my interest in the field, which led to this book.

There was, in fact, an incident where I worked which bore many similarities to the Mark Standifer story, and I visited one of my colleagues in a burn unit, where he was swathed in bandages, lying in a hospital bed, and unable to speak. He also, has since recovered and returned to his electrical career. But many are not so lucky, and the number of fatalities is still unacceptably high.

Another episode, of which I was aware, was when an engineer went to measure the voltage and current of energized electrical equipment. Here, one attaches voltage leads to the "hot" conductors, for example, putting an alligator clip around the end of a bolt and clamp-on current probes around a conductor. The current probes had an iron core, and when they were opened, the conductive iron was exposed, and an arc occurred to an energized conductor, causing severe burns, and sending the engineer to the hospital.

A third example was the case of the motor control centers (MCCs). Here, an experienced engineer and a newly hired engineer were doing troubleshooting of some low voltage, 480 V, MCCs. This work involved taking measurements with a digital voltmeter (DVM) of the voltages on the equipment. The lead engineer had to step out for a minute to answer a phone call, and the younger engineer continued working, taking more measurements. When the first engineer returned to the room, he found the other engineer knocked down on the floor, and severely burned. This was because the next MCC was a 4160 V high voltage unit. This tragedy should never have happened. The first mistake was inadequate preparation and planning. The tasks should have been clearly laid out, the equipment to be worked on identified, and safety procedures put in place. All personnel who work on specific equipment are required to be trained in that equipment in addition to their general safety training. This training did not occur for the high voltage MCC, because it was not part of the work scope. When the first engineer left the room, all work should have stopped. The rule is never to work alone on electrical equipment. No measurements should be taken on any equipment unless the expected voltage level is known, and the appropriate test equipment is used. In this case, VOMs should never be used on high voltage circuits.

Electrical accidents have been relatively common in the industry. Incidents such as these are readily preventable, but it takes knowledge and organization to provide effective protection. With the advent of more comprehensive safety programs, safer equipment, greater awareness, and improved arc flash protection, they are fortunately becoming rarer.

Electrical safety is an often-neglected area of electrical engineering. There has been a wide-ranging and pervasive set of changes taking place in attitudes toward electrical safety. Beginning with the Institute of Electrical and Electronics Engineers (IEEE) annual Electrical Safety Workshops, and with new and updated safety standards, the process of changing the electrical safety culture has been changing the world. The earlier attitude toward electrical safety was that industrial production took priority and that if it was necessary to take risks by working on live equipment, this went with the job. This was compounded by a lack of safe work procedures, inadequate safety equipment, and unawareness or indifference to the terrible human cost of industrial accidents. It has become clear now that electrical injuries are not acceptable. People's lives and health should not be sacrificed for the sake of production. An occupational health and safety policy (AIHA, 2012) should commit the organization to "protection and continual improvement of employee health and safety." The ultimate goal of electrical safety is "prevention by design," which is designing or redesigning equipment and systems such that they are safe to work with in the first place.

The word electricity is derived from the Greek "elektron," for amber. This substance, a fossil tree resin, produces static electricity when rubbed on cloth or fur. Everybody is familiar with the "tingle" of electricity when touching a household conductor, 120 V or higher. Children have been electrocuted while playing with electrical outlets and sticking objects into the openings. Electricity has been the cause of innumerable fires, in homes and elsewhere, which are surely also electrical accidents.

Electricity is a hazardous substance, just as arsenic is hazardous, or any of hundreds of other materials which cause injury on exposure, contact, or ingestion. This has not always been considered to be the case, because electricity is invisible, odorless, and colorless. Electricity travels through solid materials, as well as through gas and liquids, and even vacuum. Furthermore, electricity consists of two parts, a physical flow of charged particles and a nonphysical flow of energy in force fields. So the entire concept of electricity as a substance is nebulous. But it is a substance which has its own precise definition, its characteristics, and its very definite hazards. Exposure to electricity can cause injury and death just as surely as exposure to more conventional hazardous substances. The same methodology of hazard analysis and risk assessment, preventive and protective measures should be followed with electricity as with other dangerous materials (Mitolo, 2009a). The complexity and ubiquitous nature of electromagnetic phenomena, however, put them in a different category than other hazards, and warrant their special treatment.

Electricity has always been known to be hazardous and to have significant biological effects. The early experiments of Volta with frog's legs are known to all. The muscular contractions caused by the flow of electricity through living tissue are a significant cause of injury and death. The reaction from somebody touching an energized conductor can cause them to jerk their arm and be bruised or cut or throw them across the room. Internal muscular contractions can cause invisible injuries which only show up much later or they can cause cessation of breathing or of the heartbeat, resulting in immediate death. Protection against contact with energized electrical conductors is an essential safety practice.

The well-known experiments of Franklin showed that lightning is the flow of electricity in the air, and the electrical energy can be collected for scientific analysis and human use. Lightning has been the major source of electrical injury and death throughout all of human history. Lightning has first of all and most dramatically caused death by direct strike to the person. A direct strike will first of all kill by the flow of a large current, often thousands of amperes, through the body. At this level of current, muscular contractions are not an issue. The flow of current causes heating, as it does in any conductor, causing severe burns, both internal and external. While there have been many stories of miraculous escapes, lightning can, and does, do to people what it does to trees. Who has not seen the burned and charred remnants of a direct stroke on a tree, usually damaging only part of it, causing the trunk to split and branches to fall off? The tree may live, with partial remnants of living tissue giving continuing life to some branches. What is more rarely seen is the death and destruction of a tree. The tree is totally burned, inside and out, leaving a forlorn and blackened stick. This can and does happen to people as well as trees. Fires caused by lightning, both in forests and in human structures, have probably killed far more people than the electricity itself. Lightning contains many high frequency components, and tends to travel along the surface of objects, easily jumping from one conductor to another. Owing to the high potential, materials which are not normally good conductors of electricity will nonetheless conduct large amounts of electricity. The dangers associated with lightning are now brought to nearly every home and workplace through the ubiquitous electrical power and communication systems in place. Lightning currents flow through the earth as well, once the stroke has hit the ground. Anybody standing on the ground or touching an object may find current passing through their body. Protection against lightning and its effects is a major part of electrical safety.

Sometimes relegated to mandatory safety training courses and routine safety meetings, nothing has a greater effect on the electrical workers health and well-being than electrical safety practices. Electricians, power line workers, electrical and electronic technicians, laboratory scientists and researchers, students and instructors in teaching laboratories, field engineers, and many others are exposed to the hazards of electricity on a daily basis. This extends to nonelectrical workers who may be exposed because the electrical work is in or close to their work area. The general public is also at risk, both in their homes and with outdoor conductors such as power and communication lines. The results of an electrical accident can be catastrophic, ranging from shock injury, through electrical burns to arc burns, pressure waves, and shrapnel injury. Yet this subject is rarely taught in colleges and universities, and has but a small literature. The electrical safety measures used in industry are not always applied in the electrical engineering laboratories of educational institutions. The majority of books on electrical safety are for the practicing engineer, and not of much use to the average student. It is a certainty that while standards and industry guidelines, and the law itself, are important in many ways, the foundations of electrical safety both in the electrical theory and the physical effects on the body are in science and engineering. These principles do not change, while technology and regulations are constantly changing. Electrical safety has a firm foundation in science and this should be understood. Unfortunately, it sometimes seems that a subject is not considered serious or important unless it is treated in a rigorous manner, working from the fundamentals of physics and mathematics. Safety is such an important topic that it must be a respectable subject at any level. This book will give all due respect to scientific fundamentals and their application to real life. It is the aim of this book to introduce the subject of electrical safety to a wider audience, and have it become part of the preparation of every engineer.

Bioelectricity has many effects other than the hazards discussed here. Devices such as pacemakers and defibrillators have saved countless lives. The measurement of the electrical characteristics and electrical activity of the human body have proved essential in ECG, EEG, and other techniques. The uses of electricity and electromagnetic effects in health care are immense and are only going to grow in the future. As such, this book is not intended to be an in-depth treatment of the medical and forensic aspects of electricity; these topics are more than adequately covered elsewhere.

The sections of this book are designed to provide an introduction to theory followed by a series of practical applications. Following this introduction, the second chapter provides an introduction or review of the mathematics used in analyzing electricity and magnetism dynamically in three dimensions. While three-dimensional partial differential equations are an extremely difficult concept to visualize and understand, it is crystal clear that they are essential to understanding the flow of electricity through space and the human body. Directly related to the mathematical background, and connecting to the physical world are the fundamental physical equations. Just as Newton's laws are the foundation of dynamics, providing the tools to analyze the motion of solid bodies through three-dimensional space, Maxwell's equations are the foundational description of electromagnetism in physics. As any physicist will tell you, Maxwell's equations are not a true description of electromagnetism any more due to advances in fields such as relativity and quantum mechanics. For the human scale world in which we live, their accuracy is unquestioned.

The next three chapters examine the electrical fundamentals of resistance, inductance, and capacitance as applied to the human body. Resistance, covered in Chapter 3, is the electrical analog of friction, opposition to current flow. While this may seem a simple manner, and we are all familiar with the algebraic formulation of Ohm's law: $V = IR$, when considered in a three-dimensional body with electrical and magnetic fields of varying frequency and intensity, resistance becomes a complex matter. The material in which the resistance exists is a conductor and has the property of conductivity or its inverse, resistivity. Since the human body is amorphous, unlike a well-defined electrical conductor or resistor, the flow of current is ubiquitous and changing, necessitating a more broad-based approach to resistance. In addition to current flow, resistance also concerns heating, the result of the dissipation of power, producing energy flow. This in itself will make resistance a crucial aspect of electrical safety, as we have mentioned the deleterious effects of current causing electrical burns when the body dissipates excessive electrical power.

Capacitance, covered in Chapter 4, is the capability of "space" to store electrical energy. This energy can be put into space and returned by moving electrical potentials. These potentials are usually considered as voltages on conductors. When this stored energy is returned to a conductor, the possibility of electrical injury occurs. Capacitance is the measure of the amount of electrical energy which can be stored in a given physical situation. Capacitance may exist in the storage of a physical charge as in static electricity. Capacitance also exists in relation to electrical conductors, where the charge in the conductors may be either fixed or in motion. The material, or dielectric, in which the capacitance exists, affects the amount of energy storage. The most common safety hazard of capacitance is the discharge of a discrete, fixed capacitor, causing electric shock, burns, sparking, and arcing. Capacitance exists in other electrical equipment besides fixed capacitors. Power lines and cables are an example of what is called a *distributed capacitance*, where the energy storage is spread out over a distance. What is a few microfarads per meter may result in a large and dangerous energy storage device. Capacitance must be considered in the analysis of electrical circuits which people may come in contact with. Capacitance exists within the human body and between the human body and other objects, which will affect the occurrence and magnitude of electrical shocks. Capacitance in "space" is a fundamental part of electromagnetic waves, which propagate everywhere and can have harmful effects.

Inductance, which is examined in Chapter 5, is the capability of "space" to store magnetic energy. Normally, we think of magnetism as a fixed quantity, such as produced by a bar magnet. For the magnet to produce its physical effects of attraction and repulsion, magnetic energy must travel from object to object where there may be no physical medium. Electromagnetism is magnetism produced by the flow of electrical current, both constant and changing. As with capacitance, energy can be stored in "space" and returned by currents (instead of potentials) in conductors. The material in which magnetism exists is measured by the property of permeability, or the "resistance" to the flow of magnetic energy. Inductance, the measure of how much magnetic energy can be stored, applies both to discrete inductors, and to distributed inductance, as in cables and transmission lines. The magnetism has effects on the human body, and on conductive objects which we may come in contact with, causing induced currents which may be harmful. Magnetic fields may also have microscopic effects, affecting the more elusive effects of electrical shock. As with capacitance, the magnetic energy storage in "space" is an integral part of the electromagnetic wave phenomenon, which can have harmful effects. The calculation of inductance is a difficult problem; indeed, even the definition of inductance can be problematical owing to its complex physical dimensions. Electromagnetism causes the physical motion of conductors which are exposed to it, just as does fixed magnetism. The hazardous physical effects of stored magnetic energy include the motion of conductors, such as power lines, and the destruction of objects, such as fixed inductors or coils, by the forces from excessive current flow.

Chapter 6 concerns the electrical properties of the human body, and how they affect the propensity for electrical injury. On the simplest level, the human body can be modeled as an electrical equivalent circuit, consisting of discrete resistance, inductance, and capacitance in the various human body parts. The most important and complex of these is the skin, through which electric current must pass to enter into the body. Internally, the current will spread throughout the body in proportion to its internal electrical properties. This level of analysis is used in determining, for example, the effect on the heart for electrical contact by hands, feet, and other body parts. On a deeper level, the human body is considered as a grouping of regions of varying degrees of resistivity, dielectric strength, and magnetic permeability in which electromagnetic fields interact. These interactions, in turn, can affect the operations of the internal organs.

This is followed by an analysis of the effects of current on the human body. Chapter 7 extends the electrical properties of the human body to the effects of current on the human body. This will range from an unfelt shock all the way up to death. As the saying goes "current kills." The effects are generally arranged into amount of current (mA or A) versus effect. The flow of current through the human body is analyzed, with especial emphasis on the International Electrotechnical Commission (IEC) methodology.

Safety in substation grounding is the first practical application, focusing on step and touch potentials. Chapter 8 examines the design and analysis of ground grids. These are primarily underground structures which are installed to mitigate the effects of electrical hazards for people in electrical substations and other locations where hazards exist from the flow of electricity in the earth. Electrical hazards in substations are mainly caused by ground faults, which are occur when an energized conductor comes in contact with the earth, whether by insulation failure, through arcing, or broken conductors. The ground fault current will disperse depending on the earth resistivity and the presence of underground metal objects. One form of danger is the "step potential" where the voltage drop across the earth is sufficient to cause current flow through the feet and body of a person standing on the ground. Another significant danger is the "touch potential" where a person standing on the ground touches a conductive object, such as a metal support, and is exposed to a dangerous potential between their hands and feet. Ground grids, usually a network of copper conductors in a rectangular array, reduce the resistance across space within the grid, and with remote earth through ground rods, will, if designed, installed, and maintained properly, lower the step and touch potentials, increasing worker safety.

Chapter 9 examines the effect of high fault currents, both short circuits between current-carrying conductors and ground faults, on electrical equipment and the potential hazards this presents to people in the vicinity. Typical effects include bending of busbars, shattering of insulators, and failure of protective equipment, such as circuit breakers. When high fault currents enter a ground grid, they can cause failure of connections and conductors, and the drying of soils, which increases earth resistivity.

The multigrounded distribution system and the effect of ground return currents are examined next. Chapter 10 goes into stray currents, which are usually continuous currents, not caused by high current faults and which flow outside their intended path. A current should go from the power source to the load and return through the intended insulated conductors, not exposing any persons to the current flow. Current which flows through unintended conductors such as safety grounding conductors, electrical conduits, conductive water pipes, or through the earth itself may cause harm to both people and animals, particularly in agricultural situations. These current flows, while elusive at times, and difficult to measure, may, in fact, be analyzed in many cases, and their levels and routing predicted. Remediation for stray currents is specific to each situation. Improper grounding and bonding (interconnection connection of noncurrent-carrying conductive objects), inductive and capacitive coupling, incorrect wiring of neutral circuits, and degradation of insulation systems are some of the causes of stray currents. Often, more than one factor may be at play in a particular case.

Chapter 11 covers the topic of arc flash hazards. Arc flash hazard analysis is based on industry standard procedures set forth in NFPA 70E (NFPA, 2014) and IEEE 1584–2002 (IEEE, 2002b) (IEEE, 2004). Arc flash has come into prominence as an electrical hazard in recent years, as the seriousness of the hazard has become apparent. When a short-circuit current flows through air, the initial spark can become an arc. The arc escalates to a high temperature plasma (many thousands of degree celsius). The great heat is first transferred outward through radiation. This can cause severe burns, second degree and higher, on unprotected or insufficiently protected skin. Heat will then be transferred by convection caused by expansion of the hot gases. The rapid heating of the air will produce an acoustic pressure wave capable of causing total hearing loss. The pressure of the expanding air can bend the steel walls of an electrical cabinet, blow doors off their hinges, send shrapnel in all directions, and throw people through the air for long distances. The vaporizing of the conductors where the arc occurs will result in molten metal globules traveling at high velocity, which produce severe burns and other injuries. Because of the great number of arc flash-related injuries, mostly severe burns, standards have been put in place to reduce arc flash hazards and provide improved personal protection in case one is exposed to an arc flash.

Chapters 12 through 15 deal with the effects of short-circuit currents on various portions of the electrical power system. Chapter 12 deals with the effect on protection and metering. Chapter 13 deals with the effects of high short-circuit currents on circuit breakers. When the systems which are in place to protect against short circuits fail to operate correctly, either due to measurement errors or due to equipment failure, a significant hazard is created. Failure to interrupt a short circuit can cause fires and explosions which damage persons and property. The first half of Chapter 13 covers high voltage circuit breakers and their failure modes when attempting to interrupt short-circuit currents. The second half of Chapter 13 provides a survey of a variety of international standards for the testing of low voltage circuit breakers.

The intention is to provide a narrative and coherence to the sometimes complex and difficult-to-read standards documents. It is not intended to provide all the information on all the tests, but to help the user to select the right standard and provide a brief introduction to its methods. Finally, a short section introduces the testing of high voltage circuit breakers. Chapter 14 deals with the mechanical forces caused by high short-circuit currents, particularly on substation equipment. Chapter 15 covers the effect of high short-circuit currents on transmission lines, conductors, and insulators.

Chapter 16 is an introduction to the effects of transients, in particular those caused by lightning, in electrical power systems. High transient voltages and currents are a significant cause of electrical injuries, including both shock and fire hazards.

Writing a book of this nature is a monumental task that has progressed over the course of many years. The list of people who have helped and encouraged me is very long, and I sympathize with the actors at an awards ceremony who must thank everybody who has helped make their success possible. Although I have never met him, I must begin with Mark Standifer, whose safety training video really opened my eyes and put me on the trail of explaining electrical safety to the world. It is impossible to name all of the many people I have worked with over the years that were instrumental in my learning the safety procedures and great hazards of electricity. I wish to thank Chris Eaves of General Electric for obtaining the clearance for me to work on this book, independently of my position there.

Special thanks are due to Mary Hatcher, my editor at Wiley-IEEE Press for her patience staying with me on this long process, and for taking care of all the many tasks that a project such as this one entails. Thanks are also due to the un-named peer reviewers for their many helpful comments and criticism, without which this book would have never have become what it is.

Finally, I wish to thank my family, my beloved wife Janis, my daughter Vanessa, and my mother Joyce Sutherland, for their constant support and encouragement, without which this book would not be possible.

Peter E. Sutherland

ACKNOWLEDGMENTS

The author thanks the International Electrotechnical Commission (IEC) for permission to reproduce information from its International Standard IEC 60479-1 ed.4.0 (2005). All such extracts are copyright of IEC, Geneva, Switzerland. All rights reserved. Further information on the IEC is available on www.iec.ch. IEC has no responsibility for the placement and context in which the extracts and contents are reproduced by the author, nor is IEC in any way responsible for the other content or accuracy therein.

CHAPTER 1

MATHEMATICS USED IN ELECTROMAGNETISM

1.1 INTRODUCTION

Electromagnetics is the branch of physical science which deals with electric and magnetic fields and their interactions with each other and with physical objects. We normally think of electricity as the flow of electrons in a conductor, or as static electricity, or lightning. We think of magnetism as in permanent magnets, the earth's magnetism, compass needles, electromagnets, patterns in iron filings, and the like. The physical elements, electrons and charged particles on the one hand, and magnetic materials on the other, are only a portion of the phenomenon. More powerful are the invisible forces which draw objects together or repel them. These forces allow powerful electric machines, motors, and generators, to transform inchoate energy into useful form. They allow the wireless transmission of information, and thus energy, over distances great or small. Although not treated here, information is a form of energy, and thus intimately intertwined with all electromagnetic phenomena. Electromagnetic fields do not exist without physical manifestations, or the physical manifestations without the invisible fields. In order to understand these complex interactions, a special form of advanced mathematics had to be developed. More complex than the calculus used to determine the motion of physical bodies. This mathematics covers diffuse energy distributed over space, and within confined areas. Its first development was with the study of the distribution of heat within a solid, the flow of fluids in pipes, and steam in all its manifestations. When it was discovered that electricity was very similar to a fluid in its behavior, the mathematics was then applied to electrical currents. The distribution of heat inside a solid became the distribution of electric and magnetic fields and variations in current density.

The mathematics and symbolism used in electromagnetics are often unfamiliar and daunting to the uninitiated. With long partial differential equations, hard-to-visualize fields and vectors, and unusual symbols such as ∇, ∇^2, ∇, $\nabla \times \partial$, $\oint$, $\oiint$, $\oiiint$, etc., the subject is often prickly and difficult to approach. This chapter is intended to provide a summary of some of the fundamental mathematical concepts used in electromagnetic field theory and the symbolism used to convey them. In this way, the reader who has been exposed to them in the past will

Principles of Electrical Safety, First Edition. Peter E. Sutherland.

have a refresher, while the beginner will have a grasp of what is being discussed, and all readers will have a page to turn to for reference during the course of this book. This chapter is not intended to take the place of a mathematical textbook, but only as an introduction to the concepts which may be found therein.

1.2   NUMBERS

The fundamental basis on which the calculation of physical quantities rests is the concept of a number. Beginning with the numbers used for counting, many types of numbers have been discovered, leading to an infinite complexity and array of concepts. A number may be considered the object upon which the mathematical procedure operates. Numbers are objects in their own right, not just concepts used to count or measure physical objects. This is the reason that mathematics describes reality so well; it is also reality, and follows the same rules. The mathematics and physics of electrical theory is based on numbers. It is not possible to understand this theory in full without a foundation in real and complex numbers, vectors and scalars, coordinate systems, and other fundamental objects of mathematics. Indeed, the fundamental intuition necessary for safety relies on being able to judge the magnitude of a possible hazard, whether voltage, current, temperature, or force. While there is an extensive and well-developed science of number theory, for the purposes of this introduction, only a few significant concepts and definitions are needed. The numbers are diagrammed in Figure 1.1, the number line.

Natural numbers or counting numbers are the set $\mathbb{N} = \{1, 2, 3, \ldots\}$. Natural numbers are based on the principles of similarity and multiplicity. Objects which are like each other can be grouped together and counted. Grouping together is recognizing a set of objects. Objects does not have to exist, either physically or mentally, to be counted. A number can be a number, in and of itself. The minimum degree of similarity is that the objects are identifiable. Multiplicity means that there is more than one object, thus counting is possible.

Integers are the set $\mathbb{Z} = \{\ldots, -3, -2, -1, 0, 1, 2, 3 \ldots\}$. Adding zero to the set was one of the significant discoveries of ancient times, without which mathematics as we know it could not exist.

Positive integers are the same as the natural numbers. The search for clarity can lead to multiple definitions. Here the definition is descriptive, rather than functional, as in "counting."

Negative integers are the negatives of the natural numbers $\{-1, -2, -3, \ldots\}$. The opposite of positive is negative, leading to another descriptive category.

Nonnegative integers are natural numbers plus zero, the set $\{0, 1, 2, 3, \ldots\}$. And the opposite of negative may also be nonnegative, rather than positive. There can also be "non-positive" integers.

Rational numbers are the set $\mathbb{Q}$ of numbers which can be expressed as the ratio of two integers, a/b, where $b \neq 0$. When written as decimals, the digits either

Figure 1.1   The number line.

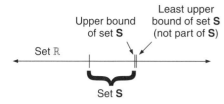

Figure 1.2   The real numbers on the number line.

terminate or are repeating. Rational numbers are the numbers used most often in computations. The longer decimals must be terminated somewhere when calculations are performed, resulting in truncation error. This is why it is best to keep ratios intact as long as possible when simplifying equations. Rational numbers are an example of the paradox of infinity. There are an infinite number of integers, extending to positive and negative infinity. Technically, one could say that there are twice infinity plus one (for zero) integers. The number of rational numbers would be determined from each integer divided by each other integer, except zero, resulting in a double infinite number of infinites.

Real numbers are the set $\mathbb{R}$ of numbers, such as the points on a line, where for every subset **S** which has an upper bound, there is a least upper bound (not a member of **S**) which is a member of $\mathbb{R}$. This means that for any set of numbers which does not increase to infinity, thus possessing an upper bound, there is one number which is not a member of the set but is larger than all members of the set, see Figure 1.2. This least upper bound is infinitely close to the largest number in the set. Thus the real numbers encompass all points on the number line.

Irrational numbers, such as $\pi, \varepsilon, \sqrt{2}$, are the set of all real numbers which are not rational numbers. They cannot be expressed as the ratio of two integers, a/b, where $b \neq 0$. When written as decimals, the digits do not terminate and do not repeat. Irrational numbers cannot be written as the quotient of two integers. In order to calculate using irrational numbers, they must be rounded to rational numbers. This is one reason why it is best to simplify an equation as much as possible before computation, so that the irrational numbers are approximated only in the final result. This results in greatest accuracy. Irrational numbers fill in the gaps on the number line between rational numbers. There is thus an infinite number of irrational numbers between each rational number. The number of infinities is basically beyond comprehension.

The *differential dx* of a number, x, is the smallest possible difference Δx as Δx is reduced toward zero. The concept of differential will be followed through different kinds of numbers, so that differentials in multiple dimensions can be defined.

Imaginary numbers are numbers whose square is a negative real number. An imaginary number is the product of a real number and the *imaginary unit*, usually called simply the "square root of minus one" expressed in mathematical notation as "i" where

$$i = \sqrt{-1} \tag{1.1}$$

A typical imaginary number would be bi, where b is a real number. Imaginary numbers are expressed as the product of a real number and the imaginary unit, not as a single symbol, unless the real number is one. In mathematical notation "i" is always placed after the number it multiplies. In electrical engineering notation, the imaginary unit is represented by "j":

$$j = \sqrt{-1} \qquad (1.2)$$

Unlike i, which is always placed after the number it multiplies, j is always placed before the number it multiplies. A typical imaginary number in electrical engineering format would be jb. When $b = 1$, the letter symbol alone is usually shown, as in $1i = i$ and $j1 = j$. The powers of j are important in many calculations:

$$j^0 = 1 \qquad (1.3)$$

$$j^1 = \sqrt{-1} \qquad (1.4)$$

$$j^2 = -1 \qquad (1.5)$$

$$j^3 = -j \qquad (1.6)$$

$$j^4 = j^0 \qquad (1.7)$$

These powers then repeat in groups of four.

Angles in geometry are formed by two rays (half lines) called the *sides* which meet at a point called the *vertex*. Angles are formally unitless, but in practice are measured in radians or degrees. Angles are measured in the counterclockwise direction from the positive real axis. Angle measurements and definitions are shown in Figure 1.3. One radian (rad) is the angle subtended by an arc having a length equal to one radius of the circle. Since the circumference of a circle of radius r is $2\pi r$, there are

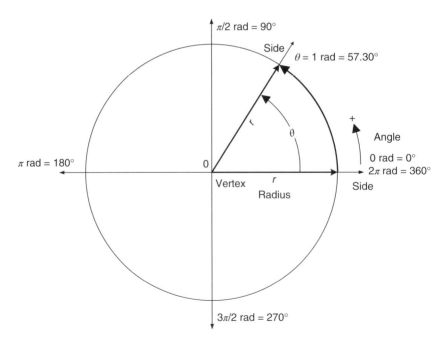

Figure 1.3   Angular measure in radians and degrees.

2π radians in a complete circle. One degree (°) is the angle subtended by one 360th part of the circumference, thus there are 360° in a complete circle. The conversion factor between degrees and radians is

$$1\,\text{rad} = \frac{1}{2\pi} = 0.159 \quad \text{circumference} = 0.159 \times 360° = 57.30° \qquad (1.8)$$

Complex numbers are the set $\mathbb{C}$ of numbers defined as the sum of one real and one imaginary number:

$$Z = a + jb \qquad (1.9)$$

When expressed in this format, complex numbers are said to be in *rectangular form*. The process of definition as the sum of a real and an imaginary may be somewhat circular. A complex number is an object which *may* be expressed as the sum of a real and an imaginary number. This does not mean that it must be, or even needs to be. The two parts of a complex number may be extracted by the functions "Re" for real and "Im" for imaginary:

$$a = \text{Re}(Z) \qquad (1.10)$$

$$b = \text{Im}(Z) \qquad (1.11)$$

Complex numbers may be plotted in the *complex plane* where the horizontal axis is the real axis and the vertical axis is the imaginary axis. The *rectangular form* of a complex number $a + jb$ is shown graphically as a point on the complex plane. Figure 1.4 shows the complex number $3 + j4$ plotted in the complex plane.

The object which is a complex number can also be expressed radially, based on a circle instead of a rectangle. The *polar form* of a complex number is represented

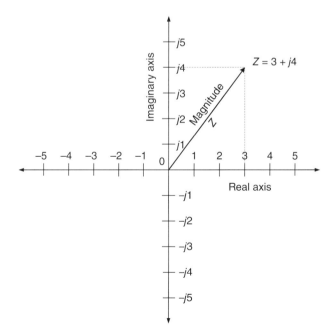

Figure 1.4   Complex numbers in rectangular form.

by a magnitude and an angle. The *magnitude* (Mag) of a complex number Z is the length of line OZ:

$$\text{Mag}(Z) = |Z| = \sqrt{a^2 + b^2} \tag{1.12}$$

The vertical bar expression is a shorthand for magnitude. This equation is the first of two equations which change the expression of a complex number from rectangular to polar form. The second one extracts the angle. The *argument* (Arg) of a complex number is the angle between the real axis and the line OZ:

$$\text{Arg}(Z) = \theta = \tan^{-1}\frac{b}{a} \tag{1.13}$$

For the example in Figure 1.4,

$$|Z| = \sqrt{3^2 + 4^2} = 5$$

and

$$\theta = \tan^{-1}\frac{4}{3} = 53.13°$$

The polar form of a complex number is written as

$$Z = |Z|\angle\theta = |Z|(\cos\theta + j\sin\theta) \tag{1.14}$$

Figure 1.5 shows the complex number $3 + j4$ in polar form as $5\angle53.13°$.

The *exponential form* of complex numbers is a variation on the polar form. It is derived from the rectangular form using Euler's formula:

$$e^{j\theta} = \cos\theta + j\sin\theta \tag{1.15}$$

where θ is the argument in radians.

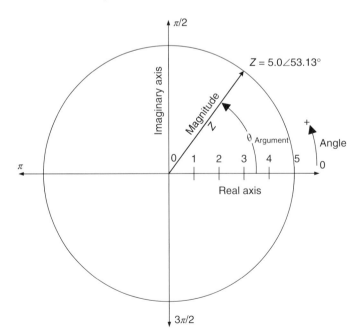

Figure 1.5   Complex numbers in polar form.

The irrational number $e = 2.718281828 \dots$ is Euler's number, the basis of the natural logarithms, named in honor of the Swiss mathematician Leonhard Euler. The expression $e^{j\theta}$ describes a circle with a radius of 1. A circular trajectory is followed as θ is increased from 0 to 2π. The natural logarithms are so called because this number is observed in nature in many forms. And yet, one cannot *measure* e to the degree of accuracy, to which it is known. The number e is its own entity, and can be calculated to any degree of precision, regardless of what is known about nature.

Then

$$z\, e^{j\theta} = a + jb \qquad (1.16)$$

and

$$a = z \cos \theta \qquad (1.17)$$
$$b = z \sin \theta \qquad (1.18)$$

Vectors are geometric quantities which possess both magnitude and direction. Vectors are primarily used for the mathematical representation of forces. A mechanical force will have a certain magnitude and direction. The gravitational force vector points to the center of mass of the object, with the magnitude of weight. The vector of a hammer's force will be down onto the head of the nail. The vector of wind velocity will have a direction by the compass $(N-S-E-W)$ and a magnitude of speed. Electromagnetic vectors are the same. An electric field vector will have a magnitude of volts per meter and direction to (or from) the point charge. A magnetic field vector will go from north to south with a magnitude of tesla. An electron or moving charge will have a force vector perpendicular to both the electric and magnetic fields, with a magnitude proportional to their product. This is what produces generator or motor action.

Vectors belonging to the set $\mathbb{R}^n$ exist in n dimensions. The *unit vector* of the one-dimensional rectangular coordinate system $\mathbb{R}^1$ is the vector $\hat{\mathbf{i}}$ (designated in lowercase with a cap) of magnitude 1 in the direction of the x-axis, shown in Figure 1.6. Vectors are designated as uppercase bold face letters. A one-dimensional vector can be written as

$$\mathbf{A} = a\hat{\mathbf{i}} \qquad (1.19)$$

While complex numbers exhibit characteristics also belonging to vectors, they are not, strictly speaking, the type of vectors we are talking about. This is because geometrical vectors do not contain imaginary numbers. In the wider mathematical sense, one can, and does, have vectors containing imaginary numbers. *Phasors* are the vector representations of complex numbers. A phasor is drawn as a directed line on the complex plane from $0 + j0$ to the complex number Z at point $a + jb$. The phasor representation is shown as the arrow in Figures 1.4 and 1.5. Phasors are used to represent the magnitude and phase angle of sinusoidal waveforms, such as voltage and currents. The angle θ in $Z = z\angle\theta$ is replaced by the angle ωt, where $\omega = 2\pi f$, where f is the frequency of the sinusoidal wave.

Figure 1.6   One-dimensional vectors.

Scalar is another name for a real number, especially as in contrast to a vector.

The *norm* $\|\mathbf{A}\|$ of the vector $\mathbf{A}$ in the set $\mathbb{R}^1$ is defined as the magnitude or length of the vector. For the one-dimensional vector $\mathbf{A}$, the norm is a. This appears to be redundant, saying that the magnitude of a number is the number itself. In mathematics, the definitions must be precise and even apply to the cases of the obvious.

The *unit vectors* of the two-dimensional rectangular coordinate system $\mathbb{R}^2$ are the vectors $\hat{\mathbf{i}}$ and $\hat{\mathbf{j}}$ (designated in lowercase with a cap) of magnitude 1 in the direction of the axes x and y. Rectangular coordinates are often called *Cartesian coordinates* in honor of their discoverer René Descartes. A two-dimensional vector can be written as the ordered pair (a, b) or as the sum of vectors $a\hat{\mathbf{i}} + b\hat{\mathbf{j}}$. Thus the unit vectors can be written as:

$$\hat{\mathbf{i}} = (1, 0) \tag{1.20}$$

$$\hat{\mathbf{j}} = (0, 1) \tag{1.21}$$

The *origin* of a two-dimensional Cartesian coordinate system is the point $(0, 0)$. The origin of an n-dimensional Cartesian coordinate system is the n-dimensional point $(0, 0, \ldots, 0)$. The origin may seem trivial, but the placement of the origin in a logical location will greatly simplify the solution of many a geometrical problem.

The *norm* $\|\mathbf{A}\|$ of the vector $\mathbf{A}$ in the set $\mathbb{R}^2$ is defined as the magnitude or length of the vector. This is comparable to the magnitude of a complex number. The magnitude of vector (a, b) is

$$A = \|\mathbf{A}\| = \sqrt{a^2 + b^2} \tag{1.22}$$

The magnitude of vector $(4, 5)$ is calculated as

$$A = \sqrt{4^2 + 5^2} = 6.40$$

Figure 1.7 shows the vector $\mathbf{A} = (4, 5)$. In the rectangular plane, a *differential length vector* $\mathbf{dl}$ may be derived from the differential lengths of the axes dx and dy as

$$\mathbf{dl} = dx\hat{\mathbf{i}} + dy\hat{\mathbf{j}} \tag{1.23}$$

In the rectangular plane, a *differential area vector* $\mathbf{dS}$ may be derived from the differential lengths dx and dy as

$$\mathbf{dS} = dx\hat{\mathbf{i}}\, dy\hat{\mathbf{j}} \tag{1.24}$$

The differential area vector of dx and dy will point in the vertical, or z-direction.

While logical and easy to analyze, rectangular formulations are inconvenient for many geometrical arrangements. In order to make calculations easier, or even possible, coordinate systems often need to be adjusted to suit the physical arrangement being investigated. Much geometry is circular, such as the cross-section of a wire or of a rotating machine, motor, or generator. It is therefore much simpler to consider the fields and vectors based on radius and angle than in x and y. The *polar form* of a two-dimensional vector which would be (a, b) in Cartesian coordinates is defined as

$$\mathbf{A} = (r, \phi) \tag{1.25}$$

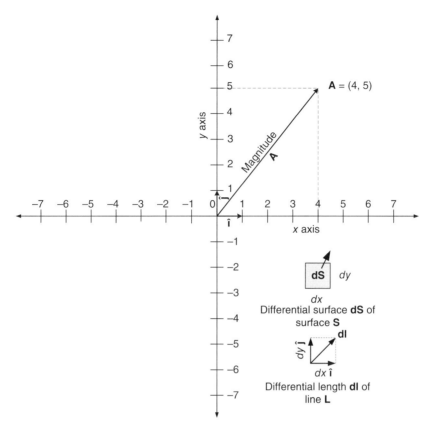

Figure 1.7   Two-dimensional vectors in rectangular form.

where r is the radius and ϕ is the angle of the vector relative to the positive x-axis. Angles are measured in the counterclockwise direction from the positive x-axis. The polar coordinate system is shown in Figure 1.8.

The unit vectors of the polar coordinate system are $\hat{\mathbf{r}}$ and $\hat{\boldsymbol{\phi}}$, but unlike the Cartesian unit vectors, they are not on fixed axes, but are located at the endpoint of a vector $\mathbf{A}$ whose starting point is the origin. Thus their location changes with magnitude $\mathbf{A}$ and angle ϕ. The radial unit vector $\hat{\mathbf{r}}$ touches the end point of the vector $\mathbf{A}$ and is normal (perpendicular) to a circle C with its center in the origin. The angular unit vector $\hat{\boldsymbol{\phi}}$ is normal to the vector $\mathbf{A}$, with its starting point at the end of $\mathbf{A}$ and is tangential to the circle C, in the direction of increasing angle. The polar unit vectors can be expressed in terms of the Cartesian unit vectors:

$$\hat{\mathbf{r}} = \cos\phi\hat{\mathbf{i}} + \sin\phi\hat{\mathbf{j}} \tag{1.26}$$

$$\hat{\boldsymbol{\phi}} = -\sin\phi\hat{\mathbf{i}} + \cos\phi\hat{\mathbf{j}} \tag{1.27}$$

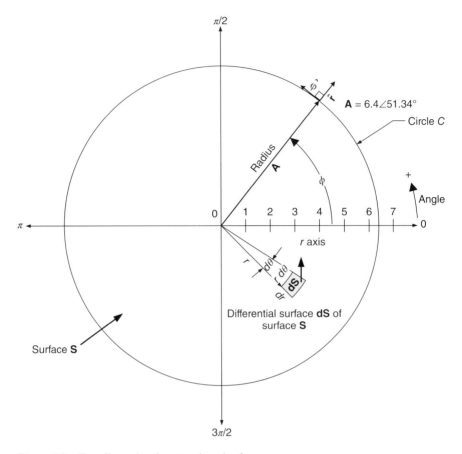

Figure 1.8   Two-dimensional vectors in polar form.

The components of the vector (a, b) converted to polar coordinates (r, ϕ) are

$$r = \sqrt{a^2 + b^2} \tag{1.28}$$

$$\phi = \tan^{-1}\frac{b}{a} \tag{1.29}$$

The Cartesian unit vectors are expressed in polar form as

$$\hat{\mathbf{i}} = \cos\phi\hat{\mathbf{r}} - \sin\phi\hat{\boldsymbol{\phi}} \tag{1.30}$$

$$\hat{\mathbf{j}} = \sin\phi\hat{\mathbf{r}} + \cos\phi\hat{\boldsymbol{\phi}} \tag{1.31}$$

The components of the vector (r, ϕ) converted to rectangular coordinates (a, b) are

$$a = r\cos\phi \tag{1.32}$$

$$b = r\sin\phi \tag{1.33}$$

The vector $\mathbf{A} = (4, 5)$, which has been plotted in Cartesian coordinates in Figure 1.7, has the following polar coordinates:

$$r = \sqrt{4^2 + 5^2} = 6.40$$
$$\phi = \tan^{-1}\frac{5}{4} = 51.34°$$

A plot of the vector $\mathbf{A} = (6.40, 51.34°)$ in polar coordinates is shown in Figure 1.8.

In the polar plane, *differential area* $\mathbf{dS}$ may be defined from the differential length dr and the differential angle $d\theta$ as

$$\mathbf{dS} = r dr d\theta \tag{1.34}$$

As with the rectangular formulation, $\mathbf{dS}$ is a unit vector in the vertical or z-direction.

Cartesian coordinates in three dimensions are members of the set $\mathbb{R}^3$ and are similar to those in two dimensions, with the addition of the vertical z-axis and the unit vector $\hat{\mathbf{k}}$ in the positive z-direction. We have already entered into the third dimension with the differential surface vectors. The unit vectors can be written as

$$\hat{\mathbf{i}} = (1, \ 0, \ 0) \tag{1.35}$$
$$\hat{\mathbf{j}} = (0, \ 1, \ 0) \tag{1.36}$$
$$\hat{\mathbf{k}} = (0, \ 1, \ 0) \tag{1.37}$$

The Cartesian coordinates of a three-dimensional vector (a, b, c) are defined as $B = (a, b, c)$ where a and b are the x and y dimensions of the projection of the vector $\mathbf{B}$ on the x–y plane and c is the height of the end of the vector $\mathbf{B}$ above the x–y plane. The vector $\mathbf{B} = (4, 5, 6)$ is shown in Figure 1.9, the vector $\mathbf{A} = (4, 5)$ is the projection of the vector $\mathbf{B}$ on the x–y plane. The magnitude of a vector $\mathbf{B} = (a, b, c)$ in three-dimensional Cartesian coordinates is

$$\mathbf{B} = \|\mathbf{B}\| = \sqrt{a^2 + b^2 + c^2} \tag{1.38}$$

The magnitude of the vector $\mathbf{B} = (4, 5, 6)$ is

$$\mathbf{B} = \sqrt{4^2 + 5^2 + 6^2} = 8.775$$

In the three-dimensional Cartesian coordinate system, a *differential length vector* $\mathbf{dl}$ may be defined from the differential lengths dx, dy, and dz as

$$\mathbf{dl} = dx\hat{\mathbf{i}} + dy\hat{\mathbf{j}} + dz\hat{\mathbf{k}} \tag{1.39}$$

In the three-dimensional Cartesian coordinate system, *three differential area vectors* $\mathbf{dS}_x$, $\mathbf{dS}_y$, and $\mathbf{dS}_z$, may be defined from the differential lengths dx, dy, and dz as

$$\left.\begin{array}{l} \mathbf{dS}_x = dy\,dz \\ \mathbf{dS}_y = dx\,dz \\ \mathbf{dS}_z = dx\,dy \end{array}\right\} \tag{1.40}$$

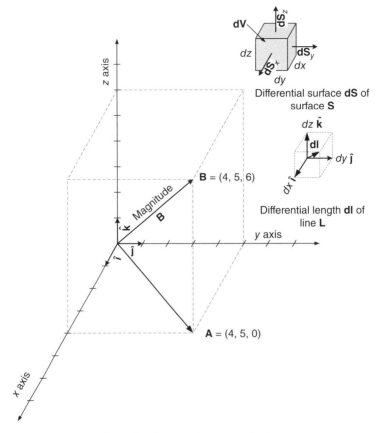

Figure 1.9   Three-dimensional vectors in rectangular form.

Each surface vector will point in the subscripted dimension, exactly as the differential surface vectors in a plane. A differential volume **dV** may be defined as

$$\mathbf{dV} = dx\,dy\,dz \qquad (1.41)$$

There are two versions of the polar format for three dimensions, cylindrical and spherical.

The *cylindrical coordinate* system is simply a vertical extension of the polar coordinate system into three dimensions. The cylindrical coordinates of a three-dimensional vector $\mathbf{B} = (a,\,b,c)$ in Cartesian coordinates are defined as $\mathbf{B} = (r,\phi,k)$ where r is the magnitude of **B**, ϕ is the angle of the projection of the vector on the $x-y$ plane relative to the positive x-axis and k is the height of the end of the vector above the $x-y$ plane. It can be seen that r and ϕ are the same as the two dimensions of polar coordinates. Angles are measured in the counterclockwise direction from the positive x-axis, looking downward from the positive z-axis. The cylindrical coordinate system is shown in Figure 1.10.

The unit vectors of the cylindrical coordinate system are $\hat{\mathbf{r}}, \hat{\boldsymbol{\phi}},$ and $\hat{\mathbf{k}}$, but, except for $\hat{\mathbf{k}}$, which is identical to the Cartesian unit vector $\hat{\mathbf{k}}$, they are not on fixed axes but

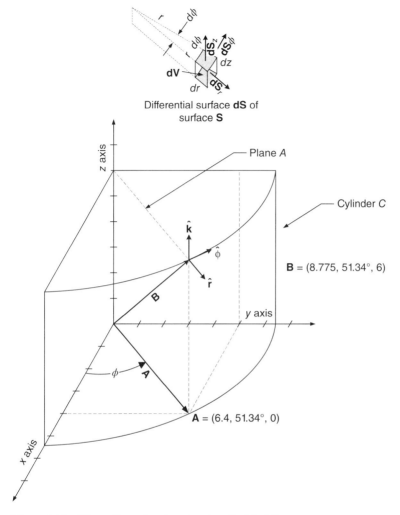

Figure 1.10   Three-dimensional vectors in cylindrical form.

are located at the endpoint of a vector whose starting point is the origin. Thus their location changes with radius r, angle ϕ, and height k. The radial unit vector $\hat{\mathbf{r}}$ is parallel to the $x-y$ plane and normal to the surface of cylinder C and normal to and pointing away from the z-axis. The angular unit vector $\hat{\boldsymbol{\phi}}$ is normal to the plane A, defined by the vector $\mathbf{B}$ and the vertical z-axis. The angular unit vector $\hat{\boldsymbol{\phi}}$ is tangential to the cylinder C, in the direction of increasing angle. The cylindrical unit vectors can be expressed in terms of the Cartesian unit vectors:

$$\hat{\mathbf{r}} = \cos\phi\hat{\mathbf{i}} + \sin\phi\hat{\mathbf{j}} \qquad (1.42)$$

$$\hat{\boldsymbol{\phi}} = -\sin\phi\hat{\mathbf{i}} + \cos\phi\hat{\mathbf{j}} \qquad (1.43)$$

$$\hat{\mathbf{k}} = \hat{\mathbf{k}} \qquad (1.44)$$

The components of the vector (a, b, c) converted to cylindrical coordinates (r, ϕ, k) are

$$r = \sqrt{a^2 + b^2 + c^2} \tag{1.45}$$

$$\phi = \tan^{-1}\frac{b}{a} \tag{1.46}$$

$$k = c \tag{1.47}$$

The cylindrical coordinates of the example vector of $\mathbf{B} = (4, 5, 6)$, shown in Figure 1.10 are

$$r = \sqrt{4^2 + 5^2 + 6^2} = 8.775$$

$$\phi = \tan^{-1}\frac{5}{4} = 51.34°$$

$$k = 6$$

The Cartesian unit vectors are expressed in cylindrical form as

$$\hat{\mathbf{i}} = \cos\phi\hat{\mathbf{r}} - \sin\phi\hat{\boldsymbol{\phi}} \tag{1.48}$$

$$\hat{\mathbf{j}} = \sin\phi\hat{\mathbf{r}} + \cos\phi\hat{\boldsymbol{\phi}} \tag{1.49}$$

$$\hat{\mathbf{k}} = \hat{\mathbf{k}} \tag{1.50}$$

The components of the vector $\mathbf{B} = (r, \phi, k)$ converted to Cartesian coordinates (a, b, c) are

$$a = r\cos\phi \tag{1.51}$$

$$b = r\sin\phi \tag{1.52}$$

$$c = k \tag{1.53}$$

In the three-dimensional cylindrical coordinate system, a *differential length vector* **dl** may be defined from the differential lengths dr, $d\phi$, and dz as

$$\mathbf{dl} = dr\hat{\mathbf{r}} + rd\phi\hat{\boldsymbol{\phi}} + dz\hat{\mathbf{k}} \tag{1.54}$$

In the three-dimensional cylindrical coordinate system, *three differential area vectors* $\mathbf{dS}_r$, $\mathbf{dS}_z$, and $\mathbf{dS}_\phi$, may be defined from the differential lengths dr, dz, and $d\phi$ as

$$\left.\begin{array}{l} \mathbf{dS}_r = r\,d\phi\,dz \\ \mathbf{dS}_z = r\,d\phi\,dr \\ \mathbf{dS}_\phi = dr\,dz \end{array}\right\} \tag{1.55}$$

Each surface vector will point in the subscripted dimension, exactly as the differential surface vectors in a plane. A differential volume **dV** may be defined as

$$\mathbf{dV} = rdrd\phi dz \tag{1.56}$$

The *spherical coordinates* of a three-dimensional vector $\mathbf{B} = (a, b, c)$ in Cartesian coordinates are defined as $\mathbf{B} = (r, \theta, \phi)$ where r is the magnitude of $\mathbf{B}$, θ is the angle of the vector on the $r-z$ plane relative to the positive z-axis and ϕ is the angle of the projection of the vector on the $x-y$ plane relative to the positive x-axis. Angles are measured in the counterclockwise direction from the positive x-axis.

The unit vectors of the spherical coordinate system are $\hat{\mathbf{r}}$, $\hat{\boldsymbol{\theta}}$, and $\hat{\boldsymbol{\phi}}$, but unlike the Cartesian unit vectors, they are not on fixed axes, but are located at the endpoint of a vector whose starting point is the origin. Thus their location changes with radius r, angle θ, and angle ϕ. The radial unit vector $\hat{\mathbf{r}}$ is always radially pointing away from the origin and normal to the sphere S. The angular unit vector $\hat{\boldsymbol{\theta}}$ is normal to the cone C, generated by sweeping the vector $\mathbf{B}$ around the vertical z-axis, and pointing away from the vertical z-axis. The angular unit vector $\hat{\boldsymbol{\theta}}$ is normal to the vector $\mathbf{B}$. The angular unit vector $\hat{\boldsymbol{\phi}}$ is identical to the same vector in the cylindrical coordinate system, normal to the plane A, defined by the vector $\mathbf{B}$ and the vertical z-axis. The angular unit vector $\hat{\boldsymbol{\phi}}$ is tangential to the sphere S, in the direction of increasing angle, and normal to the other two unit vectors. The spherical coordinate system is shown in Figure 1.11. The spherical unit vectors can be expressed in terms of the Cartesian unit vectors:

$$\hat{\mathbf{r}} = \sin\theta\cos\phi\hat{\mathbf{i}} + \sin\theta\sin\phi\hat{\mathbf{j}} + \cos\theta\hat{\mathbf{k}} \tag{1.57}$$

$$\hat{\boldsymbol{\theta}} = \cos\theta\cos\phi\hat{\mathbf{i}} + \cos\theta\sin\phi\hat{\mathbf{j}} - \sin\theta\hat{\mathbf{k}} \tag{1.58}$$

$$\hat{\boldsymbol{\phi}} = -\sin\phi\hat{\mathbf{i}} + \cos\phi\hat{\mathbf{j}} \tag{1.59}$$

The components of the vector (a, b, c) converted to spherical coordinates (r, θ, ϕ) are

$$r = \sqrt{a^2 + b^2 + c^2} \tag{1.60}$$

$$\theta = \cos^{-1}\frac{c}{r} \tag{1.61}$$

$$\phi = \tan^{-1}\frac{b}{a} \tag{1.62}$$

The cylindrical coordinates of the example vector of $\mathbf{B} = (4, 5, 6)$ shown in Figure 1.11 are

$$r = \sqrt{4^2 + 5^2 + 6^2} = 8.775$$

$$\theta = \cos^{-1}\frac{6}{8.775} = 46.86°$$

$$\phi = \tan^{-1}\frac{5}{4} = 51.34°$$

The Cartesian unit vectors are expressed in spherical form as

$$\hat{\mathbf{i}} = \sin\theta\cos\phi\hat{\mathbf{r}} + \cos\theta\cos\phi\hat{\boldsymbol{\theta}} - \sin\phi\hat{\boldsymbol{\phi}} \tag{1.63}$$

$$\hat{\mathbf{j}} = \sin\theta\sin\phi\hat{\mathbf{r}} + \cos\theta\sin\phi\hat{\boldsymbol{\theta}} + \cos\phi\hat{\boldsymbol{\phi}} \tag{1.64}$$

$$\hat{\mathbf{k}} = \cos\theta\hat{\mathbf{i}} - \sin\theta\hat{\mathbf{j}} \tag{1.65}$$

The components of the vector (r, θ, ϕ) converted to rectangular coordinates (a, b, c) are

$$a = r\sin\theta\cos\phi \tag{1.66}$$

$$b = r\sin\theta\sin\phi \tag{1.67}$$

$$c = r\cos\theta \tag{1.68}$$

In the three-dimensional spherical coordinate system, a *differential vector* $\mathbf{dl}$ may be defined from the differential lengths dr, $d\theta$, and $d\phi$ as

$$\mathbf{dl} = dr\,\hat{\mathbf{r}} + rd\theta\hat{\boldsymbol{\theta}} + r\sin\theta d\phi\hat{\boldsymbol{\phi}} \tag{1.69}$$

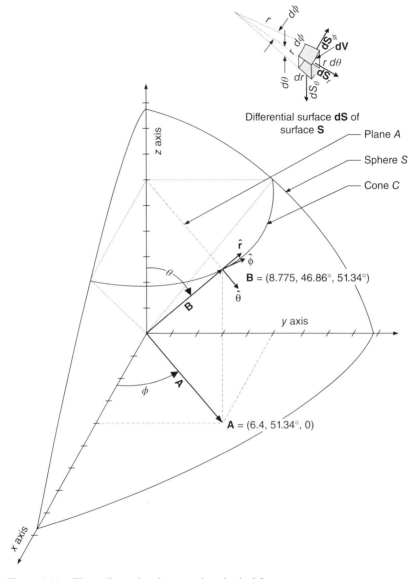

Figure 1.11   Three-dimensional vectors in spherical form.

In the three-dimensional spherical coordinate system, *three differential area vectors* $\mathbf{dS}_r$, $\mathbf{dS}_\theta$, and $\mathbf{dS}_\phi$, may be defined from the differential lengths dr, $d\theta$, and $d\phi$ as

$$\left.\begin{aligned}
\mathbf{dS}_r &= r^2 \sin\theta \, d\theta \, d\phi \\
\mathbf{dS}_\theta &= r \sin\theta \, dr \, d\phi \\
\mathbf{dS}_\phi &= r \, dr \, d\theta
\end{aligned}\right\} \tag{1.70}$$

Each surface vector will point in the subscripted dimension, exactly as the differential surface vectors in a plane. A differential volume $\mathbf{dV}$ may be defined as

$$\mathbf{dV} = r^2 \sin\theta dr d\theta d\phi \tag{1.71}$$

A *matrix* is an extension of the concept of a vector, where the elements can be arranged in rows and columns in two or more dimensions. Matrices are used to simplify the notation when a calculation is made with several nearly identical equations, such as one for each dimension in a system. Matrices are also a type of number in their own right. The 3×3 matrix $\mathbf{M}$ is written as

$$\mathbf{M} = \begin{bmatrix} m_{11} & m_{12} & m_{13} \\ m_{21} & m_{22} & m_{23} \\ m_{31} & m_{32} & m_{33} \end{bmatrix} \tag{1.72}$$

The magnitude of a matrix is the *determinant*. For a 3×3 matrix,

$$|\mathbf{M}| = \begin{vmatrix} m_{11} & m_{12} & m_{13} \\ m_{21} & m_{22} & m_{23} \\ m_{31} & m_{32} & m_{33} \end{vmatrix}$$

$$= m_{11}(m_{22}m_{33} - m_{23}m_{32}) + m_{12}(m_{21}m_{33} - m_{23}m_{31}) + m_{13}(m_{21}m_{32} - m_{22}m_{31}) \tag{1.73}$$

1.3 MATHEMATICAL OPERATIONS WITH VECTORS

Vector *addition* and *subtraction* are carried out term by term in the rectangular coordinate system. For vectors $\mathbf{A} = (a, b, c)$ and $\mathbf{B} = (d, e, f)$:

$$\mathbf{A} + \mathbf{B} = (a + d,\ b + e, c + f) \tag{1.74}$$

$$\mathbf{A} - \mathbf{B} = (a - d,\ b - e, c - f) \tag{1.75}$$

The product of a *vector* $\mathbf{A}$ and a *scalar* α is the product of the scalar and each element of the vector:

$$\alpha\mathbf{A} = (\alpha a,\ \alpha b, \alpha c) \tag{1.76}$$

The scalar may be positive or negative. If the scalar is negative, the direction of the vector will be reversed. Note that while the magnitude of a vector is always positive, the vector may have any direction.

The *dot product* or *inner product* of two vectors $\mathbf{A} = (a, b, c)$ and $\mathbf{B} = (d, e, f)$ separated by angle α is a scalar quantity defined as

$$\mathbf{A} \cdot \mathbf{B} = ad + be + cf \tag{1.77}$$

or

$$\mathbf{A} \cdot \mathbf{B} = AB \cos\alpha \tag{1.78}$$

A two-dimensional example of a dot product is shown in Figure 1.12. The dot product is commutative:

$$\mathbf{A} \cdot \mathbf{B} = \mathbf{B} \cdot \mathbf{A} \tag{1.79}$$

The *cross product* or *outer product* of two vectors $\mathbf{A}$ and $\mathbf{B}$ separated by angle α is a vector defined as

$$\mathbf{A} \times \mathbf{B} = \hat{\mathbf{i}}_{\mathbf{n}} AB \sin\alpha \tag{1.80}$$

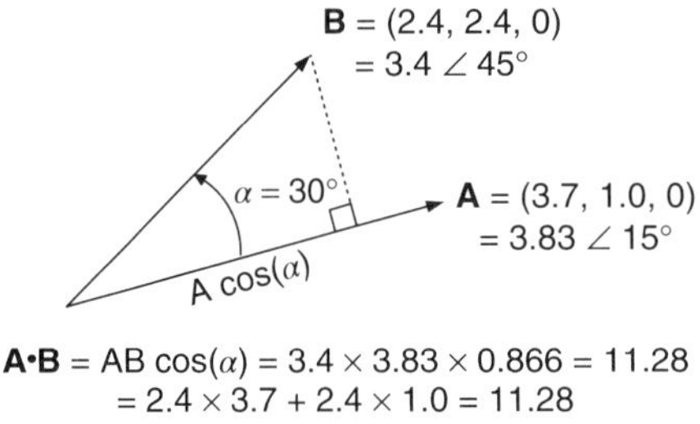

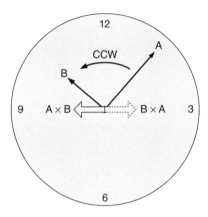

$\mathbf{B} = (2.4, 2.4, 0)$
$= 3.4 \angle 45°$

$\alpha = 30°$

A cos(α)

$\mathbf{A} = (3.7, 1.0, 0)$
$= 3.83 \angle 15°$

$\mathbf{A \cdot B} = AB \cos(\alpha) = 3.4 \times 3.83 \times 0.866 = 11.28$
$= 2.4 \times 3.7 + 2.4 \times 1.0 = 11.28$

Figure 1.12   Dot product of two vectors.

12

CCW

A

B

9   A × B   B × A   3

6

Figure 1.13   Right-hand rule for the cross product of two vectors.

The unit vector $\hat{\mathbf{i}}_n$ is normal (perpendicular) to the two vectors $\mathbf{A}$ and $\mathbf{B}$ in the direction indicated by the "right-hand rule." The right-hand rule can be visualized as a clock face with two hands A and B, Figure 1.13, where A is to the right of B. A vector emerging perpendicular to the face of the clock toward the observer represents the direction of the cross product. An example of a cross product is shown in Figure 1.14. Because of the right-hand rule, the cross product is not commutative:

$$\mathbf{B} \times \mathbf{A} = -\mathbf{A} \times \mathbf{B} \tag{1.81}$$

The vector $\mathbf{B} \times \mathbf{A}$ is also shown in the figures, in the direction opposite to $\mathbf{A} \times \mathbf{B}$. The cross product, like the dot product, may also be calculated using the elements of each vector:

$$\mathbf{A} \times \mathbf{B} = \begin{vmatrix} \hat{\mathbf{i}} & \hat{\mathbf{j}} & \hat{\mathbf{k}} \\ a & b & c \\ d & e & f \end{vmatrix} = \hat{\mathbf{i}}(bf - ce) + \hat{\mathbf{j}}(af - cd) + \hat{\mathbf{k}}(ae - bd) \tag{1.82}$$

1.4   CALCULUS WITH VECTORS — THE GRADIENT

A *scalar field* is a function which assigns a scalar to each point within a region of a space. A scalar field may be written as $f(x, y, z)$ in a three-dimensional space. Examples of a scalar field are gravitational fields and electric fields.

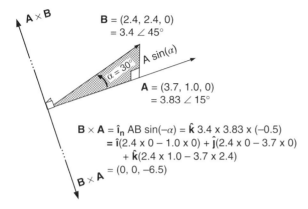

$$\mathbf{A} \times \mathbf{B} = \hat{\mathbf{i}}_n \, AB \sin(\alpha) = \hat{\mathbf{k}} \, 3.4 \times 3.83 \times 0.5$$
$$= \hat{\mathbf{i}}(1.0 \times 0 - 2.4 \times 0) + \hat{\mathbf{j}}(3.7 \times 0 - 1.0 \times 0)$$
$$+ \hat{\mathbf{k}}(3.7 \times 2.4 - 1.0 \times 2.4)$$
$$= (0, 0, 6.5)$$

$\mathbf{A} \times \mathbf{B}$

$\mathbf{B} = (2.4, 2.4, 0)$
$= 3.4 \angle 45°$

$A \sin(\alpha)$

$\alpha = 30°$

$\mathbf{A} = (3.7, 1.0, 0)$
$= 3.83 \angle 15°$

$$\mathbf{B} \times \mathbf{A} = \hat{\mathbf{i}}_n \, AB \sin(-\alpha) = \hat{\mathbf{k}} \, 3.4 \times 3.83 \times (-0.5)$$
$$= \hat{\mathbf{i}}(2.4 \times 0 - 1.0 \times 0) + \hat{\mathbf{j}}(2.4 \times 0 - 3.7 \times 0)$$
$$+ \hat{\mathbf{k}}(2.4 \times 1.0 - 3.7 \times 2.4)$$
$$= (0, 0, -6.5)$$

$\mathbf{B} \times \mathbf{A}$

Figure 1.14   Cross product of two vectors.

The *differential* of a scalar field is

$$df(x, y, z) = \frac{\partial f}{\partial x} dx + \frac{\partial f}{\partial y} dy + \frac{\partial f}{\partial z} dz \tag{1.83}$$

The *del operator* (∇) is used to denote the taking of partial derivatives in each direction:

$$\nabla = \hat{\mathbf{i}} \frac{\partial}{\partial x} + \hat{\mathbf{j}} \frac{\partial}{\partial y} + \hat{\mathbf{k}} \frac{\partial}{\partial z} \tag{1.84}$$

The del operator is the three-dimensional equivalent to the symbol for taking the derivative with no function specified, $\frac{d}{dx}$. The del operator applied to a three-dimensional scalar field is called the *gradient* (grad) of the field:

$$\mathbf{grad} \, f(x, y, z) = \nabla f(x, y, z) = \frac{\partial f}{\partial x} \hat{\mathbf{i}} + \frac{\partial f}{\partial y} \hat{\mathbf{j}} + \frac{\partial f}{\partial z} \hat{\mathbf{k}} \tag{1.85}$$

The gradient in cylindrical coordinates is

$$\nabla f(r, \phi, z) = \frac{\partial f}{\partial r} \hat{\mathbf{r}} + \frac{1}{r} \frac{\partial f}{\partial \phi} \hat{\boldsymbol{\phi}} + \frac{\partial f}{\partial z} \hat{\mathbf{z}} \tag{1.86}$$

The gradient in spherical coordinate is

$$\nabla f(r, \theta, \phi) = \frac{\partial f}{\partial r} \hat{\mathbf{r}} + \frac{1}{r} \frac{\partial f}{\partial \theta} \hat{\boldsymbol{\theta}} + \frac{1}{r \sin \theta} \frac{\partial f}{\partial \phi} \hat{\boldsymbol{\phi}} \tag{1.87}$$

The *scalar field* of a set of points on the r-axis may be written as $f(r)$. An example of a scalar field is the electric potential $V(r)$ of a point charge q:

$$V(r) = \frac{q}{4\pi\varepsilon_0 r} \text{V} \tag{1.88}$$

where,

$V(r) =$ is the potential in volts (V) at a point r meters (m) from the point charge,

$q =$ is the charge in coulombs (C), and

$\varepsilon_0 = \frac{1}{4\pi \times 10^{-7}} =$ is the permittivity of free space in farad per meter (F/m).

Work is required to move a charge in an electric field. If a charge of 1 C is moved across a potential difference of 1 V, 1 J (joule) of work will be expended. If there are two charges of opposite signs which are naturally attracted to each other, work is required to separate them. If the two charges are of the same polarity, the force will be required to move them toward each other. The force in newtons (N) exerted by the charge q on the charge q' (and by the charge q' on the charge q) at a distance x along the direction of the vector $\hat{\mathbf{i}}$ is given by *Coulomb's law*:

$$\mathbf{F} = \frac{qq'}{4\pi\varepsilon_0 r^2}\hat{\mathbf{i}} N \tag{1.89}$$

A *vector field* differs from the scalar field in that a vector exists at each point in the Euclidean space. An example of this is the *electric field* (V/m). The electric field established by a charge q is defined where the charge q' (at distance x from the charge q) becomes infinitely small, so that only the charge q is being considered.

$$\mathbf{E} = \frac{\mathbf{F}}{q'} = \frac{q}{4\pi\varepsilon_0 r^2}\hat{\mathbf{i}}\;\frac{V}{m} \tag{1.90}$$

The electric field vector is the gradient of the scalar electric field.

$$\mathbf{E} = -\nabla V \tag{1.91}$$

The electric field of a point charge calculated by the gradient definition:

$$\mathbf{E} = -\nabla V = -\frac{\partial V}{\partial r}\hat{\mathbf{r}} = -\frac{\partial}{\partial r}\frac{q}{4\pi\varepsilon_0 r}\hat{\mathbf{r}} = \frac{q}{4\pi\varepsilon_0 r^2}\hat{\mathbf{r}} \tag{1.92}$$

Figure 1.15 shows the field produced by a charge of 1 nanocoulomb (nC or 10^{-9} C) over the range of 2–5 m. At point a, 2 m, the field is 4.5 V, and at point b, 5 m, the field is 1.8 V. Thus, if the 1 nC charge q at the origin is positive, and the charge q' at point a is -1 nC, then it will take 2.7 nJ to move this charge to point b. This can be calculated more formally is we consider the differential distance $\mathbf{dl}$ along the path a to b, where

$$\mathbf{dl} = dr\hat{\mathbf{r}} \tag{1.93}$$

and

$$d\mathbf{V} = \nabla V \cdot \mathbf{dl} \tag{1.94}$$

If we integrate from a to b,

$$\int_a^b d\mathbf{V} = \int_a^b \nabla V \cdot \mathbf{dl} = V(b) - V(a) = \frac{q}{4\pi\varepsilon_0}\left(\frac{1}{a} - \frac{1}{b}\right) \tag{1.95}$$

For example, the scalar field

$$f(x, y) = 0.2x^2 + 0.5y$$

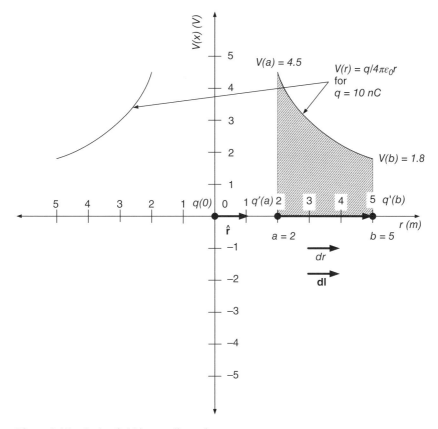

Figure 1.15   Scalar field in one dimension.

represents a specific numerical value at each point in the x, y space. The differential of the field is

$$df(x, y) = 0.4xdx + 0.5dy$$

and the gradient of the field is

$$\nabla f = 0.4x\hat{\mathbf{i}} + 0.5\hat{\mathbf{j}}$$

The angles of the gradient are

$$\left.\begin{array}{c} \phi_g = \tan^{-1}\left[\dfrac{0.5}{0.4\,x}\right] = \tan^{-1}\left[\dfrac{1.25}{x}\right] \\[4mm] \theta_g = \tan^{-1}\left[\dfrac{\sqrt{0.16\,x^2 + 0.25}}{0}\right] = \dfrac{\pi}{2} \end{array}\right\}$$

The gradient has a magnitude representing the slope of the field in the x- and y-directions, pointing in the direction of the greatest change in slope from its origin at a point (x, y). Recalling the distance vector **dl**, the differential of the scalar field may be expressed as

$$df(x, y, z) = \left(\frac{\partial f}{\partial x}\widehat{\mathbf{i}} + \frac{\partial f}{\partial y}\widehat{\mathbf{j}} + \frac{\partial f}{\partial z}\widehat{\mathbf{k}} \right) \cdot (dx\widehat{\mathbf{i}} + dy\widehat{\mathbf{j}} + dz\widehat{\mathbf{k}}) = \frac{\partial f}{\partial x}dx + \frac{\partial f}{\partial y}dy + \frac{\partial f}{\partial z}dz$$

(1.96)

since $\widehat{\mathbf{i}} \cdot \widehat{\mathbf{i}} = \widehat{\mathbf{j}} \cdot \widehat{\mathbf{j}} = \widehat{\mathbf{k}} \cdot \widehat{\mathbf{k}} = 1$. The angles of $\mathbf{dl}$ are

$$\left. \begin{array}{l} \phi_{dl} = \tan^{-1}\left[\dfrac{dy}{dx} \right] \\[1.5em] \theta_{dl} = \tan^{-1}\left[\dfrac{\sqrt{dx^2 + dy^2}}{dz} \right] \end{array} \right\}$$

(1.97)

Then

$$df = \boldsymbol{\nabla} f \cdot \mathbf{dl} = |\boldsymbol{\nabla} f| \, |\mathbf{dl}| \, \cos \theta = |\boldsymbol{\nabla} f| dl \cos \theta_{df}$$

(1.98)

Where θ_{df} is the angle between $\boldsymbol{\nabla} f$ and $\mathbf{dl}$. In the example, the angles of $\mathbf{dl}$ are (with $dz = 0$)

$$\left. \begin{array}{l} \phi_{dl} = \tan^{-1}\left[\dfrac{dy}{dx} \right] = \tan^{-1}\left[2\dfrac{df}{dx} - 0.8x \right] = \tan^{-1}0 = 0 \\[1.5em] \theta_{dl} = \tan^{-1}\left[\dfrac{\sqrt{dx^2 + dy^2}}{0} \right] = \dfrac{\pi}{2} \end{array} \right\}$$

Since the example is in two dimensions, $\theta_g = \theta_{dl} = \pi/2$. Because of the definition of the function $f(x, y) = 0.2x^2 + 0.5y$, $\phi_{dl} = 0$. Then $\theta_{df} = \phi_g$ and $df(x, y)$ may be calculated as

$$df(x, y) = \sqrt{0.16x^2 + 0.25} \, dl \cos\left[\tan^{-1}\left(\frac{1.25}{x} \right) \right]$$

It is apparent that this is equal to the same quantity calculated previously:

$$df(x, y) = (\widehat{\mathbf{i}}0.4x)(dx\,\widehat{\mathbf{i}}) + (\widehat{\mathbf{j}}0.5)(dy\,\widehat{\mathbf{j}}) = 0.4xdx + 0.5dy$$

The *gradient theorem* may be defined for a line between two points A and B:

$$\int_A^B df = \int_A^B \boldsymbol{\nabla} f \cdot \mathbf{dl} = f(B) - f(A)$$

(1.99)

The line integral is dependent only on the end points and the function f, and does not change if the length of the line is longer or shorter, or if the line takes different routes from A to B. The gradient theorem is also called the *fundamental theorem of calculus for line integrals*. In Figure 1.16, the starting point $A = (5, 1)$ and the end point $B = (2, 3)$ making $f(A) = 5.2$ and $f(B) = 2.3$. Three different paths are shown in the figure, all with the same line integral. Thus:

$$\int_A^B df = f(B) - f(A) = 2.3 - 5.2 = -3.2$$

A closed line integral, where $A = B$, is symbolized with a circle over the integral sign:

$$\oint df = f(A) - f(A) = 0$$

(1.100)

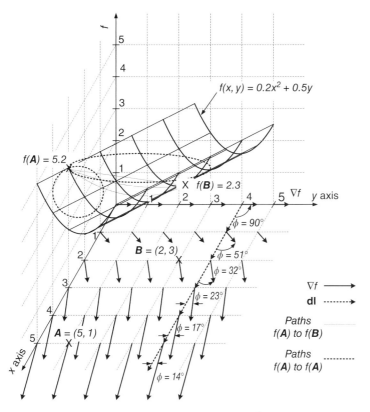

Figure 1.16   Two-dimensional scalar field $f(x,y) = 2x^2 + 5y$ and its gradient grad f.

A corollary to the gradient theorem is that *the line integral of a gradient around a closed loop is always zero*. In Figure 1.16, lines $f(A)$ to $f(A)$ have a line integral of zero.

1.5   DIVERGENCE, CURL, AND STOKES' THEOREM

When a boundary is established by a closed surface S in a three-dimensional space which contains a vector field A, the net amount of flow through the space is called the *flux*.

$$\Phi = \oint_S A \cdot dS \qquad (1.101)$$

Because the flux will also be different at each point in space, the derivative of the flux at any point in space is defined as a quantity called the *divergence*:

$$\nabla \cdot A = \left(\hat{i}\frac{\partial}{\partial x} + \hat{j}\frac{\partial}{\partial y} + \hat{k}\frac{\partial}{\partial z} \right) \cdot (a\hat{i} + b\hat{j} + c\hat{k}) = \frac{\partial a}{\partial x} + \frac{\partial b}{\partial y} + \frac{\partial c}{\partial z} \qquad (1.102)$$

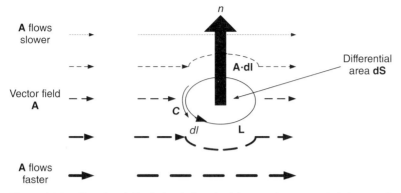

Figure 1.17   Closed path **L** of circulation C with normal vector **n** defining the direction of **A·dl** in accordance with the right-hand rule.

The *divergence theorem* relates the flux within a volume **V** composed of many infinitesimal differential volumes **dV** with the surface **S** composed of many infinitesimal differential areas $d\mathbf{S}$.

$$\Phi = \oint_{\mathbf{S}} \mathbf{A} \cdot d\mathbf{S} = \int_{\mathbf{V}} \nabla \cdot \mathbf{A} \, d\mathbf{V} \qquad (1.103)$$

The circulation C of a fluid in a closed path **L** against a vector field **A** is shown in Figure 1.17. The normal vector **n**, established by the right-hand rule, establishes the direction of **A·dl**. The circulation is expressed by

$$C = \oint_{\mathbf{L}} \mathbf{A} \cdot d\mathbf{l} \qquad (1.104)$$

The *curl* is a vector operation defined as

$$\text{curl } \mathbf{A} = \nabla \times \mathbf{A} = \begin{vmatrix} \hat{\mathbf{i}} & \hat{\mathbf{j}} & \hat{\mathbf{k}} \\ a & b & c \\ d & e & f \end{vmatrix} = \hat{\mathbf{i}}\left(\frac{\partial c}{\partial y} - \frac{\partial b}{\partial z}\right) + \hat{\mathbf{j}}\left(\frac{\partial a}{\partial z} - \frac{\partial c}{\partial x}\right) + \hat{\mathbf{k}}\left(\frac{\partial b}{\partial x} - \frac{\partial a}{\partial y}\right)$$
$$(1.105)$$

The circulation for an infinitesimally small loop **L**, shown in Figure 1.17 as an eddy in flowing water, in differential form is:

$$C = (\nabla \times \mathbf{A}) \cdot d\mathbf{S} \qquad (1.106)$$

Stokes' theorem links together these results for a finite-sized open surface **S** enclosed by a line **L**:

$$\oint_{\mathbf{L}} \mathbf{A} \cdot d\mathbf{l} = \int_{\mathbf{S}} (\nabla \times \mathbf{A}) \cdot d\mathbf{S} \qquad (1.107)$$

Recalling the gradient theorem, it is clear that the line integral around a closed loop must be zero, then using Stokes' theorem:

$$\oint_{\mathbf{L}} \nabla f \cdot d\mathbf{l} = \int_{\mathbf{S}} (\nabla \times \nabla f) \cdot d\mathbf{S} = 0 \qquad (1.108)$$

Using the definitions of divergence and curl, a useful identity for vector analysis is

$$\nabla \cdot (\nabla \times \mathbf{A}) = 0 \qquad (1.109)$$

1.6   MAXWELL'S EQUATIONS

The Scottish physicist James Clerk Maxwell (1831–1879) codified the empirically derived physical laws discovered by Faraday, Ampère, and Gauss along with some necessary corrections into a set of 20 equations which define electromagnetic field theory. The original versions of these equations have been updated by Oliver Heaviside (1850–1925) using the concepts of vector calculus outlined above into four equations, which are today generally known as "Maxwell's Equations."

Faraday's law, states that a changing magnetic field induces a corresponding electric field:

$$\nabla \times \mathbf{E} = -\frac{\partial \mathbf{B}}{\partial t} \qquad (1.110)$$

This can be thought of as an electric circuit composed of a loop of wire in combination with a time-varying magnetic field produced by a spinning magnet, the combination being an electric generator, Figure 1.18. In integral form, the moving magnetic field provides the work to move an electric charge along a specific path.

$$\oint_{L} \mathbf{E} \cdot \mathbf{dl} = -\frac{d}{dt} \int_{S} \mathbf{D} \cdot \mathbf{dS} \qquad (1.111)$$

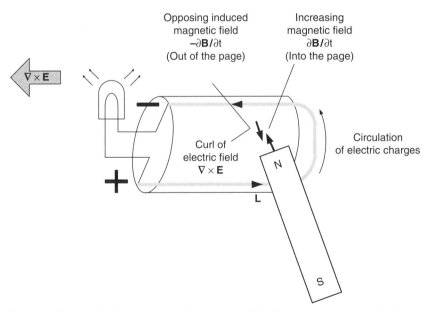

Figure 1.18   Faraday's law: the moving magnetic field induces an electric field which produces a current that induces a magnetic field opposing that which caused it.

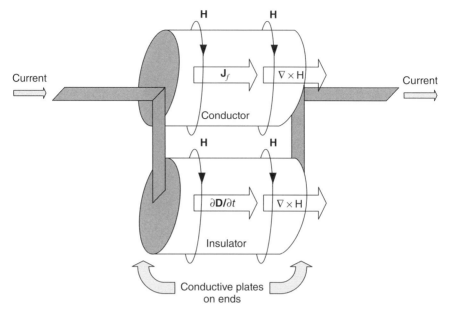

Figure 1.19   Illustration of Ampère's law for a resistor and capacitor.

Ampère's law with Maxwell's displacement current correction states that a changing electric field produces a corresponding magnetic field, as in Figure 1.19,

$$\mathbf{\nabla} \times \mathbf{H} = J_f + \frac{\partial \mathbf{D}}{\partial t} \qquad (1.112)$$

In point form, you imagine a moving electric charge producing a magnetic field according to the right-hand rule. In integral form, the work required for a magnet to move along a path $\mathbf{L}$ is provided by electric charges moving across the boundaries of a volume.

$$\oint_{\mathbf{L}} \mathbf{H} \cdot \mathbf{dl} = \int_{\mathbf{S}} J_f \cdot \mathbf{dS} + \frac{d}{dt} \int_{\mathbf{S}} \mathbf{D} \cdot \mathbf{dS} \qquad (1.113)$$

Gauss's law for electric fields in volume form is, "The electric flux through any closed surface is proportional to the enclosed electric charge." A "Gaussian surface" is an arbitrary closed three-dimensional surface (such as a sphere) which is chosen to ease the computation of volume integrals. This is illustrated in Figure 1.20, where the electric field diverges from the positive point charge and converges on the negative point charge.

$$\mathbf{\nabla} \cdot \mathbf{D} = \rho_f \qquad (1.114)$$

If Gauss's law for electric fields is expressed in integral form, it becomes

$$\oint_{\mathbf{S}} \mathbf{D} \cdot \mathbf{dS} = \int_{\mathbf{V}} \rho_f d\mathbf{V} \qquad (1.115)$$

Where the surface encloses some electrical charges as in the figure, and if you make the surface such that all field lines cross it perpendicularly, then the dot product becomes multiplication and the math is easier.

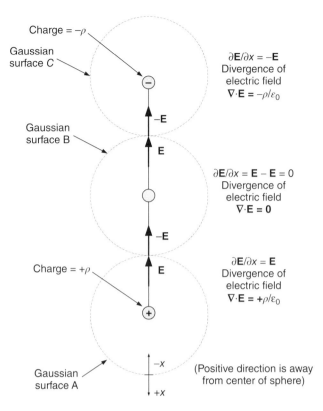

Charge = $-\rho$

Gaussian
surface C

$\partial \mathbf{E}/\partial x = -\mathbf{E}$
Divergence of
electric field
$\nabla \cdot \mathbf{E} = -\rho/\varepsilon_0$

$-\mathbf{E}$

Gaussian
surface B

$\mathbf{E}$

$\partial \mathbf{E}/\partial x = \mathbf{E} - \mathbf{E} = 0$
Divergence of
electric field
$\nabla \cdot \mathbf{E} = 0$

$-\mathbf{E}$

$\partial \mathbf{E}/\partial x = \mathbf{E}$
Divergence of
electric field
$\nabla \cdot \mathbf{E} = +\rho/\varepsilon_0$

Charge = $+\rho$

$\mathbf{E}$

$-x$

(Positive direction is away
from center of sphere)

Gaussian
surface A

$+x$

Figure 1.20   Illustration of Gauss's law for the field around a point charge.

Gauss's law for magnetic fields implies that there is no magnetic current point charge:

$$\nabla \cdot \mathbf{B} = 0 \tag{1.116}$$

All magnetic sources are dipoles, having both a north and a south pole. Unlike electric fields, this, while they may have dipoles, may also have both independent positive and independent negative charges as sources. This can be seen in Figure 1.21, where no surface can enclose a net divergence or convergence of lines of flux. The integral form is

$$\oint_S \mathbf{B} \cdot d\mathbf{S} = 0 \tag{1.117}$$

This means that whatever three-dimensional enclosure you construct, it will have as many magnetic field lines entering as leaving, as can be seen in the figure.

Maxwell's equations are written in both derivative and integral forms, with many variations based on the physical situation which they are being used to evaluate. The derivative forms have more simplicity and compactness, but the integral form may be easier to visualize as a three-dimensional object is involved, even though it is only a fictitious Gaussian surface. Whether visualizing points or surfaces, the three-dimensional characteristics of invisible, and usually moving, force fields are

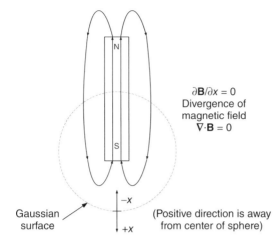

$\partial \mathbf{B}/\partial x = 0$
Divergence of
magnetic field
$\nabla \cdot \mathbf{B} = 0$

Gaussian
surface

−x

+x

(Positive direction is away
from center of sphere)

Figure 1.21   Illustration of
Gauss's law for magnetic fields.

more difficult to picture in mathematical than in physical terms. The terms used here
are listed for reference:

$\mathbf{E} =$ electric field intensity (volts per meter)

$\mathbf{B} =$ magnetic field density (tesla or webers per square meter)

$\mathbf{H} =$ magnetic field intensity (amperes per meter)

$\mathbf{J}_f =$ free current density (in a conductor) (amperes per square meter)

$\mathbf{D} =$ electricdisplacement field (coulombs per square meter)

Additionally, Maxwell's equations include the conservation of charge, which states
that charge can neither be created nor destroyed:

$$\nabla \cdot \mathbf{J}_f + \frac{\partial \rho_f}{\partial t} = 0 \tag{1.118}$$

The additional variable is

$\rho_f =$ free charge density (in a conductor) (coulombs per cubic meter) (not to be
confused with electrical resistivity also designated by ρ)

Maxwell's equations are supplemented by the constitutive laws, which are necessary for their application to physical materials:

$$\mathbf{D} = \varepsilon \mathbf{E} + \mathbf{P} \tag{1.119}$$

$$\mathbf{B} = \mu \mathbf{H} \tag{1.120}$$

The additional variables are

$\varepsilon =$ permittivity (farad per meter)

$\mu =$ permeability (henry per meter)

$\mathbf{P} =$ portion of the electric field due to polarization in matter

These equations define how much electric field strength it requires to impose a charge
on a material, and how much magnetizing intensity it takes to magnetize a material.

Ohm's law for fixed and moving conductors is also a necessary part of Maxwell's equations:

$$J_f = \sigma(\mathbf{E} + \mathbf{v} \times \mathbf{B}) \tag{1.121}$$

The additional variables are

σ = conductivity (siemens)

$\mathbf{v}$ = velocity (meters per second)

The current density in a material includes both current imposed by an electric field dependent on the conductivity of the material and current induced by movement of the material in a magnetic field.

CHAPTER 2

ELECTRICAL SAFETY ASPECTS OF THE RESISTANCE PROPERTY OF MATERIALS

2.1 INTRODUCTION

Electrical shock is caused by the passage of electrical current through the body. The amount of shock can vary from a small tingling sensation to severe burns, temporary paralysis, cardiac arrest, and death. Often, surprisingly small amounts of current can cause severe damage, depending where the current flows through the body. Electrical burns can be among the most severe because the current passes through the body and internal organs. The amount of current that flows through the body is estimated in many sources as the ratio of the applied voltage to the resistance to current flow presented by the human body. This effect varies with the characteristics of the applied voltage in terms of duration, frequency of repetition, whether the voltage is alternating current or direct current, frequency of alternating current, rise and fall times of transient waves, and other factors. Similarly, the electrical resistance of the human body is not a single value, but has different values depending on the path of the current, the area of contact, the properties of skin and internal organs, the electrical characteristics of the voltage as described above, and of the corresponding current, the presence, condition, and characteristics of clothing, footwear, gloves, headgear and the like. Thus, for example, the *IEEE Guide for Safety in Substation Grounding* (IEEE, 2000), which is concerned with sinusoidal AC systems, despite an extensive discussion of the various factors affecting electrical shock, ends with an assumed body resistance of $1000\,\Omega$, and a duration factor of $1/\sqrt{t}$, where t is the exposure duration. This section will examine the phenomenon of body resistance and its inverse, body conductance. While such assumptions provide a starting point, the resistance of the human body is a complex, nonlinear phenomenon which deserves more accurate modeling, especially in cases where human life and health are at stake.

The second area where resistance affects electrical safety is in the resistance properties of the electrical circuits and components with which a person may come in contact with. In the case of shock hazards, circuit resistance in combination with body resistance will help to determine the magnitude and duration of the exposure to voltage and current. Electrical resistance is also a significant cause of electrical

Principles of Electrical Safety, First Edition. Peter E. Sutherland fires, of burns caused by coming in contact with an electrically heated object, and other possible injuries. When electrical systems and equipment are designed, these hazards of resistance itself, along with the potential hazards of coming in contact with energized resistors both in terms of electrical shock and thermal burns must be taken into account, so that it is inherently safe to work with and use the equipment.

2.2   HAZARDS CAUSED BY ELECTRICAL RESISTANCE

In electrical engineering, resistivity is a material property which affects the behavior of an electrical component. When the material takes a specific form, the resistance is determined from the resistivity of the material and its size and shape. Resistance is an example of what is called a "lumped circuit parameter," where a property of an extensive physical object is expressed as if it were a single object with a single property. For example, electrical conductors have an intrinsic degree of resistance, which could be considered as a form of opposition to their assigned task of being forced to carry electrical current. A copper wire contains resistance which, in the case of a small degree of current flow, will cause heating of the metal. This is due to the atomic structure not being entirely uniform, resulting in collisions between freely moving electrons and more stationary metal atoms. Heat is produced by the motion of atoms and molecules, so the copper wire will heat up as the current is increased. As the conductor becomes warmer, increased atomic motion results in more collisions between atoms and electrons, causing the resistance of the conductor to increase. This can be calculated using the "temperature coefficient of resistivity," α for that material. While this quantity may sometimes be presented as a constant, it is actually a function of temperature also, and any particular value of α which is given, in a reference table, for example, should be accompanied by a note at what temperature it was measured. In the vast majority of commonly used conductor materials, α will increase with temperature. When resistance is the property which is desired, as in an electric heating device, the conductor is usually made of a higher resistivity material than copper, such as nichrome, iron, and steel. Resistance is also used in lighting, which began with Edison's carbon filament lamp, but is nowadays most recognized in the tungsten lamp.

At a certain point, the conductor's temperature will exceed the ambient temperature, and it may begin to feel warm or hot to the touch. Now, one would not touch a bare copper wire which is electrically energized, but if the wire is sufficiently well insulated and the potential of the wire is low enough, one can touch the insulation. Considering devices using iron wire or similar materials, which are meant to produce heat, as in an electric toaster, stove, oven, space heater, or radiant floor, to name a few, one may feel the heat rising from the conductors. Undesired heat in a conductor may be caused by an overload, for example, connecting too many lamps to an extension cord or outlet, or by a short circuit, where two conductors come into contact. If the heat is sufficient to melt or burn through the insulation allowing an arc to form between the two conductors, then a fire can readily start. An internal short circuit which suddenly decreases the resistance of the device, resulting in a high current flow, may also cause overheating leading to a fire. Heating may also be caused by poor connections, because any junction between two conductors must encompass a certain surface area at a certain pressure in order to ensure good conductivity. A high resistance junction, such as a plug which is loose in a socket, may cause sufficient heating to start a fire. When wires and other conductors are being inspected as part of an electrical maintenance procedure, such as at a power plant, along a power line, or in a large commercial or industrial facility, infrared thermometers and infrared video cameras are often used to locate "hot spots" where the conductivity is poor and the conductors are heating up. This allows maintenance to be scheduled before a catastrophic failure such as an electrical fire can occur.

The detailed application of electrical cables and conductors is a subject in itself, requiring an analysis of both the heating effects of electrical current on conductors and insulation, and the thermal resistivity of the materials between the conductor and the ambient (such as air) which causes cooling of the conductor. In addition, the characteristics of protective devices, such as fuses and circuit breakers must be considered. Tables for general use are provided in electrical code books, such as the National Electrical Code (NFPA, 2014) in the United States, and IEC 60364-5-52 (IEC, 2009a) in other countries. More detailed evaluation requires the use of mathematical modeling, such as the Neher-McGrath equations (Neher and McGrath, 1957) and the IEEE Power Cable Ampacity Tables (IEEE, 1994). Alternative calculation methods such as finite element analysis may also be used.

A National Fire Protection Association Report (Hall, 2012) estimates that of the 44,800 reported structural blazes involving homes, 12,000 had electrical causes. The electrically caused fires resulted in 320 deaths and over 1000 injuries (excluding firefighters and other emergency responders). While the vast majority of electrical fires were due to arcing short circuits, a significant number, perhaps one-third, were caused by heating of combustible materials by electrical sources. In many cases, the insulation of the wire or cable was ignited, perhaps by the overcurrent caused by a short circuit or by overloading of a circuit.

Electrical insulation for conductors, such as cables, cords, and the like, usually has a temperature rating. The circuit should be designed such that during normal loading, this temperature is never exceeded. Most fuses and circuit breakers have what is called an "overload characteristic," which is selected to operate at a lower temperature than the insulation of the conductors that it is protecting. Conventional "thermal-magnetic" circuit breakers are used in most homes. The current passes through a thermal element consisting of a bimetallic strip that bends as it is heated, causing the circuit breaker to open above a certain temperature. Thus the characteristics of the protective device work on a similar heating principle to the device being protected, and the protection is compatible with the protected device. At high short-circuit currents, faster action is required and the circuit breaker's magnetic element will quickly force the contacts apart. Thermal protection provided by fuses also works by heating of a wire, but in this case the wire will melt, causing the current to be interrupted. Fuses are normally designed with slow-acting regions for thermal protection and fast-action regions for short-circuit protection. These differences can be caused by changes in the width or thickness of the fusible element, or by holes punched in a fusible strip, or by making a fuse of two or more dissimilar materials.

Heating of conductors which are ductile also may cause them to stretch or lengthen, if they are suspended, as between two poles in a power line. A sagging power line may arc to some nearby person or object, resulting in electrical shock or an arc starting a fire in a combustible object such as a building or a tree. Even if protection systems cause fuses or circuit breakers to open, the damage may have already been done. A typical current-limiting fuse may operate in one–one hundredth of a second (10 ms), but a circuit breaker may take from two to five power cycles of the alternating current. This amounts to 33-83 ms in a 60-cycle system or 40-100 ms in a 50-cycle system. This is more than enough time to kill a human being or to start a large fire. In many cases, power line conductors are uninsulated, and even in cases where they are insulated, a failure of some sort may cause the insulation to melt or break. Safety regulations, such as the National Electrical Safety Code (IEEE, 2012a) require certain clearances, depending on the operating voltage between power lines and the ground and between power lines and other structures such as buildings. This clearance is not only affected by sag due to heating, but by motion due to the wind, and by loading from the accumulation of ice. Sag, of course, is affected by ambient temperature and by cooling due to wind, rain, and ice and snow accumulation. Electric utility companies manage the flow of energy through power lines in accordance with many economic and operational factors, but they are always constrained by the load limits due to sag.

Resistors which are electrical components may be made of a volume or sheet of material, such as carbon, which does not conduct very well. That will provide a convenient way to manufacture a specific value of resistance. The now obsolete carbon composition resistors, encased in a plastic shell, with a color code of painted bands indicating their resistance in ohms, and two axial leads, are the classic example of this type of resistor. This could be compared to a body part such as a finger or leg, which is a cylinder of poorly conducting material. The resistor has a number of properties besides resistance. The second most important is wattage, that is, how much power can be dissipated due to I^2R heating without damaging or destroying the resistor. The current-carrying capacity or current rating of the resistor will be directly related to the wattage. A resistor which is improperly applied or which experiences current flow which causes its rated wattage to be exceeded will heat, causing the encapsulating material to discolor, possibly to burn and eventually disintegrate. Many electrical fires have been caused by overheating resistors. The resistor will also have a maximum voltage at which the resistor can carry a given current without overheating, above which the voltage drop ($V = IR$) across the resistor may become so high that arcing will occur across the surface of the resistor or through the air around it. Again, this can cause electrical fires, along with failure of the resistor.

Figure 2.1 illustrates some common types of fixed resistors used in electronic devices, ranging from computers and televisionsets to industrial equipment. The obsolete carbon composition resistor (Figure 2.1a), discussed above, is marked with a color code that is well-known in the industry, which gives its resistance value and tolerance. The wattage is indicated by the size of the resistor. Film resistors (Figure 2.1b) come in many varieties, carbon or metallic film, thick or thin film, being the most common. The carbon-film resistors shown come in the same general size

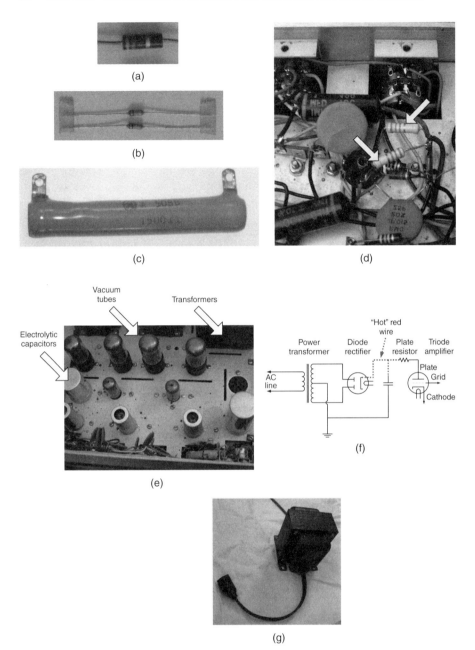

Figure 2.1 Resistors used as electronic components. (a) Carbon composition resistor, axial lead, ½ W, 470 W, 10% tolerance. (b) Carbon film, axial lead, 1/4 Watt, 10 kW, 5% tolerance. (c) Vitreous enamel-coated wire-wound power resistor, 1500 W. (d) Plate resistors in vacuum tube amplifier. (e) Vacuum tube amplifier where plate resistors would be used. (f) Schematic of vacuum tube amplifier where plate resistors would be used. (g) Isolation transformer (115 V, 250 VA) for servicing "transformerless" electronic equipment.

and shape as the carbon composition resistors, with the same markings. The carbon film is coated on a ceramic tube, and then a spiral is etched away, leaving a helical track of carbon which creates the resistance needed, depending on its composition, length, width, and thickness. Metal-film resistors are similar in construction, but use a metallic rather than a carbon-based film. Flameproof resistors are available, which eliminate the hazard of overheating mentioned above. When higher wattages are required, the wire-wound resistor (Figure 2.1c) is often used. This consists of a ceramic cylinder wound with resistance wire (iron or nichrome, e.g.), inside a heat-resistant coating, vitreous enamel being a common choice. Here, color-coded bands are not used and the resistor data are simply printed on the enamel. If the ratings of the wire-wound resistor are exceeded, it may heat enough to cause nearby materials to burst into flame, but it will not burn itself. The enamel may blacken, the ceramic tube may crack, and the wire may burn through, resulting in device failure. The composition and film resistors are of the "axial lead" type, with the leads in line with the axis of the cylindrical form. The wire-wound resistor is shown with solder lugs for connections. These are extending radially from the surface of the cylinder. In other cases, a "radial lead" construction may be used, with the connection leads extending at right angles to the surface of the cylinder, and parallel to each other.

Components can be named by their location in a circuit as well as the type of construction. An example is the plate resistor used in vacuum tube (valve) circuits, as in Figure 2.1d, is not to be confused with the plate resistors used in industrial systems, which are discussed below. The plate resistor is usually a smaller carbon film or composition resistor, usually 0.5-1 W, used to limit the current in the plate circuit of a vacuum tube amplifier, Figure 2.1e. This resistor also can pose a safety hazard for energized work, because it is at a voltage in the range of 100-1000 V and more in high power radio transmitters. The high voltage or "hot" lead is shown with dotted lines in Figure 2.1f. The hot circuit will go from the power transformer, shown in the photo of Figure 2.1e and the schematic diagram of Figure 2.1f, to a vacuum tube rectifier, electrolytic capacitors, and then to the plate circuits of the amplifier tubes, where the plate resistors are. The capacitors, with their stored charge, in fact, provide the greatest hazard in this circuit. Any repair shop which works on vacuum tube electronics, such as those contained in vintage equipment, amateur radio gear, or guitar amplifiers should observe the appropriate safety precautions for the voltage involved. When "transformerless" circuits are used, with the tubes connected directly to the AC line voltage, the available fault current may be significant, and extra care is required, such as the use of an isolation transformer, Figure 2.1g.

High power, high current resistors for industrial applications are shown in Figure 2.2. These resistors are used for a variety of applications including

1. Neutral grounding resistors
2. Harmonic filter resistors
3. Snubber resistors
4. Braking resistors used in electric trolleys
5. Starting resistors for DC Motors
6. Load banks for testing generators.

(a)

(b)

(c)

(d)

(e) Plate rheostat (GE)   (f) Schematic of plate rheostat (GE)

Figure 2.2   Resistors used in electric power systems. (a) Plate resistor. (b) Ribbon resistor. (c) Noninductive resistor. (d) Ribbon-wound resistor. (e) Plate rheostat (GE). (f) Schematic of plate rheostat (GE).

The resistor in Figure 2.2a is the plate resistor, consisting of flat sheets of steel which are spaced to allow for ventilation, and are supported on each side by spacer bars, with the metal section for each plate junction separated by insulating barriers from the adjacent plates. The plates will thus have an arrangement similar to the schematic symbol for a resistor ----/\/\/\/\/----. At each junction, a tab may be placed to provide a tapped resistance value.

The resistor in Figure 2.2b is a ribbon resistor. Here, a strip of steel is bent in the shape of a ribbon, similar to a piece of ribbon candy. It can also have taps made of tabs at the end of a bend. The resistor in Figure 2.2c is a noninductive resistor. This is a film-type resistor on a ceramic tube, but unlike the standard carbon-film resistor, it does not have an etching in the form of a spiral. Such a spiral shape creates the form of a coil, and thus has significant inductance. The noninductive resistor has a resistive coating over the entire surface of the tube, except at the ends where contacts are made. The diameter is generally about 2 in. and length about 2 ft. The noninductive resistor will have the very low inductance characteristics which may be seen in a large diameter tubular conductor. The high power resistor shown in Figure 2.2d is the ribbon-wound resistor, where a strip of steel is in the form of a ribbon with the flat surface parallel to the ends of the cylinder. This allows for good ventilation to occur between the turns. In the figure, a group of ribbon-wound resistors are combined in series–parallel to provide a particular value of resistance, with the possibility of tap connections at the junctions. One resistor in the group with taps attached to the ribbons can be used for fine-tuning.

The hazards associated with high power industrial resistors are primarily due to their open construction, which is necessary for cooling. The exposed conductors which make up the resistors can be not only a shock hazard but also a thermal burn hazard. The resistors must be mounted on insulators suitable for the working voltage and any harmonics or transients which may raise the voltage. Where there is the possibility of contact, such as when mounted on a concrete pad on the ground, or on top of a transformer, they should be enclosed in a grounded metal cage. There is little possibility of these resistors catching fire themselves, as no flammable materials are used in their construction or mounting. However, if flammable material such as some types of cable insulation is connected to them, this may present a possible hazard. Therefore, the connections to high power resistors are usually bare conductors.

Another type of power resistor is a variable resistor, in the form of the "plate rheostat" (where rheostat is another name for a variable resistor) pictured in Figure 2.2e, with a representative schematic in Figure 2.2f, named because of its shape like a dinner plate. Plate rheostats (General Electric Company, undated) are rated in the hundreds up to over a thousand watts, with plate sizes of 6-12 in. The rated currents are up to 26 A, and the resistance values vary from 20 to 5000 Ω. Plate rheostats were used for

- "Speed control of dc motors. Use rheostat to control field excitation.
- Power-factor control of synchronous motors. Use rheostat to control field excitation.
- Voltage control of generators and exciters. Use rheostat to control field excitation.

- Adjustment of voltage in control circuits. Use rheostat to vary impressed voltage.
- Speed control of wound-rotor motors. Use rheostat to vary resistance in motor secondary; not over 15 HP."

They have largely been replaced by solid-state devices. Their application is described as

> A rheostat is a resistor provided with a ready means for varying its resistance. The usual application of rheostats is in the field circuits of motors or generators for the control of speed or voltage, or in control circuits. For most applications the size of the rheostat is determined by the characteristics of the load it must control. Other applications require a certain number of steps which determine the rheostat size.

The principle safety hazards with plate rheostats are

- exposure to high voltages at rheostat terminals;
- burns from coming in contact with heated plates.

2.3   RESISTANCE AND CONDUCTANCE

The fundamental definition of resistance (R) assumes that an object, such as the cylinder in Figure 2.3, with a defined length (ℓ) and constant cross-sectional area (A) has a uniform material property which we call *resistivity* (ρ):

$$R = \rho \frac{\ell}{A} \ \ \Omega \tag{2.1}$$

where

$R =$ resistance in ohms (Ω)

$\rho =$ resistivity in ohm-meter (Ω-m)

$\ell =$ length in m

$A =$ cross-sectional area in m^2

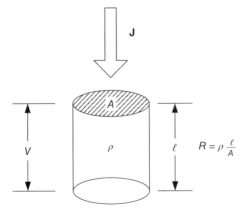

Figure 2.3   Resistance, R, is derived from the resistivity, ρ, of an object.

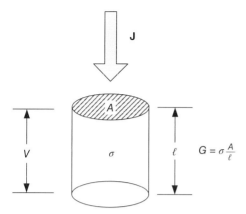

Figure 2.4   Conductance, G, is derived from the conductivity, σ, of an object.

The reciprocal of resistance is conductance (G). When applied to a uniform body such as the cylinder in Figure 2.4, it is based on the material property of conductivity (σ). Conductivity is the reciprocal of resistivity.

$$G = \sigma \frac{A}{\ell} \qquad (2.2)$$

where

$G = 1/R =$ conductance in siemens (S)

$\sigma = 1/\rho =$ conductivity in siemens/meter (S/m)

When a voltage is applied to an object possessing the property of resistance, an electric current will flow through the object. In the case of metallic conductors, the current flow is accomplished by the motions of electrons, although other particles such as ions may be the means of current flow in other media, as we shall see later, such as when we consider the flow of current through the human body. It will be assumed for now that the diffusion of current throughout the object is uniform, although, in actuality, this is rarely the case. Current flow in a wire (I) has been compared to the flow of water in a pipe (Φ_V), which is measured in cubic meters per second (m^3/s) or in cubic feet per second (ft^3/s). Volumes of electrons are measured in the amount of electrical charge possessed by the aggregate of electrons. A "volume" of electrons is not a fixed quantity of space, but a fixed number of charged particles. It is as if volumes of water were measured by a count of water molecules instead of the volume which they occupy (at a given temperature and pressure, of course). The unit of charge is the coulomb (C), which is the amount of electrical charge which is transported by 1.0 ampere (A) of current during a time period of 1.0 second (s). The charge on an electron (e) is $1.602176487 \times 10^{-19}$ C. The number of electrons (N_e) in a coulomb is

$$N_e = \frac{C}{e} = 1/1.602176487 \times 10^{-19} = 6.2415 \times 10^{18} \text{ electrons} \qquad (2.3)$$

There are 6.2415×10^{18} electrons flowing in any given electrical conductor during every second when a current of 1.0 A is flowing.

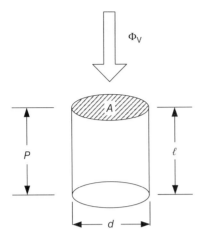

Figure 2.5   Water pipe analogy.

Voltage is the electrical analog of pressure in a water pipe (Figure 2.5). The pressure (P) is measured in pascals (Pa) where apascalis a pressure of one newton (N) per square meter (m^2). Thus $1.0\,\text{Pa} = 1.0\,\text{N/m}^2$. In English units, pressure is measured in pounds per square inch (lb/in.^{2}). Since water can be pressurized without flowing, pressure is a form of potential energy. The amount of potential energy is the volume of water times the pressure. Energy is measured in joules (J), thus $1.0\,\text{Pa} = 1.0\,\text{J/m}^3$. When the water flows, the energy transferred is the pressure difference between the ends of the pipe times the internal volume of the pipe. Thus $1.0\,\text{J} = 1.0\,\text{Pa} \times \text{m}^3$. Thermal energy is dissipated by the flow of water under pressure in a pipe of a given diameter (d). The property of pipes to dissipate thermal energy due to the flow of water is called head loss. The head loss of a pipe is inversely proportional to the diameter of the pipe.

Now consider an electrical system. Field strength is exerted on charges, which attempts to make the units of charge (such as electrons) move through a medium (such as a cylindrical object). Field strength is measured in volts, with one volt corresponding to a force of one newton-meter (N-m) per coulomb (C). Thus $1.0\,\text{V} = 1.0\,\text{N-m/C}$. Since electricity can be pressurized without flowing, volts are a form of potential energy. The amount of potential energy is the charge (coulombs) times the volts. Energy is measured in joules (J), thus $1.0\,\text{V} = 1.0\,\text{J/C}$. When the charge flows, the energy released is the voltage difference between the ends of the object times the charge in coulombs. The rate at which the work of charges flowing is performed is measured in watts (W) where $1.0\,\text{W} = 1.0\,\text{J/s} = 1.0\,\text{V} \times 1.0\,\text{C/s} = 1.0\,\text{V} \times 1.0\,\text{A}$. The thermal energy dissipated by the flow of one ampere under the pressure of one volt is one watt. The property of materials to dissipate thermal energy due to the flow of electric current is called *resistance*, which is measured in Ohms (Ω). The electrical conductivity of an object is analogous to the diameter of a water pipe. Thus one ohm of resistance with one ampere flowing through it will have a voltage drop of one volt and a thermal energy of one watt. If this continues for one second, one joule of energy will be used.

TABLE 2.1   Ohmic Conductivity of Some Common Materials (at Room Temperature)

Material	σ (S/m)
Pure water	4.0×10^{-6}
Carbon	7.3×10^{-4}
Sea water	4.0
Lead	0.5×10^{7}
Tin	0.9×10^{7}
Zinc	1.7×10^{7}
Gold	4.2×10^{7}
Pure Copper	5.9×10^{7}
Silver	6.3×10^{7}
Aluminum	2.7×10^{8}

TABLE 2.2   Ohmic Conductivity of Some Body Parts (at Power Frequency)

Body Part	σ (S/m)
Bone	0.013
Fat	0.04
Liver	0.13
Muscle (perpendicular)	0.076
Lung	0.092
Blood	0.60
Muscle (parallel)	0.52
Sweat (0.3% saline)	1.4

The linear relation of current to voltage in an object is the familiar Ohm's law. When one first considers a conductive object, such as a copper wire, we can see the simplest form of Ohm's law:

$$\mathbf{J} = \sigma E \tag{2.4}$$

Typical conductivity values are shown in Tables 2.1 and 2.2. Considering resistivity as the inverse of conductivity,

$$\mathbf{J} = \frac{E}{\rho} \tag{2.5}$$

where

$\mathbf{J} =$ current density in amperes/meter^2 (A/m^2)

$E =$ electric field in volts/meter (V/m)

Ohm's law for lumped circuit components, as shown in Figure 2.6, is expressed in the form

$$I = \frac{V}{R} \tag{2.6}$$

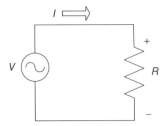

Figure 2.6   Circuit model of a resistive circuit.

where

I = current in amperes (A)

V = voltage in volts (V)

Using conductance instead of resistance,

$$I = GV \tag{2.7}$$

2.4   EXAMPLE — TRUNK OF A HUMAN BODY

Given the trunk of a human body, represented by an elliptical shape as in Figure 2.7, the area

$$A = \pi ab \tag{2.8}$$

while the circumference may be found approximately using

$$C_e \approx 2\pi \sqrt{\frac{a^2 + b^2}{2}} \tag{2.9}$$

The exact calculation of the circumference of an ellipse requires the use of the elliptical integral, and is not necessary for this example. Given, for example, a person with a 1 m waist, then

$$a^2 + b^2 = \frac{1}{2\pi^2} = 0.0507 \, \text{m}^2$$

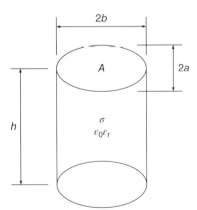

Figure 2.7   Model of body trunk with elliptical cross section.

If the ratio $a/b = 2$, then:

$$a = \frac{1}{\pi \sqrt{2 \times 1.25}} = 0.201 \, \text{m}$$
$$b = 0.101 \, \text{m}$$

and the area is

$$A = \pi ab = 0.064 \, \text{m}^2$$

A comparable circle with the same circumference will have

$$r = \frac{C}{2\pi} = 0.159 \, \text{m}$$
$$A = \pi r^2 = 0.080 \, \text{m}^2$$

showing the importance of using appropriate shapes in the model.

If the height of the trunk is 0.5 m, and the conductivity of internal organs is 0.1 S/m, then the resistance can be calculated as follows:

$$R = \frac{h}{\sigma A} = \frac{0.5}{0.1 \times 0.064} = 78 \, \Omega$$

If the circular assumption had been used, we have

$$R = \frac{h}{\sigma A} = \frac{0.5}{0.1 \times 0.08} = 62 \, \Omega$$

which is 20% too small.

2.5   EXAMPLE — LIMB OF A HUMAN BODY

Given a limb, Figure 2.8, where $2b = 0.10 \, \text{m}$, $2a = 0.05 \, \text{m}$, $h = 0.5 \, \text{m}$, containing a bone with $r = 0.02 \, \text{m}$.

$$A = \pi ab = \pi \times 0.05 \times 0.025 = 3.93 \times 10^{-3} \, \text{m}^2$$
$$B = \pi r^2 = \pi \times 0.02^2 = 1.26 \times 10^{-3} \, \text{m}^2$$

The area of flesh is

$$F = A - B = 2.67 \times 10^{-3} \, \text{m}^2$$

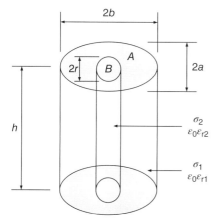

Figure 2.8   Model of limb with elliptical cross section and circular bone at the center.

The resistance of flesh is based on the longitudinal conductivity of muscle, $\sigma = 0.52\,\text{S/m}$.

$$R_F = \frac{h}{\sigma F} = \frac{0.5}{0.52 \times 2.67 \times 10^{-3}} = 360\,\Omega$$

The resistance of bone is based on the conductivity of bone, $\sigma = 0.013\,\text{S/m}$.

$$R_B = \frac{h}{\sigma B} = \frac{0.5}{0.013 \times 1.26 \times 10^{-3}} = 30\,\text{k}\Omega$$

The total resistance is

$$R = \frac{1}{\dfrac{1}{R_F} + \dfrac{1}{R_B}} = 356\,\Omega$$

If a current of 1 A flowed through the limb, 0.988 A would flow through the flesh and 0.012 A would flow through the bone. Total power dissipation would be 356 W, with 351.4 W in the flesh and 4.4 W in the bone.

2.6 POWER AND ENERGY FLOW

For a uniform material,

$$E = \frac{V}{\ell} \tag{2.10}$$

In electromagnetic field theory, power density is defined in terms of the Poynting vector (Sutherland, 2007):

$$\mathbf{S} = \mathbf{E} \times \mathbf{H} \tag{2.11}$$

where

$\mathbf{S} =$ power density in watts per square meter (W/m^2)

$\mathbf{E} =$ electric field vector in volts per meter (V/m)

$\mathbf{H} =$ magnetic field intensity in amperes per meter (A/m)

Over a closed surface S:

$$P_{in} = -\oint_S (\mathbf{E} \times \mathbf{H}) \cdot dS \tag{2.12}$$

Consider the cylindrical body made of a material whose resistivity has a uniform value of ρ, whose length is ℓ, and whose radius $r = a$, such that area $A = \pi a^2$. When an electric field is applied uniformly across each end of the cylinder, there is a constant axial voltage drop across the length ℓ. The magnetic field will consist of circular lines of force, such that the Poynting vector will be radial, pointed inward. Energy which enters the volume, but does not leave, will produce heating. Energy which enters at one end of area A, will leave by the other end, contributing no net energy. Energy entering the sides of the cylinder contributes to heating. The surface area of the sides of the cylinder is the circumference times the length.

Assuming perfect conductivity and uniform current distribution, the electric field cannot sustain itself within the conductor, and is present only as radial lines leaving the surface of the conductor. The electric field strength, E_r, of these radial vectors is

$$E_r = \frac{\lambda}{2\pi\varepsilon_0 r} \qquad (2.13)$$

where λ is the line charge, ε_0 is the permittivity, and r the distance from the conductor center. The radius of the conductor does not affect this result. In the case of electric power cables, a grounded electrostatic shield may encase each individual conductor, confining the electric field to within the cable insulation.

In a uniform cylindrical conductor, the magnetic field is divided into the part within the conductor, and that outside the conductor.

$$B_\phi = \begin{cases} \dfrac{\mu_0 I}{2\pi r}, & r > a \\[2mm] \dfrac{\mu_0 I r}{2\pi a^2}, & r < a \end{cases} \qquad (2.14)$$

where

$$B_\phi = \mu_0 H \qquad (2.15)$$

$B =$ magnetic flux density, tesla (T)

$\mu_0 =$ permeability of free space $= 4\pi \times 10^{-7}$ H/m

Assuming perfect conductivity and uniform current distribution, the electric field cannot sustain itself within the conductor, and is present only as radial lines leaving the surface of the conductor. The electric field strength, E_r, of these radial vectors is

$$E_r = \frac{\lambda}{2\pi\varepsilon_0 r} \qquad (2.16)$$

where λ is the line charge, ε_0 is the permittivity, and r the distance from the conductor center. (Figure 2.9).

The radius of the conductor does not affect this result. In the case of electric power cables, a grounded electrostatic shield may encase each individual conductor, confining the electric field to within the cable insulation.

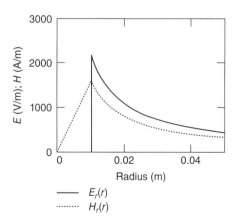

Figure 2.9   Electric and magnetic fields around a circular conductor.

The magnetic field, H_t, on the contrary, is present both inside and outside a single conductor:

$$H_t = \begin{cases} \dfrac{Ir}{2\pi a^2}, & r < a \\[2ex] \dfrac{I}{2\pi r}, & r > a \end{cases} \tag{2.17}$$

where

$\quad a =$ radius of the conductor and

$\quad I =$ current in the conductor.

The magnetic field vectors are tangential. The resultant vector direction of P_{in} is thus away from the source and toward the load. If the electric flux lines, **E**, are radial and the magnetic flux lines, **H**, are concentric, the magnitude of the power density as a function of distance from the conductor center is the product $\mathbf{E} \times \mathbf{H}$ shown in Figure 2.10.

Most of the power transmission occurs within a relatively small radius of the conductors. The power density is substantially less in most of the space between the conductors. Here, 50% of the power is transmitted within the first 10% of the distance between the conductors. Power is transmitted in the same direction by both the outgoing and returning current flow.

If we take the closed surface S over the outside of the cylinder in Figure 2.11,

$$P_{in} = 2\pi a\ell EH$$

$$= 2\pi a\ell \frac{V}{\ell} \frac{I}{2\pi a} = VI \tag{2.18}$$

Ohm's law for electric fields is expressed as

$$\mathbf{J} = \sigma\mathbf{E} \tag{2.19}$$

where

$\quad \mathbf{J} =$ current density in A/m^2 and

$\quad \sigma =$ conductivity in siemens (S).

$$\sigma = \frac{1}{\rho} \tag{2.20}$$

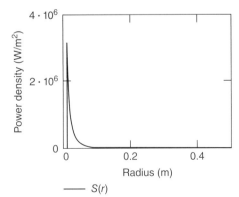

Figure 2.10   Power density is highest closest to the conductors.

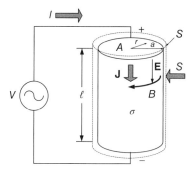

Figure 2.11 Cylindrical conductor with applied voltage.

where

$\rho =$ resistivity in ohm-meters (Ω-m).

Multiplying by m^2, the expression is converted to circuit units, as in Figure 2.11, resulting in the familiar Ohm's law,

$$I = \frac{V}{R} \qquad (2.21)$$

and the power equation,

$$P = VI = \frac{V^2}{R} \qquad (2.22)$$

Resistivity is thus a measure of the ability of a material to convert electromagnetic energy to thermal energy.

2.7 SHEET RESISTIVITY

An object, such as skin, which consists of a sheet of material, can be modeled considering that the contact electrode current (I_C), divides into an area-proportional current (I_P) and a spreading current (I_S), as shown in Figure 2.12, with an equivalent circuit in Figure 2.13. The depth current operates on the principle of sheet resistivity. Let R_P be the resistance seen by the uniform vertical current I_P over length ℓ, width w, and depth d:

$$R_P = \frac{\rho}{\ell} \frac{d}{w} = \frac{\rho}{A_0} \frac{d}{n_s} \qquad (2.23)$$

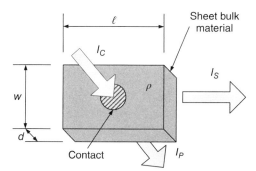

Figure 2.12 Sheet resistivity.

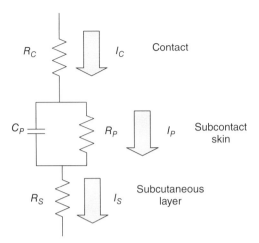

Figure 2.13   Simplified equivalent circuit of skin.

Where

$$A_0 = 1\,\text{m}^2$$

and $n_s = \dfrac{\ell w}{A_0}$ = number of squares. Typically, $n_s > 1$.

Then

$$\rho_P = \rho d \quad \Omega\text{-m}^2 \tag{2.24}$$

so that

$$R_P = \frac{\rho_P}{\ell w} = \frac{\rho_P}{A_0 n_s} \; \Omega \tag{2.25}$$

2.8   EXAMPLE — SQUARE OF DRY SKIN

Consider a square of dry skin area $1\,\text{cm}^2$, with a resistivity of $1000\,\text{k}\Omega\text{-cm}^2 = 100\,\Omega\text{-m}^2$. The resistance from an electrode also of area $1\,\text{cm}^2 = 10^{-4}\,\text{m}^2$ to the underlying body is

$$R_P = \frac{\rho_P}{A_0 n_s} = \frac{10^2}{10^{-4} \times 1} = 10^6\,\Omega = 10^3\,\text{k}\Omega$$

The thickness of the skin is in the range $10–100\,\mu\text{m}$. If the actual thickness is $50\,\mu\text{m}$, the resistivity of the skin may be calculated:

$$\rho = \frac{\rho_P}{d} = \frac{100}{50 \times 10^{-6}} = 2 \times 10^6\,\Omega \cdot \text{m}$$

2.9   SPREADING RESISTANCE

Let R_S be the resistance seen by the uniform horizontal current I_S over length ℓ, width w, and depth d:

$$R_S = \frac{\rho}{d} \frac{\ell}{w} \; \Omega \tag{2.26}$$

expresses the resistivity on ohm-meters as in the formula for bulk resistivity. However, if we consider only the squares of surface area, then

$$\rho_S = \frac{\rho}{d} \, \Omega/\text{square} \tag{2.27}$$

and

$$R_S = \frac{\rho}{d}\frac{\ell}{w} = \rho_s n_s \, \Omega \tag{2.28}$$

where $n_s = \frac{\ell}{w}$ is the number of squares. Typically, $n_s > 1$. At high frequencies, where the current only travels in the surface of the material, to a depth called the "skin depth" δ (by analogy to, but not the same as the biological skin),

$$\rho_S = \frac{\rho}{\delta} \, \Omega/\text{square} \tag{2.29}$$

2.10   EXAMPLE — CIRCLE OF DRY SKIN

In Figure 2.14, the $1 \, \text{cm}^2 = 10^{-4} \, \text{m}^2$ circular contact area has a radius of

$$r = \frac{\sqrt{A}}{\pi} = \frac{0.01}{\pi} = 3.18 \times 10^{-3} \, \text{m}$$

This is at the center of a circular area of skin with an area of $10 \, \text{cm}^2 = 10^{-3} \, \text{m}^2$, radius

$$r = \frac{\sqrt{A}}{\pi} = \frac{\sqrt{10^{-3}}}{\pi} = 10.1 \times 10^{-3} \, \text{m}$$

The ring of skin not covered by the contact has a width of $(10.1 - 3.2) \times 10^{-3} \, \text{m} = 6.9 \times 10^{-3} \, \text{m}$, and an area of $9 \, \text{cm}^2 = 0.9 \times 10^{-4} \, \text{m}^2$. Assuming $\rho = 2.0 \times 10^6 \, \Omega\text{-m}$ for a skin layer of $50 \, \mu\text{m}$.

The surface resistivity is

$$\rho_S = \frac{\rho}{d} = \frac{2 \times 10^6}{50 \times 10^{-6}} = 4 \times 10^{10} \, \Omega/\text{square}$$

while the subcutaneous layer composed of muscle and fat has a much lower resistivity. Assuming $\sigma = 0.05 \, \text{S/m}$ for a subcutaneous layer of $2.0 \, \text{mm}$:

$$\rho_{SC} = \frac{\rho}{d_{SC}} = \frac{20}{2.0 \times 10^{-3}} = 10 \times 10^3 \, \Omega/\text{square}$$

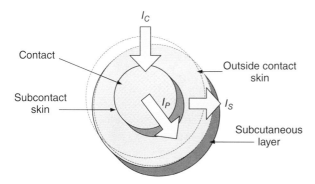

Figure 2.14   Surface current flows.

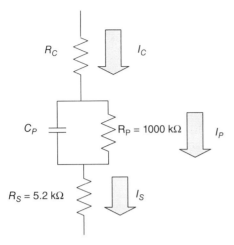

Figure 2.15   Example of skin impedance.

TABLE 2.3   **Surface Resistivity of Wet and Dry Skin (at Power Frequency)**

Body Part	$R_P + R_S$ (kΩ-cm^2)
Skin (dry)	100–1000
Skin (wet)	<100

The width of the exposed skin area is 6.89×10^{-3} m. The circumference of the contact is 20×10^{-3} m. The circumference of the total skin area is 63.2×10^{-3} m. The average value of the contact and the skin areas is 41.6×10^{-3} m. The area of a square is thus approximately $(41.6 \times 10^{-3})^2$ m^2= 1.73×10^{-3} m^2. The number of squares is also approximately $n_s = 0.9/1.73 = 0.52$.

Then

$$R_S = \rho_s n_s = 10 \times 10^3 \times 0.52 = 5.2 \times 10^3 \, \Omega$$

The resulting skin model is shown in Figure 2.15, with typical values in Table 2.3. The capacitance will be calculated in a later section.

2.11   PARTICLE CONDUCTIVITY

The current we have discussed to this point is due to charges moving in response to the energy provided by the Poynting vector, directly as a result of energy flow in the circuit. There is a second type of current that must be considered, which is current as a result of the mechanical or chemical motion of charged particles. The electrochemical equilibrium of the human cell is disturbed by the flow of electrical current, and can have long-lasting effects. This is why every incident of electrical shock has a permanent effect, and medical attention should be sought. The field version of Ohm's law is then extended:

$$\mathbf{J}_{\pm} = \pm\sigma_{\pm}\mathbf{E} - D_{\pm}\nabla\rho_{\pm} \tag{2.30}$$

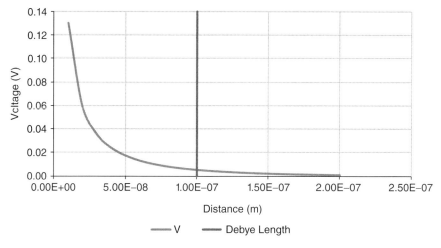

Figure 2.16   Voltage around a charged particle and Debye length.

where

$D\pm$ = diffusion coefficient in square meters per second (m²/s)

ρ = electric charge density in coulombs per cubic meter (C/m³).

A charged particle, Q, has a voltage field around it, which decreases expo-nentially over a small volume, shown in Figure 2.16. For a typical material at room temperature, this length, called the Debye length, l_d, is approximately 10^{-7} m. The voltage magnitude is

$$V = \frac{Q}{4\pi\varepsilon r}e^{-\frac{r}{l_d}} \text{ V} \tag{2.31}$$

Suppose that $r = l_d$, then,

$$V = \frac{Q}{4\pi\varepsilon l_d}e^{-1} = \frac{1.602 \times 10^{-19}}{4\pi \times \dfrac{10^{-9}}{36\pi} \times 10^{-7}} = 5.3 \times 10^{-3} \text{ V}$$

This field strength is in the order of millivolts, and the Debye length an order of magnitude greater than the thickness of a cellular membrane (10^{-8} m). The movement of charges through a cellular membrane is shown in Figure 2.17. The membrane has different equilibria on each side, which results in the potential difference shown in Figure 2.18. The potential difference across a cellular membrane is defined by the Nernst equation, (Reilly, 1998, p. 78) given here in simplified form for

$$V_m = Z61\log_{10}\frac{[S]_o}{[S]_i} \text{ mV} \tag{2.32}$$

where

V_m = Nernst potential

$[S]_o$ = ionic concentration outside the membrane in micromoles per cubic centimeter (μM/cm³)

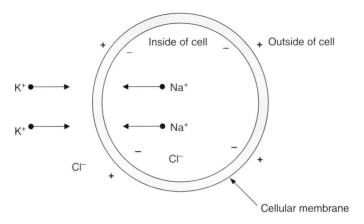

Figure 2.17   Cellular membrane with ionic movements.

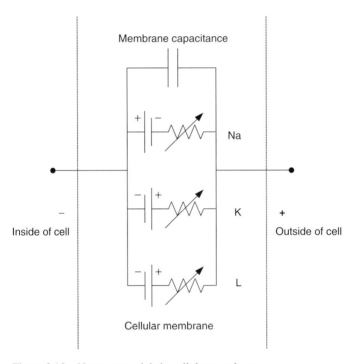

Figure 2.18   Nernst potentials in cellular membrane.

$[S]_i$ = ionic concentration inside the membrane, and

Z = ionic polarity.

The Nernst potential is given with different constants instead of 61 V in different sources, depending on the assumptions made, the choice of base for the logarithm, and the temperature. It is given as 58 V at 17 °C with the logarithm to the base 10 or 25 V with a natural logarithm in (Plonsey and Barr, 2007), also as 26 V at "room temperature" with a natural logarithm in (Christensen, 2009, p. 178). The membrane voltage V_m is in the range of ± 100 mV. The potentials are associated with variable conductances, g, which model the biological action of the cell membrane in allowing the transit of ions in one direction or another.

2.12   EXAMPLES — POTASSIUM, SODIUM, AND CHLORINE IONS

With the potassium ion K^+, let

$$[S]_o = 4 \ \mu M/cm^3$$
$$[S]_i = 155 \ \mu M/cm^3$$
$$V_m = +61 \log \frac{4}{155} = -97 \, mV$$

With the sodium ion Na^+, let

$$[S]_o = 145 \ \mu M/cm^3$$
$$[S]_i = 12 \ \mu M/cm^3$$
$$V_m = +61 \log \frac{145}{12} = 66 \, mV$$

With the chlorine ion Cl^-, let

$$[S]_o = 120 \ \mu M/cm^3$$
$$[S]_i = 4 \ \mu M/cm^3$$
$$V_m = -61 \log 4 = 90 \, mV$$

When cells are exposed to external electrical fields, the ionic equilibria are disturbed, as the external fields move the ions across the cellular membrane in opposition to the normal cellular mechanisms. Even small electrical shocks can cause permanent injuries that are not always apparent initially.

2.13   CABLE RESISTANCE

Electrical cables play a major part of the electrical power system. The impedance of power cables is a major factor which limits the severity of short circuits and their related hazards (Sutherland, 2012). The resistance values for cables are the DC resistance of the conductor adjusted for skin effect at the frequency of the transients (Paul, 2008). Resistance per unit length for a conductor with a circular cross section can be calculated in the ideal case as

$$r_{dcw} = \frac{\rho}{A} = \frac{1}{\sigma \pi r_w^2} \ \Omega/m \tag{2.33}$$

where

$$r_{dcw} = \text{DC resistance of the conductor in } \Omega/\text{m}$$
$$A = \text{cross-sectional area of the conductor in m}^2$$
$$r_w = \text{radius of the conductor in m}$$
$$\rho = 1.673 \times 10^{-8}\,\Omega\,\text{m} = \text{resistivity of copper at } 20\,°\text{C, and}$$
$$\sigma = 1/\rho = 5.977 \times 10^7\,\text{S/m} = \text{conductivity of copper at } 20\,°\text{C}.$$

The appropriate calculated or table value of DC resistance, r_{dcw}, should be used for the particular cable in use, taking into account such factors as the type of copper, stranding, and coating. (Southwire®, 2005) Ambient temperature correction factors should be applied as necessary.

The skin depth is

$$\delta(f) = \sqrt{\frac{1}{\pi f \mu_0 \sigma}} \; \text{m} \tag{2.34}$$

where

$$\delta(f) = \text{skin depth in m}$$
$$f = \text{frequency, in Hz, and}$$
$$\mu_0 = 4\pi \times 10^{-7}\,\text{H/m} = \text{permeability of free space.}$$

The skin depth of a conductor is inversely proportional to the square root of frequency. For a copper conductor,

$$\delta(f) = \frac{0.0655}{\sqrt{f}} \; \text{m} \tag{2.35}$$

as shown in Figure 2.19 and Table 2.4.

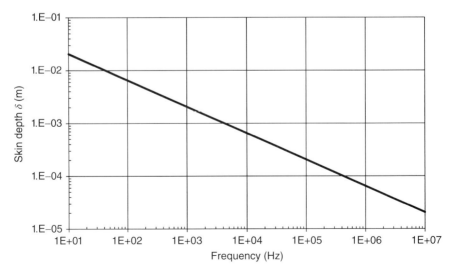

Figure 2.19   Skin depth, δ, of a copper conductor.

TABLE 2.4   Skin Depth, δ, of a Copper Conductor

Frequency, f (Hz)	Skin Depth, δ (m)	Frequency, f (Hz)	Skin Depth, δ (m)
10	2.06E−02	1E+04	6.51E−04
20	1.46E−02	2E+04	4.60E−04
50	9.21E−03	5E+04	2.91E−04
100	6.51E−03	1E+05	2.06E−04
200	4.60E−03	2E+05	1.46E−04
500	2.91E−03	5E+05	9.21E−05
1000	2.06E−03	1E+06	6.51E−05
2000	1.46E−03	2E+06	4.60E−05
5000	9.21E−04	5E+06	2.91E−05

The frequency for a given skin depth is

$$f = \frac{1}{\delta(f)^2 \pi \mu_0 \sigma} \text{ Hz} \tag{2.36}$$

The exact formula for the ratio of r_{ac} conductor AC resistance to r_{dc} conductor DC resistance is a solution for the diffusion equation involving the use of Bessel functions. However, the following approximation may be made:

$$r_{acw} \cong \begin{cases} r_{dcw} & \text{if } r_w < 2\delta \\ r_{dcw} \dfrac{r_w \sqrt{\pi \mu_0 \sigma}}{2} \sqrt{f} & \text{if } r_w > 2\delta \end{cases} \quad \Omega/\text{m} \tag{2.37}$$

For the second case, where the skin effect is operative, the AC resistance is a function of the square root of the frequency. The factor $r_{acw}/\sqrt{f}$ can be used to calculate the AC resistance:

$$\frac{r_{acw}}{\sqrt{f}} = r_{dcw} \frac{r_w \sqrt{\pi \mu_0 \sigma}}{2} \quad \text{for } r_w > 2\delta \tag{2.38}$$

The breakpoint in equation (2.37) occurs when the radius of the conductor equals twice the skin depth:

$$r_w = 2\delta(f) \text{ m} \tag{2.39}$$

A comparison of the radii of conductors and sheaths versus $\delta(f)$ and r_w is shown in Figure 2.20.

Combining (2.34) and (2.39), the conductor may be said to be at the skin-depth transition frequency

$$f_{tw} = \frac{4}{r_w^2 \pi \mu_0 \sigma} \text{ Hz} \tag{2.40}$$

The skin-depth transition frequency for copper conductors with a circular cross section is

$$f_{tw} = \frac{0.01695}{r_w^2} \text{ Hz} \tag{2.41}$$

Skin-depth transition frequencies, DC resistance, and the factor $r_{acw}/\sqrt{f}$ are shown in Table 2.5 for some common sizes of copper conductors.

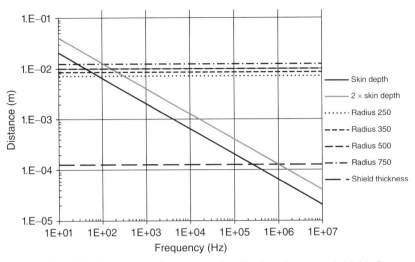

Figure 2.20   Skin depth versus frequency and radii of conductors and shields for several common cable sizes (kcmil).

TABLE 2.5   Resistance of Copper Conductors

Size (AWG/kcmil)	r_w (m)	f_{tw} (Hz)	r_{dcw} (Ω/m)	$r_{acw}/\sqrt{f}$ (Ω/m)
2	0.0036	1311.7	5.31E−04	1.47E−05
1	0.0041	1013.4	4.23E−04	1.33E−05
1/0	0.0046	802.9	3.35E−04	1.18E−05
2/0	0.0051	640.4	2.66E−04	1.05E−05
3/0	0.0058	505.7	2.11E−04	9.38E−06
4/0	0.0065	401.2	1.67E−04	8.34E−06
250	0.0071	337.7	1.41E−04	7.67E−06
350	0.0084	240.2	1.01E−04	6.52E−06
500	0.0100	168.3	7.08E−05	5.46E−06
750	0.0123	112.1	4.72E−05	4.46E−06
1000	0.0142	84.2	3.54E−05	3.86E−06

This approximation is not used in calculating the power frequency resistance of conductors for power-system studies such as short-circuit and load-flow analysis. However, it can be used for transient studies above the transition frequency.

For high frequency transients, the frequency of interest may vary between 10 kHz and several megahertz, depending on the mode of the oscillation. Once the frequency of oscillation is determined, the resistance for the simulation may be calculated using (2.38).

At high frequencies, the cable acts as a coaxial cable and the return current flows through the shield. The currents being considered are capacitive ground currents through the cable and transformer capacitances and not the load currents which are phase to phase and do not involve ground. If the shield consists of a conducting tape of fixed thickness, its DC resistance may be calculated as follows:

$$r_{dcs} = \frac{\rho}{A} = \frac{1}{\sigma \pi [(r_s + t_s)^2 - r_s^2]} \, \Omega/\text{m} \tag{2.42}$$

where

r_s = average radius of the shield in m and

t_s = thickness of the shield in m

For higher frequencies, current flows only within the inner surface of the shield. The DC resistance equals the AC resistance for all frequencies below where the skin depth equals the thickness of the shield:

$$t_s = \delta(f) \, \text{m} \tag{2.43}$$

The skin-depth transition frequency for the shield is

$$f_{ts} = \frac{1}{t_s^2 \pi \mu_0 \sigma} \, \text{Hz} \tag{2.44}$$

The skin-depth transition frequency for copper shield conductors is

$$f_{ts} = \frac{0.004238}{t_s^2} \, \text{Hz} \tag{2.45}$$

For a typical copper shield of 5 mils thickness or 0.127 mm, $f_{ts} = 263\,\text{kHz}$. When the transition to AC resistance is made, the resistance changes from being fixed by the thickness of the shield to varying with the skin depth by $\sqrt{f}$:

$$r_{acs} \cong \begin{cases} r_{dcs} & \text{if} \quad t_s < \delta \\ r_{dcs} t_s \sqrt{f \pi \mu_0 \sigma} & \text{if} \quad t_s > \delta \end{cases} \, \Omega/\text{m} \tag{2.46}$$

Then,

$$\frac{r_{acs}}{\sqrt{f}} = r_{dcs} t_s \sqrt{\pi \mu_0 \sigma} \quad \text{if} \quad t_s > \delta \tag{2.47}$$

The factor $r_{acs}/\sqrt{f}$ which can be used to calculate the AC resistance is shown in Table 2.6 for the shields of some common sizes of copper conductors.

TABLE 2.6   Resistance of 5 mil (0.127 mm) Copper Shield

Size (AWG/kcmil)	r_s (m)	f_{ts} (Hz)	r_{dcs} (Ω/m)	$r_{acs}/\sqrt{f}$ (Ω/m)
2	0.01051	2.63E+05	1.98E−03	3.87E−06
1	0.01102	2.63E+05	1.89E−03	3.69E−06
1/0	0.01153	2.63E+05	1.81E−03	3.53E−06
2/0	0.01204	2.63E+05	1.73E−03	3.38E−06
3/0	0.01267	2.63E+05	1.65E−03	3.21E−06
4/0	0.01337	2.63E+05	1.56E−03	3.05E−06
250	0.01407	2.63E+05	1.48E−03	2.90E−06
350	0.01540	2.63E+05	1.36E−03	2.65E−06
500	0.01699	2.63E+05	1.23E−03	2.40E−06
750	0.01937	2.63E+05	1.08E−03	2.10E−06
1000	0.02121	2.63E+05	9.86E−04	1.92E−06

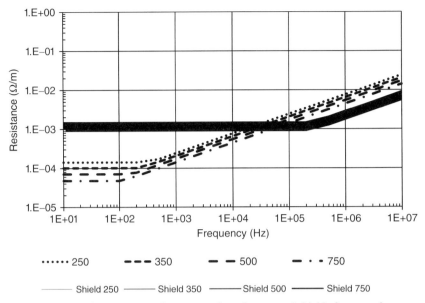

Figure 2.21   Resistance versus frequency of conductors and shields for several common cable sizes (kcmil).

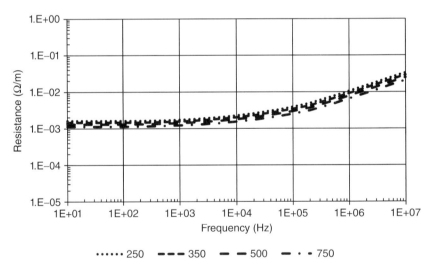

Figure 2.22   Conductor plus shield resistance.

The shield resistances are plotted in Figure 2.21, and are seen to be constant up to the shield skin-depth transition frequency. The conductor resistance goes to the load, and the shield resistance is the ground return, so that the total resistance is the sum of the two. The total resistance is plotted in Figure 2.22 and is dominated by the shield resistance.

CHAPTER **3**

CAPACITANCE PHENOMENA

3.1 FUNDAMENTALS OF CAPACITANCE

We are all familiar with the capacitance effect of the human body. The phenomenon of building up a static charge when walking on a carpet illustrates the charge-storing abilities of human body capacitance. Many electronic devices, such as touch-screen displays, touch-sensitive switches, and computer touch pads make use of this property. In terms of electrical hazards, the capacitance is a property which must be considered, along with inductance and resistance, in evaluating the effect of electrical currents on the human body. Capacitors also present a significant electrical safety hazard in their ability to store charge. Systems and equipment which use capacitors must be designed in such a way that this stored charge does not present a hazard to workers.

Capacitance is a property where an insulator opposes a change in the amount of electrical potential, due to the electric field produced by the potential. Since the human body is relatively small compared with the rate of change of power frequency potentials, the capacitance effect is relatively small. Nonetheless, it can have a significant effect at higher frequencies. However, devices which have a large capacitance may be hazardous owing to the stored charge which they may contain.

Electrical energy is stored in the electric field, and in monopoles (electrons and ions) and dipoles (polarized molecules) within the material, all of which must be factored into the total relationship between voltage and current. The dielectric losses may be considered as a bulk resistance and analyzed as previously discussed for resistors. For the initial discussion, the resistance will be assumed to be infinite, so that the capacitive effects may be isolated and analyzed separately. A real object, such as a body part, contains resistive, capacitive, and inductive components.

Capacitors as with resistors are not only a material property, but can be electronic components or power equipment. Capacitors are used in electronics for a wide variety of purposes. A few examples are shown in Figure 3.1. Capacitors may be used for blocking the flow of DC and only permitting the flow of AC; this application is called a coupling capacitor. A coupling capacitor might be applied in series with the input of an audio amplifier. Capacitors are widely used for filtering out AC components of DC voltages, where they are connected in shunt with the DC source. Bypass capacitors are small capacitors for filtering out high frequencies. Filter capacitors are large capacitors for smoothing out low frequencies, widely used in power supplies. Often, a large and a small capacitor may be applied in parallel. Capacitors are used

Principles of Electrical Safety, First Edition. Peter E. Sutherland.
© 2015 John Wiley & Sons, Inc. Published 2015 by John Wiley & Sons, Inc.

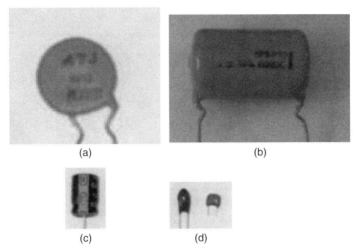

(a)   (b)

(c)   (d)

Figure 3.1   Capacitors used in electronic applications. (a) Ceramic disk capacitor. (b) Film capacitor. (c) Electrolytic capacitor. (d) Tantalum capacitors.

in a wide variety of filter circuits, such as low pass, high pass, band-pass, and many more elaborate designs. The uses of capacitors in electronics are virtually unlimited, and there are thousands of types available. Pictured here in Figure 3.1a is the ubiquitous ceramic disk capacitor, which generally has a low capacitance value, usually in the pico farad range, (this one has a capacitance of 47 pF) and can have a voltage rating from 100 V up to several kilovolts. Capacitors made of a roll of foil separated by an insulating material are used for higher capacitance values. Figure 3.1b is a 0.1 μF, 600 VDC polyester film capacitor encapsulated in a hard resin case. Earlier versions of this type of capacitor were made with paper insulation and dipped in wax for protection against moisture. When large capacitance values are needed, electrolytic capacitors are normally used as shown in Figure 2.1(e) and Figure 3.1c. These are polarized, and can carry a high capacitance value (10's to 1000's of μF) at voltages from low DC values to hundreds of volts, in a small package. Tantalum capacitors, Figure 3.1d, are widely used in computer circuits owing to their small size and low series inductance. They are available in both bypass and filter capacitor forms.

In terms of hazards to personnel, capacitors with a high capacitance value present the greatest danger due to stored charges. Even at low voltages, a discharge can produce a high current flow. Unless there is a parallel resistance to discharge the capacitor, the stored charge may last for a long time after the equipment is de-energized. Since the stored energy is $^{1}/_{2}CV^{2}$ as the voltage increases, the amount of capacitance which presents a danger decreases. Capacitors should always be discharged through a resistor, not by applying a short circuit, in order to avoid the dangers of arcing.

Capacitors used in electric power applications are typified by the rectangular cans with one or two bushings. The stack rack configuration Figure 3.2a is often used in substations. The stack rack configuration allows parallel and series combinations to be constructed, typically for application in substations. Capacitors may also

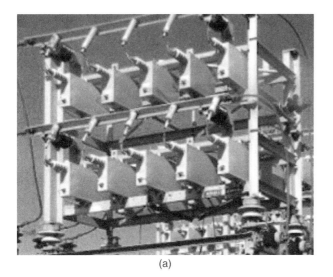

(a)

(b)

Figure 3.2   (a) Stack rack capacitors used in electric power systems.
(b) Distribution capacitor bank with switches.

be mounted on poles for applications along distribution lines, Figure 3.2b (Crudele, Short, and Sutherland, 2006; EPRI, 2005). Power capacitors are generally constructed of rolls of aluminum foil separated by polyethylene or similar plastic film, and then impregnated in oil. The rectangular cans at high voltage will contain just such rolled capacitors. At lower voltages, they may contain an array of smaller capacitors interconnected to make up the required value. Power capacitors are generally specified in kVAr and voltage, rather than in μF, although both values may be given.

These capacitors are typically used for power factor correction and harmonic filters. When capacitors are used on power distribution lines, they also provide the function of voltage support. Capacitor banks are often switched due to changes in load, power factor, or voltage. Banks can be switched as a whole, or in steps. Power capacitors are used in series with long, high voltage AC transmission lines to reduce the inductance effect. Smaller capacitors, called surge capacitors, are used to protect motors, generators, and transformers against high voltage transients. In power conversion systems, such as variable frequency drives (VFDs) for motor speed control, capacitors are used to store energy on an internal DC bus.

There are a variety of power electronic devices which provide continuous variation in capacitance at a faster speed than is possible with switching. These include the STATCOM or static compensator, the SVC or static VAR controller, the TCSR or thyristor-controlled shunt reactor which controls capacitance by varying a parallel reactor. Power capacitors are required to have internal discharge resistors (IEEE, 2002c), which discharge a capacitor of 600 V or less in no more than 1 min, and of greater than 600 V in no more than 5 min. The internal discharge resistor is not a substitute for discharging the capacitor through an external resistor before performing work on the bank.

3.2 CAPACITANCE AND PERMITTIVITY

The capacitance of the cylindrical geometry shown in Figure 3.3, is defined as

$$C = \frac{\varepsilon A}{\ell} = \frac{\varepsilon \pi r^2}{\ell} \tag{3.1}$$

where

$A =$ area of one of two identical conductive plates (m^2)

$\ell =$ distance between the surfaces of the two plates (m)

$\varepsilon =$ electrical permittivity of the material between the plates in farads/meter (F/m).

$\varepsilon =$ product of two components, $\varepsilon = \varepsilon_0 \varepsilon_r$. They are

$\varepsilon_0 =$ permittivity of free space $\approx 10^{-9}/36\pi$ F/m

$\varepsilon_r =$ relative permittivity of the material (unitless).

The energy that is stored in a capacitor can be analyzed by drawing the Poynting surface, S_1, around one of the end plates. The energy will enter one side of the plate, and exit the other side into the dielectric. Similarly, at the other plate, the energy will

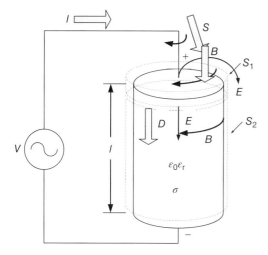

Figure 3.3   Capacitive energy storage.

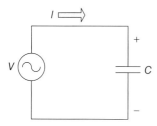

Figure 3.4   Circuit model of a lossless capacitive circuit.

exit around the other conductor. Since the conductance, σ, is zero, and the capacitor is lossless, the circuit model in figure 3.4 is very simple.

$$P_{in} = -\int_{S_1} (\mathbf{E} \times \mathbf{H}) \cdot dS_1$$
$$= -\pi a^2 EH + \pi a^2 EH = 0 \tag{3.2}$$

If the surface is taken around the dielectric,

$$P_{in} = -\int_{S_2} (\mathbf{E} \times \mathbf{H}) \cdot dS_2$$
$$= \ell\,(\pi a^2 EH + \pi a^2 EH) = \ell\, 2\pi a^2 EH = \ell\, 2A \frac{V}{\ell} \frac{I}{2A} = VI \tag{3.3}$$

In a pure capacitive volume, there is infinite resistivity, or zero conductivity, and thus no current flow. However, the electromagnetic energy must transition from one plate to another, so a vector quantity called the displacement current, $\mathbf{D}$, is postulated:

$$D = \varepsilon \mathbf{E}$$
$$= \varepsilon_0 \varepsilon_r \mathbf{E} \tag{3.4}$$

The relative permittivity of some common materials is listed in Table 3.1, and of some common body parts in Table 3.2.

TABLE 3.1 Relative Permittivity (ε_r) of Some Common Materials (at Room Temperature)

Material	ε_r
Copper	1.0
Vacuum	1.0
Teflon	2.1
Paraffin wax	2.2
Polyethylene	2.2
Plexiglas	3.4
Borosilicate glass	4.0
Ruby mica (muscovite)	5.4
Polyvinyl chloride	6.6
Pure water, sea water, fresh water	80

TABLE 3.2 Relative Permittivity (ε_r) of Some Body Parts (at Power Frequency)

Body Parts	ε_r
Skin (dry)—corneum	283
Bone	>3800
Fat	1.5×10^5
Muscle (perpendicular)	3.2×10^5
Lung	4.5×10^5
Liver	8.5×10^5
Muscle (parallel)	1.1×10^6

The displacement current is the subject of one of the most famous equations in physics, Maxwell's displacement current correction of Ampère's law, which is the cornerstone of Maxwell's equations.

$$\nabla \times \mathbf{H} = \mathbf{J}_f + \frac{\partial \mathbf{D}}{\partial t} \tag{3.5}$$

This demonstrates that while the conventional current flow produces a magnetic field, the change in displacement current also produces a magnetic field. It is changing displacement which causes energy flow through a capacitor. A capacitor is normally thought as providing energy storage in an electric field. This is true in the static case. In the dynamic case, it provides energy flow between two conducting materials, similar to a radio transmission. Each plate of the capacitor may be considered as an antenna.

When the dielectric materials contain polarized molecules (dipoles), the polarization may be caused by field cancellation from the capacitor plates. In addition, charged particles (ions) may be present within the material.

$$\mathbf{D} = \varepsilon \mathbf{E} = \varepsilon_0 \mathbf{E} + \mathbf{P} \qquad (3.6)$$

Thus the relative permittivity relates the polarizability to the electric field strength:

$$\varepsilon_r = 1 + \chi_e \qquad (3.7)$$

where

χ_e = electric susceptibility.

3.3   CAPACITANCE IN ELECTRICAL CIRCUITS

Discrete capacitors are widely used in electrical and electronic circuits, and often present safety hazards in their own right. In a simple DC circuit, capacitors can be charged through a resistor (Figure 3.5) at an exponential rate. The circuit equations are

$$V_1(t) = R \cdot I(t) + V_C(t)$$

$$\text{here} \quad I(t) = C \frac{dV_C}{dt}$$

$$\text{then} \quad V_1(t) = R \cdot C \frac{dV_C}{dt} + V_C(t) \qquad (3.8)$$

Solving the differential equation:

$$V_C(t) = V_1(1 - e^{-t/\tau})$$

$$\text{where} \quad V_C(t = 0) = 0$$

$$\text{and} \quad \tau = RC \qquad (3.9)$$

The result is the familiar capacitor charging exponential, as shown in Figure 3.6. Similarly, the discharge circuit is shown in Figure 3.7 and its curve is shown in Figure 3.8.

The equation for the discharge is calculated similarly as for the charging, and results in

$$V_C(t) = V_C(t = 0)e^{-t/RC} \qquad (3.10)$$

With an inductor in place of the resistor, the familiar series resonant circuit is constructed (Figure 3.9). The resonant frequency is

$$\omega_0 = \frac{1}{LC} \qquad (3.11)$$

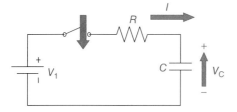

Figure 3.5   Capacitor charged from a DC source.

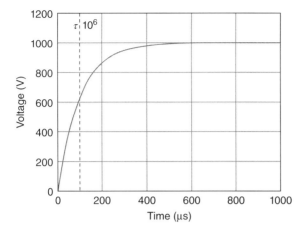

Figure 3.6   Voltage on $1.0\,\mu F$ capacitor charged through a $100\,\Omega$ resistor to $1\,kV$ with $t = 100\,ms$.

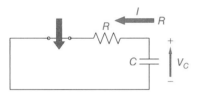

Figure 3.7   Discharge of capacitor charged from a DC source.

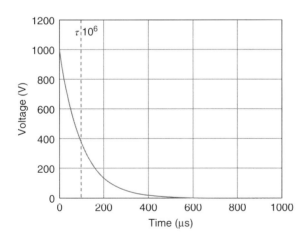

Figure 3.8   Voltage on $1.0\,\mu F$ capacitor discharged from $1\,kV$ through a $100\,\Omega$ resistor with $t = 100\,ms$.

while the voltage across the capacitor can be expressed as

$$V_C(t) = V_1(1 - \cos \omega_0 t) + V_C(t = 0)\cos \omega_0 t \qquad (3.12)$$

as shown in Figure 3.10.

   The switching of capacitive loads, as in Figure 3.11, causes transient oscillations with high di/dt.

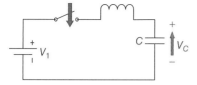

Figure 3.9   Oscillatory LC circuit.

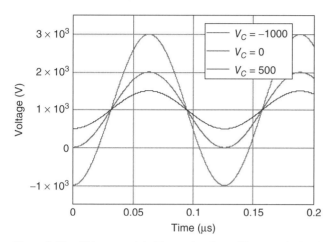

Figure 3.10   Voltage on $1.0\,\mu F$ capacitor in oscillatory LC circuit with $L = 20\,mH$ and $\omega_0 = 5.010 \times 10^{-7}$.

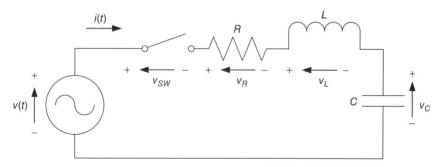

Figure 3.11   Series RLC circuit.

The voltage source is sinusoidal or

$$v(t) = V_p \sin(\omega t) \tag{3.13}$$

where the peak voltage is:

$$V_p = \sqrt{2} V_{LL} \tag{3.14}$$

Using the definitions of the voltages across resistors, inductors, and capacitors, we have the classical differential equation for a series RLC circuit.

The inductor and capacitor are energy storage devices. The current through the inductor is zero with no voltage applied. Since the current in an inductor cannot change instantaneously, the current is zero both immediately before and after time $t = 0$.

$$i_L(0^-) = i_L(0^+) = 0\,\text{A} \tag{3.15}$$

The voltage across the capacitor is zero if it is discharged before the start of the switching operation. Since the voltage in a capacitor cannot change instantaneously, the voltage is zero both immediately before and after time $t = 0$.

$$v_c(0^-) = v_C(0^+) = 0\ \text{V} \tag{3.16}$$

Using the definitions of the voltages across resistors, inductors, and capacitors, we have the classical differential equation for a series RLC circuit:

$$v(t) = v_R(t) + v_L(t) + v_c(t)$$
$$= Ri(t) + L\frac{di(t)}{dt} + \frac{1}{C}\int i(t)dt \tag{3.17}$$

This equation can be solved by a number of means, giving a generic solution of

$$i(t) = A_1 e^{s_1 t} + A_2 e^{s_2 t} \tag{3.18}$$

where

$$s_1 = -\alpha + \beta, \quad s_2 = -\alpha - \beta \tag{3.19}$$

$$\alpha \equiv \frac{R}{2L}, \quad \omega_0 \equiv \frac{1}{\sqrt{LC}}, \quad \text{and} \quad \beta \equiv \sqrt{\alpha^2 - \omega_0^2} \tag{3.20}$$

The form of the solution is determined by the relation between α and ω_0.

If $\alpha > \omega_0$, the system is described as being *overdamped*. The solution to the differential equation becomes: $i(t) = e^{-\alpha t}(A_1 e^{\beta t} + A_2 e^{-\beta t})$.

If $\alpha = \omega_0$, the system is described as being *critically damped*. The solution to the differential equation becomes: $i(t) = e^{-\alpha t}(A_1 + A_2 t)$.

If $\alpha < \omega_0$, the system is described as being *underdamped*. The solution to the differential equation becomes $i(t) = e^{-\alpha t}(A_1 \cos(|\beta|t) + A_2 \sin(|\beta|t))$ or more conveniently, $i(t) = e^{-\alpha t}A_3 \sin(|\beta|t + \phi)$

Supposing that the values of R, L, and C are

$$R = 14\ \Omega$$

$$L = 12.0\ \mu\text{H}$$

$$C = 0.08\ \mu\text{F}$$

then the resonant frequency is

$$\omega_0 \equiv \frac{1}{\sqrt{LC}} = \frac{1}{\sqrt{12 \times 10^{-6} \times 0.08 \times 10^{-6}}} = 1.021 \times 10^6$$

$$\text{giving} \quad f_0 = \frac{\omega_0}{2\pi} = 162.4\ \text{kHz}$$

Then,

$$\alpha \equiv \frac{R}{2L} = \frac{0.01 \times 87 + 21.728}{2 \times 6.4 \times 10^{-6}} = 1.755 \times 10^6$$

$$\text{and} \quad \beta \equiv \sqrt{\alpha^2 - \omega_0^2} = j|\beta| = j|\sqrt{(1.755 \times 10^6)^2 - (2.059 \times 10^6)^2}|$$

$$= j \cdot 1.077 \times 10^6$$

$\alpha < \omega_0$ because $1.755 \times 10^6 < 2.059 \times 10^6$. The system is underdamped. Initial voltages are $v(0^-) = 0\,\text{V}$ and $v(0^+) = V_m$

At time $t = 0^+$, the loop equation

$$v(t) = v_{SCR}(t) + v_R(t) + v_L(t) + v_c(t)$$

becomes

$$V_p = 0 + 0 + L\frac{di(t)}{dt}\bigg|_{0^+} + 0 \quad \text{or} \quad \frac{di(t)}{dt}\bigg|_{0^+} = \frac{V_p}{L} = \frac{5883\ \text{V}}{6.4 \times 10^{-6}\ \text{H}} = 919\ \text{A/μs}$$

At time $t = 0^+$ the underdamped solution equation is

$$i(0^+) = 0 = e^{-\alpha t}A_3 \sin(|\beta|t + \phi) = A_3 \sin(\phi)$$

Since $A_3 \neq 0$ or else there would not be a solution, and $\sin(0) = 0$, it follows that $\phi = 0$.

The derivative of the underdamped solution equation is:

$$\frac{di(t)}{dt} = A_3[e^{-\alpha t}|\beta| \cos(|\beta|t + \phi) - e^{\alpha t}\alpha \sin(|\beta|t + \phi)]$$

$$\frac{di(t)}{dt}\bigg|_{0^+} = A_3|\beta| = 2.842 \times 10^6 A_3 = 919\ \text{A/μs}$$

Thus,

$$A_3 = -\frac{919 \times 10^6}{1.077 \times 10^6} = 854$$

A plot of the solution equation

$$i(t) = 854.1e^{-1.755 \times 10^6 t} \sin(2.059 \times 10^6 t)$$

is shown in Figure 3.12.

3.4 CAPACITANCE OF BODY PARTS

Capacitance in the human body is primarily present between the layers of skin $(0.02-0.06\ \text{μF/cm}^2)$, and between the body as a whole and the earth.

3.4.1 Example — Skin Capacitance

Skin capacitance resides primarily in the dried layer of dead cells called the corneum, which is the normal surface layer. The thickness of the corneum is in the range $10-20\ \text{μm}$, and it typically consists of approximately 200 layers of cell membranes,

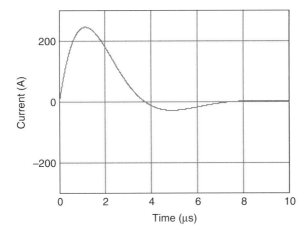

Figure 3.12   Oscillatory current transient in series RLC circuit with $L = 12\,\mu\text{H}$, $C = 0.08\,\mu\text{F}$, and $R = 14\,\Omega$.

which would each have a thickness of $0.05\,\mu\text{m}$. The capacitance of a single membrane is in the range of $2\text{–}5\,\mu\text{F/cm}^2$. This allows the relative permittivity to be calculated:

$$\varepsilon_r = \frac{\ell C}{\varepsilon_0 A} = \frac{36\pi}{10^{-9}}\frac{0.05 \times 10^{-6} \times 5 \times 10^{-6}}{10^{-4}} = 283 \ \text{F/m}$$

Using the data from the skin with an electrode, the area $A = 10^{-4}\,\text{m}^2$, and depth $\ell = 10\ \mu\text{m}$, assuming $\varepsilon_r = 283$.

$$C = \frac{\varepsilon_0 \varepsilon_r A}{\ell} = \frac{10^{-9}}{36\pi}\frac{283 \times 10^{-4}}{10 \times 10^{-6}} = 0.025 \ \mu\text{F}$$

This is in the accepted range of $0.02\text{–}0.06\,\mu\text{F/cm}^2$. At 60 Hz, the capacitive reactance is

$$X_C = \frac{1}{2\pi fC} = \frac{1}{2\pi \times 60 \times 0.025 \times 10^{-6}} = 106 \ \text{k}\Omega$$

The equivalent circuit of the skin can then be completed (Figure 3.13). For AC current, the capacitance will dominate R_P, resulting in lower skin impedance.

3.4.2   Example — Capacitance of Trunk and Limb

The trunk of the body from the previous example has a capacitance of

$$C = \frac{\varepsilon_0 \varepsilon_r A}{\ell} = \frac{10^{-9}}{36\pi}\frac{10^5 \times 0.064}{0.5} = 0.113 \ \mu\text{F}$$

At 60 Hz, the capacitive reactance is

$$X_C = \frac{1}{2\pi fC} = \frac{1}{2\pi \times 60 \times 0.113 \times 10^{-6}} = 23\,\Omega$$

The limb of the body from the previous example has a capacitance of flesh and bone. The capacitance of the bone is

$$C_B = \frac{\varepsilon_0 \varepsilon_r A}{\ell} = \frac{10^{-9}}{36\pi}\frac{3800 \times 1.26 \times 10^{-3}}{0.5} = 85\,\text{pF}$$

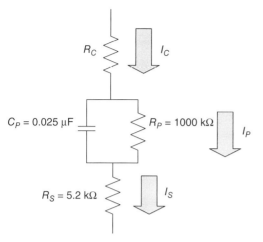

Figure 3.13   Example of skin impedance.

The capacitance of the flesh is

$$C_F = \frac{\varepsilon_0 \varepsilon_r A}{\ell} = \frac{10^{-9}}{36\pi} \frac{3.2 \times 10^5 \times 3.93 \times 10^{-3}}{0.5} = 0.022 \ \mu F$$

The total capacitance of the limb is

$$C_L = C_B + C_L = 0.022 \, \mu F$$

At 60 Hz, the capacitive reactance is

$$X_C = \frac{1}{2\pi f C} = \frac{1}{2\pi \times 60 \times 0.022 \times 10^{-6}} = 121 \ \Omega$$

3.5   ELECTRICAL HAZARDS OF CAPACITANCE

Power capacitors are used in electric power applications mainly for power factor correction and harmonic filters.

   Filter capacitors for DC power supplies act to reduce the remaining AC components in a pulsating DC voltage after the rectification process is completed. These are usually large capacitors in the hundreds to thousands of microfarads, at voltages from the single digits up to the kilovolt range. Filter capacitors cause several electrical hazards. The first, and most obvious, is the risk of electrical shock from a capacitor charged to a high voltage with a high stored energy. A filter capacitor will charge to the peak value of the sinusoidal voltage being rectified. For example, if the power transformer for a piece of vacuum-tube electronic equipment has a 500 V_{rms} center-tapped secondary, such that $E_s = E_{s1} = E_{s2} = 250 V_{(rms)}$, the full-wave rectified voltage will be

$$E_{d0} = 0.9 E_s = 0.9 \times 250 V_{rms} = 225 V_{dc}$$

However, the filter capacitor will charge to the peak value of voltage

$$V_p = \sqrt{2} \, E_s = \sqrt{2} \, 250 = 354 V_{dc}$$

If the capacitor has a size of 100 μF, then the charge on the capacitor is

$$Q = CV = 100 \ \mu F \times 354 V_{dc} = 0.035 \text{ C}$$

The energy stored in the capacitor is

$$W = \frac{1}{2}QV = \frac{1}{2}CV^2 = 6.3 \text{ J}$$

This may provide a noticeable shock. The second hazard is of high discharge currents if such a capacitor is accidentally shorted, such as by an uninsulated screwdriver or other tool.

DC Link capacitors for AC motor drives act to store energy from the AC system after it has been rectified, in order that the inverter has a constant voltage source. These are usually large capacitors in the hundreds to thousands of microfarads, at voltages from the hundreds of volts up to the kilovolt range. DC Link capacitors can cause several electrical hazards. The first, and most obvious, is the risk of electrical shock from a capacitor charged to a high voltage with a high stored energy. A filter capacitor will charge to the peak value of the sinusoidal voltage being rectified. For example, if the AC drive is powered by a 480 VAC system, and rectified by a three-phase bridge, the rectified voltage will be

$$E_{d0} = 1.35 E_L = 1.35 \times 480 V_{rms} = 648 V_{dc}$$

However, the filter capacitor will charge to the peak value of voltage

$$V_p = \sqrt{2} \ E_L = \sqrt{2} \ 480 V_{dc} = 679 \text{ V}$$

If the capacitor has a size of 4500 μF, then the charge on the capacitor is

$$Q = CV = 4500 \ \mu F \times 679 V_{dc} = 3.05 \text{ C}$$

The energy stored in the capacitor is

$$W = \frac{1}{2}QV = \frac{1}{2}CV^2 = 1037 \text{ J}$$

This is more than sufficient to provide a fatal discharge if touched, and a severe arc flash if accidentally shorted.

3.6 CAPACITANCE OF CABLES

The capacitance of a cable is a distributed quantity, and the longer the cable, the larger the capacitance (Sutherland, 2012). When cables are de-energized, they hold a charge for a long period of time unless properly discharged. It is important to know the capacitance of a cable and the voltage at which it was operated in establishing the proper safety precautions when working on it, even if de-energized. Capacitance, C, is calculated as

$$C = \frac{2\pi\varepsilon_0\varepsilon_r}{\ln r_s/r_w} \text{ F/m} \tag{3.21}$$

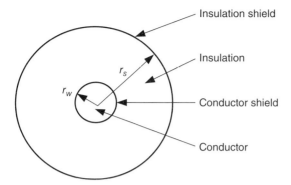

Figure 3.14   Shielded cable cross section showing radius of the wire and shield.

TABLE 3.3   Dielectric Constants of Some Cable Insulation Materials

Material	Description	Maximum Service Temperature (°C)	Dielectric Constant
PVC	Polyvinyl chloride	105	3.4
XLPE	Cross-linked polyethylene	90	5.0
MVXLPE	Medium voltage cross-linked polyethylene	90	2.3
EPR	Ethylene propylene rubber	90	2.5

where

r_s = radius of the shield

r_w = radius of the center conductor

$\varepsilon_0 = 8.857 \times 10^{-12}$ F/m, and

ε_r = (unitless) is the relative permittivity (dielectric constant) of the insulation.

The dimensions r_s and r_w are shown in Figure 3.14. Values of ε_r for some typical cable-insulating materials are shown in Table 3.3 (Southwire®, 2005). More detailed simulations will also include the permittivity and conductivity of the semi-conducting shields (Gustavsen, Martinez, and Durbak, 2005).

CHAPTER 4

INDUCTANCE PHENOMENA

4.1   INDUCTANCE IN ELECTRICAL THEORY

Inductance is a property where a conductor opposes a change in the amount of current flow, due to the magnetic field produced by the current flow. Since the human body is relatively small compared with the rate of change of power frequency currents, the inductance effect is relatively small. Nonetheless, it can have a significant effect at higher frequencies.

Inductance as a property of electrical systems, and specifically designed in a component such as an inductor, reactor, or transformer presents hazards to people who are working on the equipment or systems. The first hazard may be coming in contact with the energized conductors. Another important hazard is induced currents in conductive materials, which may cause electrical shocks, arcing, and heating. The effects of the magnetic fields produced by inductance can also affect human health, the most prominent example being biomedical devices such as pacemakers. For these and other hazards which may exist in particular circumstances, whenever an electrical equipment or system is designed, extreme care should be taken to prevent the inherent or designed inductances from causing safety hazards.

Definitions of inductance are a thorny matter, and inductance calculations can be extremely difficult. Inductance is defined as the ratio of flux linkages to current. Current is defined using Ampere's law and Maxwell's displacement current as

$$I = \int_S \mathbf{J} \cdot d\mathbf{S} + \frac{d}{dt} \int_S \mathbf{D} \cdot d\mathbf{S} = \oint_L \mathbf{H} \cdot d\mathbf{l} \tag{4.1}$$

where

$$\mathbf{D} = \varepsilon_0 \mathbf{E} + \mathbf{P} \tag{4.2}$$

is the displacement field, and $\mathbf{P}$ is the polarization. The magnetic flux is

$$\Phi = \int_S \mathbf{B} \cdot d\mathbf{S} \tag{4.3}$$

Thus,

$$L = \frac{\int_S \mathbf{B} \cdot d\mathbf{S}}{\int_L \mathbf{H} \cdot d\mathbf{l}} \tag{4.4}$$

Principles of Electrical Safety, First Edition. Peter E. Sutherland.

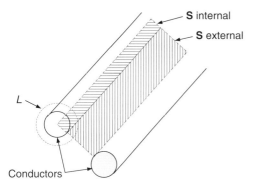

Figure 4.1   Surfaces of integration for internal and external inductance.

The Ampère's law enclosure, also labeled L, is a circle enclosing one conductor of the parallel line. The external self- and mutual inductances are defined using **S** as a rectangle bounded by the two conductors. The internal self-inductance is defined by a rectangle bounded by the axis and surface of a conductor, Figure 4.1.

For a conductor pair of separation D and radius a,

$$L = \frac{\mu_0}{\pi} \left[\ln \left(\frac{D}{a} \right) + \frac{1}{4} \right] \tag{4.5}$$

Alternatively, inductance can be calculated from magnetic energy. Inductive reactance is a measure of opposition to change in current. A tangential magnetic field **H** surrounds the axis of a conductor with a steady current I flowing. This field is divided into two parts, internal and external to the conductor. In the two-conductor system, I is the same in both conductors, but with opposing directions, and two fields are produced, $\mathbf{H}_1$ and $\mathbf{H}_2$. Any change in the current in the conductor will result in a change in the magnitude of the magnetic field everywhere. The energy stored in the magnetic field is

$$W = \frac{1}{2} \int_V \mu(\mathbf{H}_1 + \mathbf{H}_2)^2 dV \tag{4.6}$$

This will result in the familiar self and mutual inductances for a two-wire line.

$$W = \frac{1}{2}(L_1 + L_{12} + L_2)I^2 \tag{4.7}$$

The portions of the magnetic fields that produce these inductances and that which produce the Poynting vector are not exactly the same. The field inside a perfect conductor produces inductance, but does not contribute to power flow.

For these concepts to have meaning, there must be a finite amount of magnetic energy. This will be the case in all practical applications. It is sometimes stated that a linear single conductor has infinite inductance, and that finite inductance is only introduced when return conductors are considered. This is contradictory to the usual practice of referring to a discrete component called an "inductor" or "reactor," which is not a complete circuit. It is also contradictory to the practice of calculating the inductance of transmission lines, which are also components and not circuits. In the microelectronics industry, Ruehli's concept of "partial inductance" is used

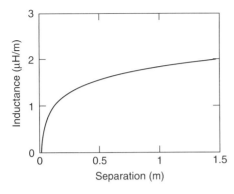

Figure 4.2 Inductance versus separation for an open-wire line.

(Ruehli, 1972). When the inductance of a power transmission line is calculated, this is, in essence, a "partial inductance."

4.2 INDUCTANCE OF WIRES

The inductance of an open-wire line increases as the conductors are separated (Figure 4.2). Infinite separation, as in a one-wire line, would mean infinite inductance. In practice, the increase in inductance with spacing is very gradual. The inductance calculation (3.5) assumes $\ell \gg D$, where ℓ is the conductor length. This is because the lines of flux are assumed to be uniform and parallel in the area S in the expression for flux (3.3). This may be a valid approximation for closely spaced conductors. However, for widely spaced conductors, the assumption of infinite length is no longer reasonable.

Using the Grover formula (Grover, 1946) for the inductance of a straight conductor does not require a return conductor:

$$L \approx \frac{\mu_0 \ell}{2\pi} \left[\ln \left(\frac{2\ell}{a} \right) - \frac{3}{4} \right] \text{ H} \tag{4.8}$$

This is an example of partial inductance. In order to model the complete circuit, the partial inductance of each element must be included, as well as the mutual inductance between each pair of elements. The partial inductances of series elements cannot be added without considering these mutual inductances.

4.3 EXAMPLE — INDUCTANCE OF A CONDUCTOR

For a conductor with radius $a = 20\,\text{mm} = 0.02\,\text{m}$, and length $\ell = 1\,\text{m}$,

$$L \approx \frac{\mu_0 \ell}{2\pi} \left[\ln \left(\frac{2\ell}{a} \right) - \frac{3}{4} \right] = \frac{4\pi \times 10^{-7} \times 1}{2\pi} \left[\ln \left(\frac{2 \times 1}{20 \times 10^{-3}} \right) - \frac{3}{4} \right] = 0.77\,\mu\text{H}$$

4.4   EXAMPLE — INDUCTANCE OF TRUNK AND LIMB

Assuming that a human trunk has a circular cross section, with radius of $a = 0.159$ m, and a length $\ell = 0.5$ m,

$$L \approx \frac{\mu_0 \ell}{2\pi} \left[\ln \left(\frac{2\ell}{a} \right) - \frac{3}{4} \right] = \frac{4\pi \times 10^{-7} \times 0.5}{2\pi} \left[\ln \left(\frac{2 \times 0.5}{0.159} \right) - \frac{3}{4} \right] = 0.109 \, \mu\text{H}$$

At a power frequency of 60 Hz, the inductive reactance is

$$X_L = 2\pi f L = 2\pi \times 60 \times 0.109 \times 10^{-6} = 41 \, \mu\Omega$$

Assuming that a limb has a circular cross section, with radius $a = 5$ cm $= 0.05$ m, and a length of $\ell = 0.5$ m, the inductance is

$$L \approx \frac{\mu_0 \ell}{2\pi} \left[\ln \left(\frac{2\ell}{a} \right) - \frac{3}{4} \right] = \frac{4\pi \times 10^{-7} \times 0.5}{2\pi} \left[\ln \left(\frac{2 \times 0.5}{0.05} \right) - \frac{3}{4} \right] = 0.225 \, \mu\text{H}$$

At a power frequency of 60 Hz, the inductive reactance is

$$X_L = 2\pi f L = 2\pi \times 60 \times 0.225 \times 10^{-6} = 85 \, \mu\Omega$$

Because the inductive reactance is several orders of magnitudes less than the resistance, it may be neglected in circuit analysis at power frequency. It is not necessary to refine the calculation for an elliptical as opposed to a circular cross section.

4.5   INDUCTORS OR REACTORS

An inductor or reactor is a construction of conductors designed to contain an induced magnetic field for temporary energy storage. The electric and magnetic fields over the inductor are parallel, resulting in no power transfer. Because power does not pass through an inductor, the Poynting vector follows the conductor through all its turns. When energy is stored in an inductor, the power flow is radial from the conductors into the magnetic field volume. The losses in an inductor, on the other hand, result from the same causes as those in a wire, and have a similar energy surface. Figure 4.3 shows examples of power reactors. The first is an outdoor current limiting reactor, with cast in concrete construction. Newer models use fiber glass epoxy construction. The second shows an indoor air core reactor with support by fiber glass sheeting.

4.6   SKIN EFFECT

The skin effect is not strictly a form of inductance, but a frequency effect on resistance. This is why demonstrations with Tesla coils of a person touching an extremely high voltage source are possible. The extremely high frequency will cause a small amount of current to flow over the surface of a person's skin and not go deep enough to cause any physical harm. Electrical current can only flow where a magnetic field

(a)   (b)

Figure 4.3   Power Reactors. (a) Cast in concrete reactor. (b) Modern fiberglass epoxy outdoor air core reactors.

is present within a material; in accordance with Ampère's law with Maxwell's displacement current correction,

$$\nabla \times \mathbf{H} = \mathbf{J}_f + \frac{\partial \mathbf{D}}{\partial t} \tag{4.9}$$

where conduction current is dominant in conductors and displacement current is dominant in a dielectric. The electrical field is related to the magnetic field through Faraday's law:

$$\nabla \times \mathbf{E} = -\frac{\partial \mathbf{B}}{\partial t} \tag{4.10}$$

When these equations are written for a wave along a plane surface, where $\mathbf{B}$ is directed along the x-axis and $\mathbf{E}$ along the y-axis, Figure 4.4, we have

$$\left.\begin{aligned} -\frac{\partial H_z}{\partial x} &= \sigma E_y + \frac{\partial E_y}{\partial t} \\ \frac{\partial E_y}{\partial x} &= -\mu \frac{\partial H_z}{\partial t} \end{aligned}\right\} \tag{4.11}$$

These can be combined to form the equation for a wave which is normal to the surface of the conducting medium.

$$\frac{1}{\mu}\frac{\partial^2 E_y}{\partial x^2} - \varepsilon \frac{\partial^2 E_y}{\partial t^2} - \sigma \frac{\partial E_y}{\partial t} = 0 \tag{4.12}$$

which reduces to

$$\frac{1}{\mu}\frac{\partial^2 E_y}{\partial x^2} - \gamma^2 E_y = 0 \tag{4.13}$$

and

$$\gamma^2 = j\omega\mu\sigma - \omega^2\mu\varepsilon \tag{4.14}$$

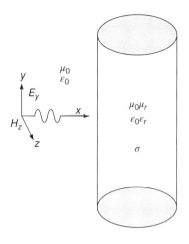

Figure 4.4   Electromagnetic wave entering a conducting medium.

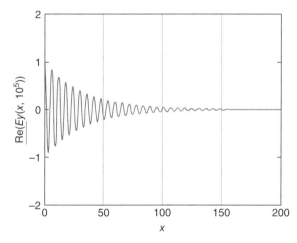

Figure 4.5   Magnitude of incident wave with $\omega = 10^5$, $E_0 = 1$, $\varepsilon_r = 10^7$, $\mu_r = 1.0$, $\sigma = 0$.

where

γ = complex propagation constant.

The wave entering the conductor will be an exponentially decaying sinusoid, Figure 4.5, of the form

$$E_y = E_0 e^{-\mathrm{Re}(\gamma)x} e^{-j\mathrm{Im}(\gamma)x} \tag{4.15}$$

where the real exponential term is the attenuation factor and the complex exponential term is the phase factor.

However, the propagation constant, γ, can be used to evaluate whether a material is a conductor, a dielectric, or a quasi-conductor at any given frequency. The ratio $\sigma/\omega\varepsilon$ is the ratio of how much of the current is conductive over how much of the current is displacement current, Figure 4.6. The dividing lines are that $\sigma/\omega\varepsilon > 100$ indicates a conductor, while $\sigma/\omega\varepsilon < 1/100$ indicates a dielectric, while the region between indicates a quasi-conductor. Biological materials can be considered conductors at power frequencies, quasi-conductors at kilohertz frequencies, and dielectrics at millihertz frequencies and above.

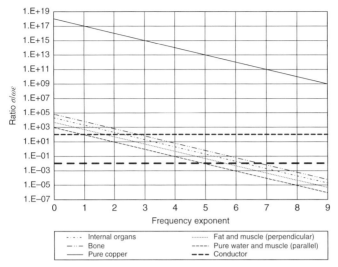

Figure 4.6   Ratio $\sigma/\omega\varepsilon$ for biological materials, copper shown for comparison. Materials are conductors above the "conductor" line, dielectrics below the "dielectric" line, and quasi-conductors in between.

The distance where the magnitude of the wave decreases to $1/e$ is called the skin depth, δ:

$$\delta = \mathrm{Re}(1/\gamma) \tag{4.16}$$

Figure 4.7 shows the skin depth for biological materials, compared to a typical radius of 0.5 m, illustrating the fact that the skin effect is not biologically significant

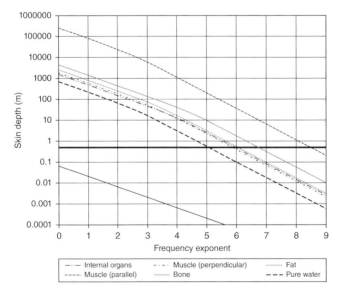

Figure 4.7   Skin depth for biological materials, compared to a radius of 0.5 m.

for frequencies of less than 0.1 mHz. This excludes commonly used power frequencies, which are less than 100 Hz. The skin effect also changes the internal inductance of a solid conductor, because the conduction occurs along the surface, making the inductance approach that of a tube rather than a cylinder. Exact calculation of skin effect resistance and inductance requires the use of the Bessel functions, which are necessary when any great precision is needed.

4.7   CABLE INDUCTANCE

Cables comprise a large portion of electrical circuits, so the inductance of cables is a major factor in limiting hazardous short-circuit current. Also, the methods used in calculating cable inductance are important in electrical engineering and can also be utilized on other types of conductors (Sutherland, 2012). For shielded power cables, as in Figure 4.8 (Southwire®, 2010), inductance is the sum of the internal inductance of the wire, the external inductance between the wire and the shield, and the internal inductance of the shield. For the capacitive ground currents being considered, the external inductance, l_e, is calculated from the dimensions of the cable:

$$l_e = \frac{\mu_0}{2\pi} \ln \left(\frac{r_s}{r_w} \right) \text{ H/m} \tag{4.17}$$

For normal phase currents, the distance between conductors would be substituted for the radius of the shield. The DC internal inductance of the wire for low frequencies is constant:

$$l_{dcw} = \frac{\mu_0}{8\pi} \text{ H/m} \tag{4.18}$$

At higher frequencies, the internal inductance l_{iw} has characteristics similar to resistance, as shown in Figures 4.9 and 4.10, exhibiting a skin effect also (Paul, 2010).

$$l_{iw} \cong \begin{cases} l_{dcw} & \text{if} \quad r_w < 2\delta \\ l_{dcw} \dfrac{1}{2r_w \sqrt{\pi \mu_0 \sigma}} \dfrac{1}{\sqrt{f}} & \text{if} \quad r_w > 2\delta \end{cases} \text{ H/m} \tag{4.19}$$

The internal inductance is generally a small percentage of the external inductance, and is often ignored. The third inductance in the cable is that of the internal inductance of

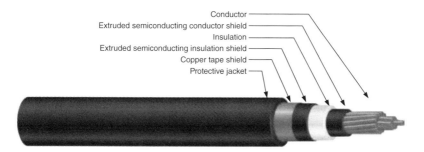

Figure 4.8   Single-conductor medium-voltage cable construction (Southwire®, 2010).

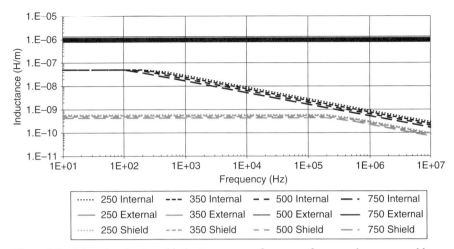

Figure 4.9   Internal and external inductance versus frequency for several common cable sizes (kcmil).

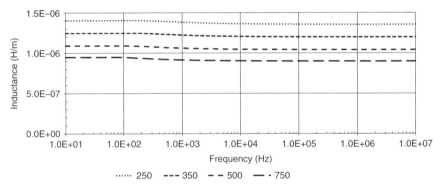

Figure 4.10   Total internal and external inductance versus frequency for several common cable sizes (kcmil).

the shield. Paul (2010) has given the calculation of this inductance by both the energy method and the method of flux linkages.

The shield transition frequency, f_{ts}, was calculated using (2.44) to be 263 kHz (Table 4.1). For frequencies above f_{ts}, the effective thickness of the shield goes from t_s to δ.

The internal DC inductance and the internal AC inductance of the shield above f_{ts} are approximated as

$$
l_{dcs} = \frac{\mu_0}{2\pi} \left[\frac{\left(r_s + t_s\right)^4 \ln\left[\dfrac{(r_s + t_s)}{r_s}\right] - (r_s + t_s)^2[(r_s + t_s)^2 - r_s^2] + \dfrac{1}{4}[(r_s + t_s)^4 - r_s^4]}{[(r_s + t_s)^2 - r_s^2]^2} \right]
$$

if $t_s > \delta$ H/m   (4.20)

TABLE 4.1   Inductance of 5 mil (0.127 mm) Copper Shield

Size (AWG/kcmil)	r_s (m)	$r_s + t_s$ (m)	f_{ts} (Hz)	l_{dcs} (H/m)
2	0.0105	0.0106	2.628E+05	8.056E−10
1	0.0110	0.0111	2.628E+05	7.686E−10
1/0	0.0115	0.0117	2.628E+05	7.346E−10
2/0	0.0120	0.0122	2.628E+05	7.035E−10
3/0	0.0127	0.0128	2.628E+05	6.682E−10
4/0	0.0134	0.0135	2.628E+05	6.335E−10
250	0.0141	0.0142	2.628E+05	6.020E−10
350	0.0154	0.0155	2.628E+05	5.498E−10
500	0.0170	0.0171	2.628E+05	4.985E−10
750	0.0194	0.0195	2.628E+05	4.371E−10
1000	0.0212	0.0213	2.628E+05	3.992E−10

$$l_{is}(f) = \frac{\mu_0}{2\pi} \left[\frac{(r_s + (f))^4 \ln\left[\frac{(r_s + (f))}{r_s}\right] - (r_s + (f))^2[(r_s + (f))^2 - r_s^2] + \frac{1}{4}[(r_s + (f))^4 - r_s^4]}{[(r_s + (f))^2 - r_s^2]^2} \right]$$

$$\text{if} \quad t_s < \delta \text{ H/m} \tag{4.21}$$

Below the transition frequency, where the skin depth is greater than the thickness of the shield, current distribution in the shield is uniform, and the DC shield inductance may be used. The shield inductance is plotted in Figure 4.9 and some of the factors used in the calculations are listed in Table 4.3. Because the internal inductances of the wire and shield are over an order of magnitude smaller than the external inductance of the wire, l_e in equation (4.16), this can be used as a good approximation for the cable inductance.

$$l_t \cong l_e = \frac{\mu_0}{2\pi} \ln \frac{r_s}{r_w} \text{ H/m} \tag{4.22}$$

4.8   SURGE IMPEDANCE

Surge impedance is the equivalent resistance of a distributed circuit, where the inductance and capacitance are defined as per unit length, rather than as discrete quantities (Sutherland, 2012). The surge impedance will determine the ratio of voltage and current for a transient wave traveling down the line. When there is an abrupt change in surge impedance, there will be reflected and refracted waves, often resulting in severe overvoltage conditions. The surge impedance of a transmission system is defined as

$$Z_0 = \sqrt{\frac{l}{c}}\,\Omega \tag{4.23}$$

TABLE 4.2 Typical Cable Resistance, Inductance, Capacitance, and Surge Impedance at 20 kHz

Size (AWG/kcmil)	R (Ω/m)	L (H/m)	C (F/m)	Z_0 (Ω)
2	4.068E−03	2.130E−06	1.296E−10	128.2
1	3.783E−03	1.967E−06	1.404E−10	118.4
1/0	3.491E−03	1.825E−06	1.512E−10	109.9
2/0	3.229E−03	1.686E−06	1.637E−10	101.5
3/0	2.982E−03	1.554E−06	1.776E−10	93.5
4/0	2.748E−03	1.430E−06	1.929E−10	86.1
250	2.576E−03	1.360E−06	2.028E−10	81.9
350	2.283E−03	1.202E−06	2.295E−10	72.4
500	2.006E−03	1.043E−06	2.643E−10	62.8
750	1.713E−03	9.010E−07	3.060E−10	54.3
1000	1.534E−03	7.973E−07	3.457E−10	48.0

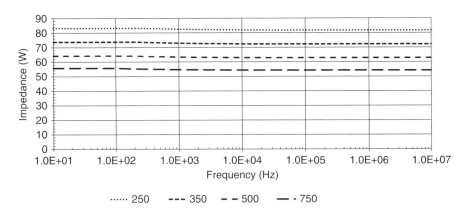

Figure 4.11   Surge impedance versus frequency for several common cable sizes (kcmil).

where

l = inductance in H/m and

c = capacitance in F/m.

Typical cable resistance, inductance, capacitance, and surge impedance at 20 kHz are shown in Table 4.2. Surge impedance versus frequency is fairly constant, Figure 4.11.

4.9   BUS BAR IMPEDANCE CALCULATIONS

The values of resistance are calculated or looked up in a reference (Copper Development Association, undated) using the area of the conductors and the resistivity of copper at 20 °C, skin and proximity effects.

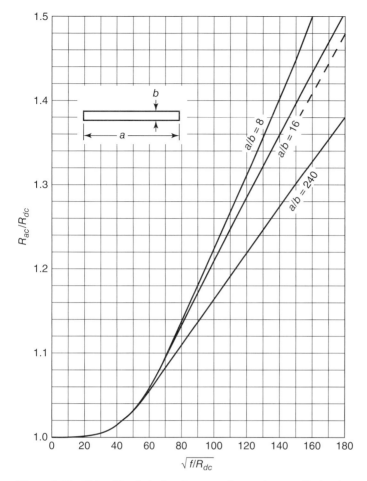

Figure 4.12   Skin effect in isolated rectangular conductors, R_{dc} in ohms per 1000 ft (Dwight, 1945, p. 225).

For example, with a resistivity of $1.72 \times 10{-}8\,\Omega\text{-m}$ @ 20 °C, a 6″ × ½″ copper bar has a DC resistance of $8.89 \times 10{-}6\,\Omega/\text{m}$ or $2.71 \times 10{-}3\,\Omega/1000\,\text{ft}$ @ 20 °C. This is divided by 2 if there are two parallel bars.

Correction for skin effect is shown in Figure 4.12 (Dwight, 1945, p. 225). the ratio $a/b = 6/0.5 = 12$, and at 60 Hz,

$$\sqrt{\frac{f}{R_{dc}}} = 149$$

By the figure in Dwight, $R_{ac}/R_{dc} = 1.4$. If two parallel ½″ conductors are modeled as a single 1″ conductor, then $a/b = 6$ and $R_{ac}/R_{dc} = 1.45$.

No formula or curve has been found in the literature for the proximity effect for rectangular bus bars. Dwight does have a curve for proximity effect for round wires,

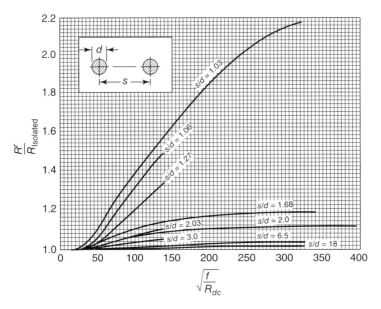

Figure 4.13   Proximity effect ratio in wires, R_{dc} in Ohms per 1000 ft (Dwight, 1945, p. 225).

reproduced here in Figure 4.13 (Dwight, 1945, p. 225). If we assume that the spacing is 90 mm or 3.54 in. and instead of the diameter, we use the thickness of 1.39″ for two conductors, then the ratio $s/d = 2.54$ and $R_{ac}/R_{dc} = 1.07$.

The values of inductance are calculated as a three-phase inductance matrix, consisting of the self- and mutual inductances of the three-phase groups of bus bars.

$$\mathbf{L} = \begin{bmatrix} L_{11} & M_{12} & M_{13} \\ M_{21} & L_{22} & M_{23} \\ M_{31} & M_{32} & L_{33} \end{bmatrix} \tag{4.24}$$

This is next converted to a reactance matrix at 60 Hz.

$$\mathbf{Z} = \begin{bmatrix} Z_{11} & Z_{12} & Z_{13} \\ Z_{21} & Z_{22} & Z_{23} \\ Z_{31} & Z_{32} & Z_{33} \end{bmatrix} \tag{4.25}$$

The following expression (Grover, 1946, p. 21) for the geometric mean radius (GMR) of a rectangular bar with sides of lengths "a" and "b," may be used in these calculations:

$$\text{GMR} = 0.2235(a + b) \tag{4.26}$$

The following more exact expression (Dwight, 1945, p. 142) for the GMR of one rectangular bar with sides of lengths "a" and "b," may be used in calculating inductance:

$$\ln(\text{GMR}) = \frac{1}{2}\ln(a^2 + b^2) - \frac{1}{12}\frac{a^2}{b^2}\ln\left(\frac{a^2 + b^2}{a^2}\right) - \frac{1}{12}\frac{b^2}{a^2}\ln\left(\frac{a^2 + b^2}{a^2}\right)$$
$$+ \frac{2}{3}\frac{a}{b}\tan^{-1}\frac{b}{a} + \frac{2}{3}\frac{b}{a}\tan^{-1}\frac{a}{b} - \frac{25}{12} \tag{4.27}$$

The following expression (Dwight, 1945, p. 143) for the geometric mean distance (GMD) between two rectangular bars with sides of lengths "a" and "b," and center-to-center spacing "s," may be used in calculating inductance:

$$
\begin{aligned}
\ln(\text{GMD}) = {} & \frac{1}{4}\left[(s+a)^2\left\{b^2 - \frac{(s+a)^2}{6}\right\} - \frac{b^4}{6}\right]\ln\{(s+a)^2 + b^2\} \\
& + \frac{1}{4}\left[(s-a)^2\left\{b^2 - \frac{(s-a)^2}{6}\right\} - \frac{b^4}{6}\right]\ln\{(s-a)^2 + b^2\} \\
& - \frac{1}{2}\left[s^2\left\{b^2 - \frac{s^2}{6}\right\} - \frac{b^4}{6}\right]\ln(s^2 + b^2) \\
& + \frac{1}{12}(s+a)^4 \ln(s+a) + \frac{1}{12}(s-a)^4 \ln(s-a) \\
& - \frac{1}{6}s^4 \ln s + \frac{1}{3}\{b(s+a)^3 - b^3(s+a)\}\tan^{-1}\left(\frac{b}{s+a}\right) \\
& + \frac{1}{3}\{b(s-a)^3 - b^3(s-a)\}\tan^{-1}\left(\frac{b}{s-a}\right) \\
& - \frac{2}{3}(bs^3 - b^3 s)\tan^{-1}\left(\frac{b}{s}\right) - \frac{25}{12}a^2 b^2 \quad\quad (4.28)
\end{aligned}
$$

The dimensions of the rectangular bus bar are shown in Figure 4.14.

For example, two parallel bus bars per phase, with length and width dimensions of $6'' \times \frac{1}{2}''$, as shown in Figure 4.15, and similar to the bus bars in the photograph of Figure 4.16, excepting that the bus bars in the photograph do not have any spacing between the two bus bars in each phase, we consider

$a = 1.39$ in. (0.0354 m) (pair of bus bars separated by 0.39 in. (0.01 m))

$b = 6$ in. $= 0.1524$ m

$s_{12} = 3.54$ in. $= 0.09$ m (two adjacent pairs)

$s_{13} = 7.09$ in. $= 0.18$ m (two outer pairs)

$\text{GMR} = 0.04197$ m (pair of bus bars separated by 0.39 in (0.01 m))

$\text{GMD}_{12} = 0.10557$ m (two adjacent pairs)

$\text{GMD}_{13} = 0.18930$ m (two outer phases)

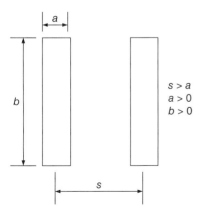

$s > a$
$a > 0$
$b > 0$

Figure 4.14   Dimensions of two parallel rectangular bus bars.

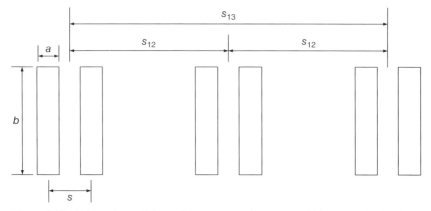

Figure 4.15   Dimensions of three-phase system with two parallel rectangular bus bars per phase.

Figure 4.16   Three-phase system with two parallel rectangular bus bars per phase.

TABLE 4.3 Sequence Impedances for the Example of Two Parallel 6″ × 1/2″ Bus Bars Per Phase

Bus Bar	Positive Sequence		Zero Sequence	
Units	R1	X1	R0	X0
Ω/m	6.24E−06	9.40E−05	1.10E−05	3.62E−04
Ω/ft	1.90E−06	2.87E−05	3.35E−06	1.10E−04

The resulting impedance matrix can be converted to a sequence impedance matrix, from which the positive and zero sequence impedances Z_1 and Z_0 are extracted to use in the calculations.

$$\mathbf{Z}_{seq} = \begin{bmatrix} Z_0 & Z_{01} & Z_{02} \\ Z_{10} & Z_1 & Z_{12} \\ Z_{20} & Z_{21} & Z_2 \end{bmatrix} \tag{4.29}$$

The results of applying equation (4.28) to the example system are shown in Table 4.3.

CHAPTER 5

CIRCUIT MODEL
OF THE HUMAN BODY

5.1 CALCULATION OF ELECTRICAL SHOCK USING THE CIRCUIT MODEL OF THE BODY

When electrical systems are being designed, the safety evaluation must include an assessment of the effect of human contact with any metallic or energized parts or conductors which may exist. Whenever possible, grounding, insulation, and physical distance should reduce or eliminate the hazards of electrical shock. During the design, the effect of contact with conductors can be evaluated using electrical circuit models of the human body. There will be many factors which affect such a model, and varying degrees of complexity of the model depending on the application and the level of potential risk which may be present. This chapter will present an introduction to this field, which is still evolving as new research is performed and more advanced mathematical modeling techniques are used.

The circuit model of a human body can then be constructed on basis of the skin model, the limbs, and the trunk. Figure 5.1 shows the combined circuit models, assuming dry skin with contact area of 1 cm^2, from the previous examples to simulate the electrical contact from hand to foot, through the trunk. Figure 5.2 shows the circuit reduction of the example body with dry skin. The circuit reduction is performed in the following steps:

Step 1. Combine duplicate models. There are two identical areas of skin contact. For the skin area, the resistances, R_P, are in series, thus $R = 2R_P$. The capacitances in series combine by reciprocals: $C = 1/(2C_P) = 0.5C_P$. The same procedure is used for the two identical limbs.

Step 2. Convert parallel circuits to series circuits, so that all resistive and capacitive elements may be summed up for all elements i.

The total admittance of the element is calculated from the conductance and susceptance.

$$\mathbf{Y}_i = G_i + jB_i = \frac{1}{R_i} + \frac{1}{j\omega C_i} = Y_i \angle \theta_i \tag{5.1}$$

where

$$\omega = 2\pi f = 2\pi \times 60 \ \text{Hz} = 377$$

Principles of Electrical Safety, First Edition. Peter E. Sutherland.
© 2015 John Wiley & Sons, Inc. Published 2015 by John Wiley & Sons, Inc.

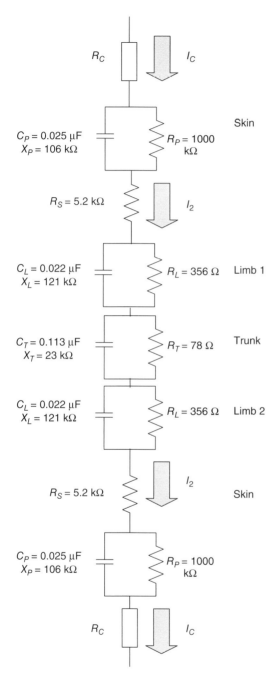

Figure 5.1   Circuit model of hand-to-foot conduction, with dry skin.

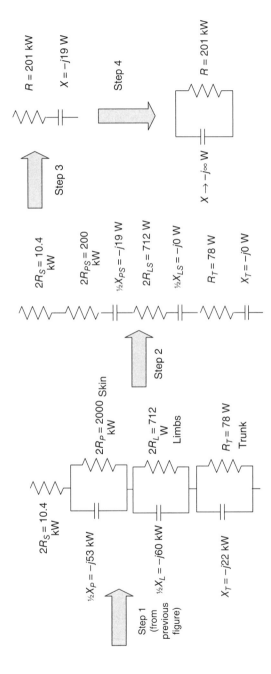

Figure 5.2   Circuit reduction of human body impedance with dry skin.

and

$$\theta_i = \tan^{-1}\left(\frac{B_i}{G_i}\right)$$

The admittance is converted to impedance.

$$\mathbf{Z}_i = \frac{1}{\mathbf{Y}_i} = \left(\frac{1}{Y_i}\right)\angle - \theta_i \qquad (5.2)$$

The impedance is broken out into two series elements.

$$R_i = \mathbf{Z}_i \cos\theta_i \qquad (5.3)$$
$$X_i = \mathbf{Z}_i \sin\theta_i \qquad (5.4)$$

Step 3.   The series elements are summed.

$$R = \Sigma R_i \qquad (5.5)$$
$$X = \Sigma X_i \qquad (5.6)$$
$$\mathbf{Z} = R + jX \qquad (5.7)$$
$$\theta = \tan^{-1}\left(\frac{X}{R}\right) \qquad (5.8)$$

Step 4.   A parallel equivalent circuit is also created.

$$Y = \left(\frac{1}{Z}\right)\angle - \theta \qquad (5.9)$$
$$G = Y\cos(-\theta) \qquad (5.10)$$
$$B = Y\sin(-\theta) \qquad (5.11)$$

This evaluation shows that the capacitive effect is minimal, as $X_C \ll R$. The model can be used to simulate the effects of electrical shock. With 120 V AC applied, the current is

$$I_{dry} = \frac{E}{Z} = \frac{120}{200 \times 10^3} = 0.6\,\text{mA}$$

This is what produces the "tingle" when house wiring is touched inadvertently.

Figure 5.3 shows the circuit reduction of the example body with wet skin, assuming that skin resistance is zero, making the only factor the internal body resistance. With 120 V AC applied, the current is

$$I_{dry} = \frac{E}{Z} = \frac{120}{790} = 152\,\text{mA}$$

Current at this level may cause ventricular fibrillation and possible death.

5.2   FREQUENCY RESPONSE OF THE HUMAN BODY

Using the circuit model developed, the inductances may be added as elements in series with the resistances, and the effect of varying frequency can be found. Continuing the previous example, inductances had been found for the limbs and the trunk:

$$L_T = 0.109 \;\; \mu\text{H}$$
$$L_L = 0.225 \;\; \mu\text{H}$$

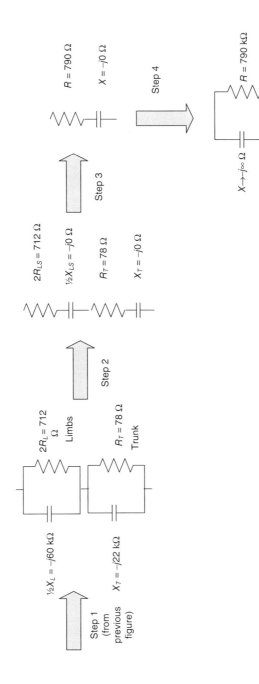

Figure 5.3   Circuit reduction of human body impedance with wet skin.

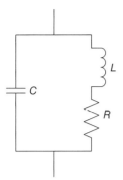

Figure 5.4   Equivalent circuit of a body part.

The resistances and capacitances were

$$R_T = 78\ \Omega \quad C_T = 0.113\ \mu\text{F}$$
$$R_L = 356\ \Omega \quad C_L = 0.022\ \mu\text{F}$$

The equivalent circuit of a body part, including inductance, is shown in Figure 5.4. The impedance as a function of frequency can be calculated as

$$Z(\omega) = \cfrac{1}{j\omega C + \cfrac{1}{R + j\omega L}}$$
$$= \frac{R + j\omega L}{1 - \omega^2(LC) + j\omega RC} \tag{5.12}$$

Separating into real and imaginary components,

$$Z(\omega) = \frac{R[1 - \omega^2(LC)] - \omega^2 RLC + j(\omega L[1 - \omega^2(LC)] - \omega R^2 C)}{[1 - \omega^2(LC)]^2 - (\omega RC)^2}$$
$$= \frac{R[1 - \omega^2(LC)] - \omega^2 RLC}{[1 - \omega^2(LC)]^2 - (\omega RC)^2} + j\frac{\omega L[1 - \omega^2(LC)] - \omega R^2 C}{[1 - \omega^2(LC)]^2 - (\omega RC)^2} \tag{5.13}$$

The resonant frequency is the frequency of maximum impedance. It is normally assumed in analysis of resonant tank circuits that R is very small, so at resonance,

$$1 - \omega^2(LC) = 0$$
$$\omega = \frac{1}{\sqrt{LC}} \tag{5.14}$$

and

$$f = \frac{1}{2\pi\sqrt{LC}} \tag{5.15}$$

Therefore, the ideal resonant frequency of the trunk is $f_T = 1.43$ MHz, and of the limb, $f_L = 2.26$ MHz.

This is not the case when R is significant. Here we need to find the minimum value of a complex function.

$$1 - \omega^2(LC) + j\omega RC = 0 \tag{5.16}$$

$$\omega = \text{Re}\left[-\frac{1}{2LC}\left[-jRC \pm \sqrt{(RC)^2 + 4LC}\right]\right]$$

$$= \text{Re}\left[\frac{jR}{2L} \pm \sqrt{-\left(\frac{R}{2L}\right)^2 + \frac{1}{LC}}\right]$$

$$= \sqrt{\frac{1}{LC} - \left(\frac{R}{2L}\right)^2} \quad \text{if} \quad \frac{1}{LC} > \left(\frac{R}{2L}\right)^2 \tag{5.17}$$

For the trunk, $1/LC = 8.12 \times 10^{13}$, and $(R/2L)^2 = 1.28 \times 10^{17}$, and resonance does not exist. For the limb, $1/LC = 2.02 \times 10^{14}$, and $(R/2L)^2 = 6.26 \times 10^{17}$, and resonance does not exist. The maximum value of resistance for which resonance exists is

$$\frac{1}{LC} = \left(\frac{R}{2L}\right)^2 \tag{5.18}$$

$$R = \frac{2L}{\sqrt{LC}} = 2\sqrt{\frac{L}{C}} \tag{5.19}$$

For the trunk $R = 1.96\,\Omega$, for the limb $R = 6.40\,\Omega$. These resistances are well below typical values for the human body. The impedance magnitude and angle plots are shown in Figures 5.5 and 5.6. For all frequencies in the kilohertz range, consisting of the power frequency and its harmonics, the resistance is the only circuit value which needs to be used. Considering the skin impedance, the skin consists of a resistance R_P

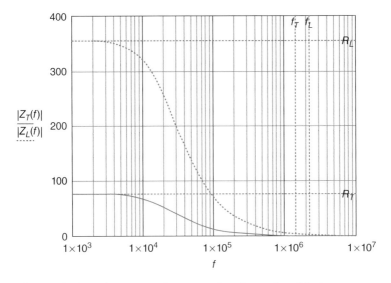

Figure 5.5   Impedance versus frequency plots for trunk and limb.

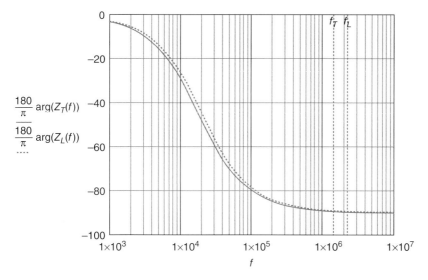

Figure 5.6   Phase angle versus frequency plots for trunk and limb.

in parallel with a capacitance C_P, along with a series resistance R_S. The impedance can be calculated as

$$Z_s(\omega) = R_S + \cfrac{1}{j\omega C_P + \cfrac{1}{R_P}}$$

$$= R_S + \frac{R_P}{1 + j\omega R_P C_P}$$

$$= R_S + \frac{R_P(1 - j\omega R_P C_P)}{1 - (\omega R_P C_P)^2}$$

$$= R_S + \frac{R_P}{1 - (\omega R_P C_P)^2} - j\frac{\omega R_P C_P}{1 - (\omega R_P C_P)^2} \qquad (5.20)$$

In the example calculation, the resistances and capacitance were

$$R_P = 1.0 \ \text{M}\Omega \quad C_P = 0.025 \ \mu\text{F}$$
$$R_S = 4.2 \ \text{k}\Omega$$

The impedance magnitude and angle plots are shown in Figures 5.7 and 5.8. For all frequencies in the kilohertz range, consisting of the power frequency and its harmonics, the resistance is the only circuit value which needs to be used. There is a circuit pole at

$$f = \frac{1}{2\pi R_P C_P} = 6.4 \ \text{Hz}$$

This is well below the normal power frequency. The circuit model for the entire body, including resistance, capacitance, and inductance can then be constructed. Because resonance is involved, the series–parallel circuit reduction method used for the resistance elements will not correctly model the frequency response. Therefore, the approach of summing the equivalent circuit models will be used.

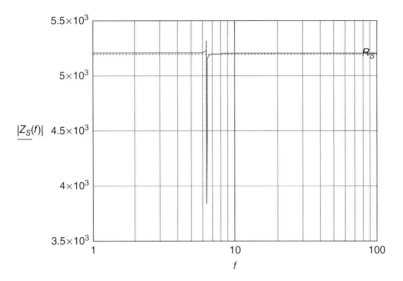

Figure 5.7   Impedance versus frequency plot for skin.

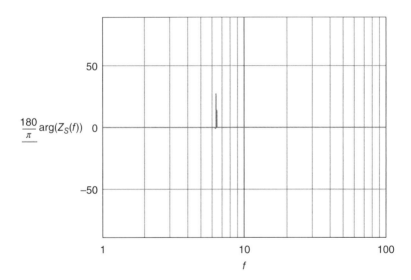

Figure 5.8   Phase angle versus frequency plot for skin.

$$Z(\omega) = \frac{R[1 - \omega^2(L_T C_T)] - \omega^2 R_T L_T C_T}{[1 - \omega^2(L_T C_T)]^2 - (\omega R_T C_T)^2} + j\frac{\omega L[1 - \omega^2(L_T C_T)] - \omega R_T^2 C_T}{[1 - \omega^2(L_T C_T)]^2 - (\omega R_T C_T)^2}$$

$$+ 2\left[\frac{R_L\left[1 - \omega^2\left(L_L C_L\right)\right] - \omega^2 R_L L_L C_L}{[1 - \omega^2(L_L C_L)]^2 - (\omega R C_L)^2} + j\frac{\omega L_L[1 - \omega^2(L_L C_L)] - \omega R_L^2 C_L}{[1 - \omega^2(L_L C_L)]^2 - (\omega R_L C_L)^2}\right]$$

$$+ 2 \left[R_S + \frac{R_P}{1 - (\omega R_P C_P)^2} - j\frac{\omega R_P C_P}{1 - (\omega R_P C_P)^2} \right] \tag{5.21}$$

The total body impedance for low frequencies is the sum of the resistances

$$R_{B\downarrow} = 2R_S + 2R_L + R_T \tag{5.22}$$

The total body impedance for high frequencies is twice the skin resistance, as the resistances in parallel with capacitors are bypassed:

$$R_{B\uparrow} = 2R_S \tag{5.23}$$

The internal body impedance for low frequencies is the sum of the resistances

$$R_{B\downarrow} = 2R_L + R_T \tag{5.24}$$

The internal body impedance for high frequencies is zero, as the resistances in parallel with capacitors are bypassed:

$$R_{B\uparrow} = 0 \tag{5.25}$$

However, this is counteracted by the skin effect, which is not included in this model, so that at high frequencies, internal body impedances are not considered. The results are shown in Figures 5.9 and 5.10. As seen by the dip in the angle plot, there are system poles for the limbs at

$$f_L = \frac{1}{4\pi L_L C_L} \left[R_L C_L \pm \sqrt{(R_L C_L)^2 - 4L_L C_L} \right] \tag{5.26}$$

For the example being considered,

$$f_L = 2.032 \times 10^4, \quad 2.518 \times 10^8$$

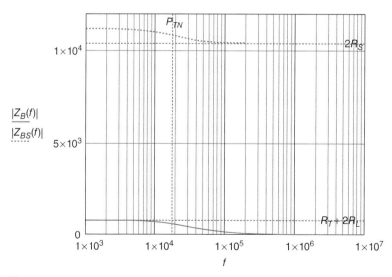

Figure 5.9   Total body impedance magnitude (Z_{BS}) and internal body impedance (Z_B) versus frequency. P_{TN} is the negative trunk pole at 18 kHz.

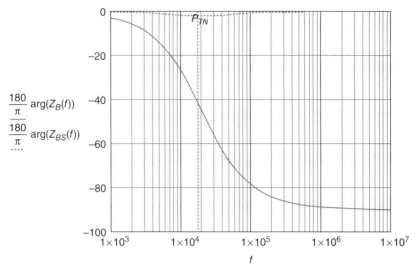

Figure 5.10   Total body impedance angle (Z_{BS}) and internal body impedance (Z_B) versus frequency. P_{TN} is the negative trunk pole at 18 kHz.

There are system poles for the trunk at

$$f_T = \frac{1}{4\pi L_T C_T} \left[R_T C_T \pm \sqrt{\left(R_T C_T\right)^2 - 4L_T C_T} \right] \qquad (5.27)$$

For the example being considered,

$$f_T = 1.806 \times 10^4, \quad 1.139 \times 10^8$$

CHAPTER 6

EFFECT OF CURRENT ON THE HUMAN BODY

6.1   INTRODUCTION TO ELECTRICAL SHOCK

The effects of electric currents flowing through a human body vary from little or no perceptible effect, to the sensation of shock, to severe injury or death by electrocution. The effect of the current flow is a function of magnitude and duration of the current. The magnitude, by Ohm's law, depends on the applied voltage divided by the electrical impedances of the body and other available current paths. The chances that an electric current will exceed the level necessary for human or animal perception increase with contact duration and the voltage and decrease with the impedance of the body. Greater time duration of exposure results in heating of body tissues, with the attendant physical damage and destruction. Because the human body impedance is nonlinear, voltage is a significant consideration in determining whether or not an energized object will have the ability to deliver current through a body. As with human body impedances, the effects of electrical shock on the human body are essential in the design of inherently safe electrical systems.

Modeling the response of the human body to electrical stimuli is a complex task. In the simplest model, two types of impedances are used: (i) skin and (ii) internal. The skin is a layered structure, with both resistance and capacitance. The resistance is nonlinear in voltage and time. The impedance of the interior is largely resistive, comparable to a similar volume of saline at human body concentration. Different types of internal tissue, muscle, and bone will have differing electrical characteristics, which must be taken into account. The skin capacitance causes impedance to decrease with frequency, as AC current shunts the high skin resistance.

Data on human and animal body electrical resistance under various conditions is widely available in the literature, and consists of both original research and numerous compilations of data for various applications. The most detailed combination of original research with a literature survey is in Reilly (Reilly, 1998). Power line effects are discussed in (IEEE, Working Group on Electrostatic and Electromagnetic Effects, 1978), while substation grounding has its own standard (IEEE, 2000). Agricultural stray voltages are analyzed in (USDA, 1991), with the main emphasis being on cattle, but including a chapter on effects on humans. Residential concerns, especially swimming pools, are discussed in an EPRI guidebook for technicians investigating

Principles of Electrical Safety, First Edition. Peter E. Sutherland.
© 2015 John Wiley & Sons, Inc. Published 2015 by John Wiley & Sons, Inc.

complaints of shocking (EPRI, 1999) and a technical brief on swimming pools (EPRI, 2000). Major papers on electrical shock hazards and their effects by C.F. Dalziel (Dalziel and Lee, 1969) (Dalziel, 1972) and R.H. Lee (Lee, 1971), presented much of the important data still used today. Kouwenhoven conducted important research at John's Hopkins University on electric shocks from the 1930s through the 1950s. Electrical safety information is also compiled in the *IEEE Yellow Book* (IEEE, 1998), and the *Electrical Safety Handbook* (Cadick, Capelli-Schellpheiffer, and Neitzel, 2000). A very useful standard, which provides compilations of much important data, is provided in *IEC 60479-1* (IEC, 2005). The *UL Standard C101* (UL, 2002) on *Leakage Current for Appliances* provides some current limits and test methods.

A key consideration with energized objects is that in order for electric current to flow through the body, the body must become part of an electrical circuit. This chapter supplies considerable detail on how currents can pass though different parts of the body and each parameter that can impede or limit that flow of current. For example, the same person touching an energized object may experience a shock if they were barefoot and may not if they were wearing tennis shoes. Similarly, the perception would be minimized if the skin were dry instead of wet, and further; there would be a different perception if the contact mode were "foot-to-foot" versus "hand-to-foot" or "hand-to-hand." The following chapter provides a very detailed description of the physiological effects of currents on the human body and details the conditions that pose the greatest dangers for severe shock and/or electrocution. The information presented indicates that the most likely scenarios for a dangerous condition would be that when a human subject comes in contact with an energized metallic object and the subject is barefoot, and the ground is moist or wet as are the subject's hands and or feet. The hazards may be less under other conditions, such as a dry day, rubber- or leather-soled shoes, or dry hands, but are not eliminated.

6.2   HUMAN AND ANIMAL SENSITIVITIES TO ELECTRIC CURRENT

Human sensitivity to electric current is classified in an increasing scale dependent on current magnitude and duration. Table 6.1 lists some of the experimental results from the literature for the following exposure levels.

Perception levels are based on touch or grip of dry skin. Perceptions levels do not present a danger under normal circumstances.

Startle currents do not produce harmful effects in themselves, but can cause injury due to reactions, such as fear and sudden movements. This is the level used to set leakage current standards for appliances and power tools. The 0.5 mA UL leakage current standards for appliances (UL, 2002) are set below the reaction currents.

The let-go current threshold is the point beyond which the subject can no longer release their grasp on a conductor. Currents above this level cause involuntary muscle contractions. If the current level is decreased, the person will be able to release their grasp, and may suffer no ill effects.

TABLE 6.1   Current Thresholds (mA) for 60 Hz Exposure

Threshold	Women		Men			Children
Percentile	0.5%	50%	0.5%	50%	99.5%	0.5%
Perception for touch	0.07	0.24	0.10	0.36	NA	NA
Perception for grip	0.28	0.73	0.40	1.10	NA	NA
Startle	NA	2.2–3.2	NA	NA	NA	NA
Let-go current	6.0	10.5	9.0	10.0^a–16.0	NA	NA
Respiratory paralysis	NA	15	NA	2–30^a	NA	NA
Fibrillation	NA	NA	$75^{a,\,b}$–$100^{c,\,d}$	NA	$250^{a,\,b}$	$35^{e,\,d}$
Heart paralysis	NA	NA	NA	4000 (2)	NA	NA
Tissue burning	NA	NA	NA	$\geq$5000 (2)	NA	NA

[a] Adult, 68 kg.

[b] Exposure time, 5 s.

[c] Adult, 70 kg.

[d] Exposure time, 3 s.

[e] Child, 20 kg.

[f] NA: Data not available.

Sutherland, Dorr, and Gomatom (2009) Figure 11. © 2009a IEEE.

Respiratory paralysis (or respiratory tetanus) can be produced by prolonged exposure to currents at the let-go current threshold and above. Respiratory paralysis can cause severe pain, exhaustion, and death from asphyxiation. Some reports indicate that currents as little as 50% above the let-go current threshold produce difficulty in breathing.

Ventricular fibrillation is a stoppage of the heart caused by electric current. It is usually fatal within a few minutes, and is the most severe consequence of electric shock. The current necessary to produce fibrillation is related to exposure time by an I^2T constant (Dalziel and Lee, 1969), the so-called electrocution equation:

$$I = \frac{K}{\sqrt{t}} \tag{6.1}$$

where

$I =$ current in amperes

$t =$ exposure time in seconds, and

$K =$ electrocution constant.

Using animal data, Lee and Dalziel extrapolated values for constant K at 0.5% percentiles for 50 kg adults and 18 kg children (Figure 6.1). This represents 67-107 mA for 3 s for adults and 30-40 mA for children. The data is valid over a range of 0.083 (5 cycles @ 60 Hz) to 5 s (300 cycles @ 60 Hz). At the minimum time, the current range is 403-642 mA for adults and 180-240 mA for children.

Heart paralysis and severe burning can result from higher currents than necessary to cause fibrillation. At these high currents, fibrillation may not occur, or may be reversed.

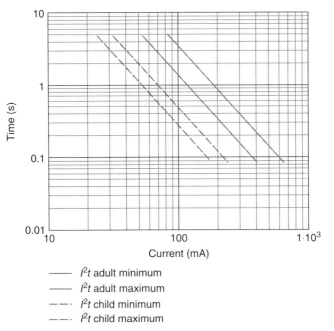

Figure 6.1 Electrocution threshold current ranges for 0.5% percentile of 50 kg adults (solid) and 18 kg children (dashed). © IEEE (2009a).

6.3 HUMAN BODY IMPEDANCE

The impedance of the human body can be broken down into the impedances of the various body parts, resulting in an equivalent circuit for the electrical path through the body. The various components are indicated in Figures 6.2 and 6.3. Human body impedance is characterized as that of a bulk medium, which combines the elements of inductance, resistance, and capacitance in one solid.

The skin resistance is dependent on the portion of the body (thicker or thinner skin), the wetness of the skin, and the area and pressure of contact. Skin resistance is divided into two parts, the surface layer resistance, R_P, in parallel with the capacitance and the spreading resistance R_S. The spreading resistance covers the zone where the current spreads out into the body from the contact area. Skin resistivity $(R_P + R_S)$ is in a range of $60–1200 \, k\Omega\text{-cm}^2$ (Reilly, 1998, p. 27). Skin resistance decreases with time of contact. After approximately 20 min, the resistivity of dry and wet skin becomes equal, at about 2/3 of the wet skin resistivity.

Skin is also subject to dielectric breakdown at an applied voltage between 150 and 250 V, and at as low as 50 V for exposures lasting several minutes. Considering the area of a foot as $200 \, cm^2$, the skin resistance of a single foot ranges from 0.3 to $6 \, k\Omega$. The skin capacitance varies from 0.02 to $0.06 \, \mu F/cm^2$ (Reilly, 1998, p. 280). The resistance in series with the skin capacitance, R_{P2}, is often neglected, while the parallel resistance, R_P, is the steady-state impedance.

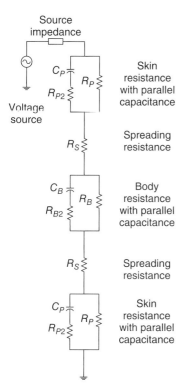

Figure 6.2   Human body impedance components.
© IEEE (2009a).

The internal body impedance is usually divided into regions, each of which is assigned a percentage of the total (Reilly, 1998, pp. 37–41), as shown in Figure 6.4. The resistance of the various body parts is proportional to the bulk resistance of an equal size and shape of saline solution with a resistivity of 80 Ω-cm. On this basis, the impedance of children is larger than that of adults.

Statistical rankings of adult total body impedance given by the IEC (IEC, 2005) are shown in Table 6.2. For a typical 125 V exposure, the 5th percentile is 1125 Ω, the 50th 1625 Ω, and the 95th 2875 Ω. These values are for hand-to-hand conduction with dry contacts.

The IEC results show impedance decreased by 10-25% for fresh water contacts, and by up to 50% for salt water. This effect is more pronounced at applied voltages less than 150 V.

Data on hand-to-foot and foot-to-foot contact shows these values are reduced by 10–30% (IEC, 2005 p. 31). The values shown in Table 6.2 are 70% of the hand-to-hand values.

Resistance to direct currents (DC) is higher than for alternating currents (AC) because the skin capacitance removes the parallel skin resistance R_{P2} from the circuit, as shown in Figure 6.2.

Skin impedance is the most complex component of human body impedance. In addition to the linear components shown in Figure 6.2, skin impedance is nonlinear with respect to both voltage and time. When skin is exposed to increasing current densities over time, the properties of the skin itself will change (IEC, 2005, p. 29).

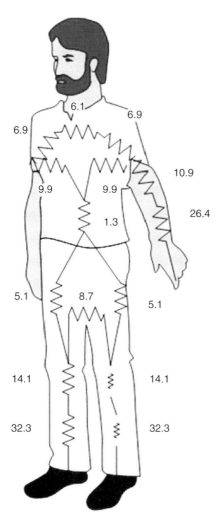

Figure 6.3   Human body impedance percentages. © IEEE (2009a).

These are divided by the IEC into four zones, ranging from no alterations (Zone 0) through carbonization (Zone 3), and plotted as a current density versus time curves. The upper limit of Zone 0 is 10 mA/cm^2 at 5 s to about 5 mA/cm^2 in the steady state. Skin resistance will decrease with time of exposure, leveling off after 20–30 min (Reilly, 1998, pp. 29–31). Short-term variations may result in distortion of sinusoidal current waveforms.

The Stevens equation (Reilly, 1998, p. 29) describes the nonlinear voltage–direct current relationship in dry skin as

$$I = aV + bV^2 \tag{6.2}$$

The nonlinearity becomes dominant above the range of 2–4 V. Above 450 to 1000 V, skin breakdown becomes significant. The parallel resistance R_P was found to be the major source of nonlinearities, which is expected from Stevens' DC results.

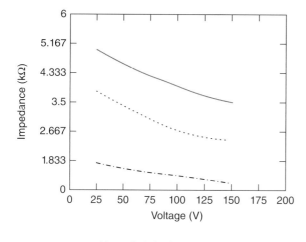

———— Upper limit (dry)
······ Average
–·–·– Lower limit (wet)

Figure 6.4   Total body
impedance ranges for
hand-to-hand or hand-to-foot
contacts. © IEEE (2009a).

TABLE 6.2   Adult Total Body Impedance (Ω) Including Skin Resistance, at Power Frequencies for Contact of Large Surface Area, as a Function of Exposure Voltage

Voltage Percentile	5%		50%		95%	
Conduction Path	Hand-to-Hand	Hand-to-Foot	Hand-to-Hand	Hand-to-Foot	Hand-to-Hand	Hand-to-Foot
25	1750	1225	3250	2275	6100	4270
50	1450	1015	2625	1838	4375	3063
75	1250	875	2200	1540	3500	2450
100	1200	840	1875	1313	3200	2240
125	1125	788	1625	1138	3875	2713
220	1000	700	1350	945	2125	1488
700	750	525	1100	770	1550	1085
1000	700	490	1050	735	1500	1050
Asymptotic	650	455	750	525	850	595

Hand to Foot Impedances are Given as 70% of Hand to Hand.

IEC (2005) Table 1.

6.4   EFFECTS OF VARIOUS EXPOSURE CONDITIONS

6.4.1   Bare Feet, Wet Conditions, and Other Variations

Resistance of footwear has a considerable impact on total body resistance. Typical shoe resistances (Lee, 1971 Figure 1) are $5-10\,k\Omega$ for wet leather soles and $100-500\,k\Omega$ for dry. Rubber soles have resistances of over $20\,M\Omega$.

Skin resistance will decrease by as much as 50% under wet conditions. This is particularly significant when barefoot contact is made. Comparison of a number of tests (Reilly, 1998) shows decreases in resistance being much greater at voltages less than 125 V. The resistance is less with saline as opposed to tap water.

The effect of pressure, for instance a hand holding a pipe, is also to decrease the resistance. For example, a dry palm touch will have a resistance of $3-8 \text{ k}\Omega$, while a hand grasp of a 1.5 in. pipe will have a resistance of $1-3 \text{ k}\Omega$. These values are reduced to $1-2 \text{ k}\Omega$ and $0.5-1.5 \text{ k}\Omega$, respectively, under wet conditions.

Human body impedance, as illustrated in Figure 6.4, varies with applied voltage (Reilly, 1998, pp. 36–37). This nonlinearity is mainly due to skin impedance, and so varies with wetness and contact area. This data can be fit to an exponential function:

$$Z(V) = ae^{bV} + c \qquad (6.3)$$

where

$$Z = \text{k}\Omega, V \text{ in volts and}$$

a, b, and c = constants.

6.4.2 Shoes and Other Insulated Objects and the Earth

The resistance of the earth can be calculated assuming the foot is a conducting disk (IEEE, 2000, p. 20). The usual assumption is of a 0.08 m radius and a footprint area of 200 cm^2. The resistance R_F of one footprint is

$$R_F = \frac{\rho}{4 \cdot r} \qquad (6.4)$$

where

ρ = resistivity of the material and

r = radius of the footprint.

If the surface material is in a thin layer, additional de-rating factors apply. The contact resistance for a single foot varies from approximately $100 \,\Omega$ to $3 \text{ k}\Omega$ for concrete and $75 \,\Omega$ to $3 \text{ M}\Omega$ for gravel. These values can be compared with a typical $1 \text{ k}\Omega$ for the human body (neglecting footwear).

6.5 CURRENT PATHS THROUGH THE BODY

The pathways electric current takes through the body are extremely dependent on the points of contact. The most dangerous pathways involve the heart. Tests using a current transformer around the heart of a cadaver (Reilly, 1998, p. 41) whose results are summarized in Table 6.3 show an average of 6.7% and a maximum of 8.5% of total current passed through the heart for right hand-to-feet conduction. For foot-to-foot conduction, the average was negligible, while the maximum was 0.4%. Similar results were found in tests on dogs. It could be assumed that current pathways through the body similarly affect respiratory paralysis. The IEC shows a heart-current factor where the current from left hand to either or both feet is taken as 100% (Table 6.4). The heart current required to cause damage is a function of time, as will be discussed

TABLE 6.3   Typical Ranges of Current Which Flow Through the Heart

Current Path	Minimum	Average	Maximum
Left hand to feet	1.2	3.3	5.1
Hand to hand	1.9	3.3	4.4
Right hand to feet	4.8	6.7	8.5
Foot to foot	<0.1	—	0.4
Head to feet	4.8	5.5	5.9

Reilly (1998), Table 2.4 with kind permission from Springer Science+Business Media B.V.

TABLE 6.4   IEC Heart-Current Factor

Current Path	Heart-Current Factor
Left hand to feet	1.0
Hand to hand	0.4
Right hand to feet	0.8
Foot to foot	0.02
Head to foot	0.8

IEC (2005), Table 12.

in a later section. Experimental results on animal hearts show variation with electrode area and placement as well (Reilly, 1998, pp. 230–237). A current flowing vertically (hand to foot) is more likely to cause fibrillation than the same current flowing horizontally (hand to hand). Very little fibrillating current flows through the heart for foot-to-foot contacts.

Body size relationships are different for fibrillating current than for shocking current (Reilly, 1998, pp. 220–225). One conclusion favored by researchers is that fibrillation depends on current density in the heart, rather than the total current magnitude. A current flowing through a child would have greater current density in the heart than the same current in an adult, and thus be more likely to cause fibrillation. However, as a child's impedance can be up to four times greater than an adult, the total current through a child would be less than an adult for exposure to the same voltage.

Electrical current is applied to the heart when an automatic electric defibrillator (AED) is used. Defibrillation is performed by applying a series of high current pulses, with magnitudes of several amperes, to a wide area of the heart. This synchronized stimulation can reverse the effects of fibrillation if performed within a short time after the onset of fibrillation.

A model (Figure 6.5) of human body impedance was constructed, based on Reilly (1998, p. 40). A typical foot-to-foot human body resistance of 802 Ω is assumed. With a 120 V, 60 Hz source and 200 Ω of footing resistance, the current through the victim is 0.12 A. This is just at or above the threshold of fibrillation listed in Table 6.1. In the case where almost none of the current passes through the chest, the greater risk is of paralysis of the limbs. Tables 6.5–6.7 show how the resistances

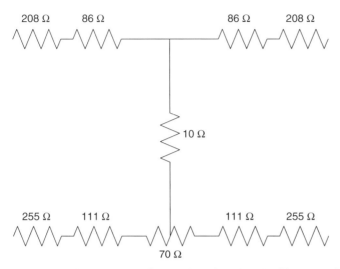

Figure 6.5   Circuit model of human body based on the fifth percentile of hand-to-foot conduction. © IEEE (2009a).

TABLE 6.5   Typical Ranges of Resistance (Ω) for Hand to Foot Conduction

Name	%Ω	5th Percentile	50th Percentile	95th Percentile
Fore arm	26.4	208	300	716
Upper arm	10.9	86	124	296
Shoulder to trunk	9.9	78	113	269
Trunk	1.3	10	15	35
Leg to trunk	5.1	40	58	138
Upper leg	14.1	111	160	383
Lower leg	32.3	255	368	876
Total	100	788	1138	2713

TABLE 6.6   Typical Ranges of Resistance (Ω) for Foot to Foot Conduction

Name	%Ω	5th Percentile	50th Percentile	95th Percentile
Lower leg	32.3	255	368	876
Upper leg	14.1	111	160	383
Leg to leg	8.7	70	99	236
Upper leg	14.1	111	160	383
Lower leg	32.3	255	368	876
Total	101.5	802	1155	2754

TABLE 6.7   Typical Ranges of Resistance (Ω) for Hand to Hand Conduction

Name	%Ω	5th Percentile	50th Percentile	95th Percentile
Fore arm	26.4	297	429	1023
Upper arm	10.9	123	177	422
Shoulder to chest	6.9	78	112	267
Chest	6.1	69	99	236
Shoulder to chest	6.9	78	112	267
Upper arm	10.9	123	177	422
Fore arm	26.4	297	429	1023
Total	94.5	1063	1536	3662

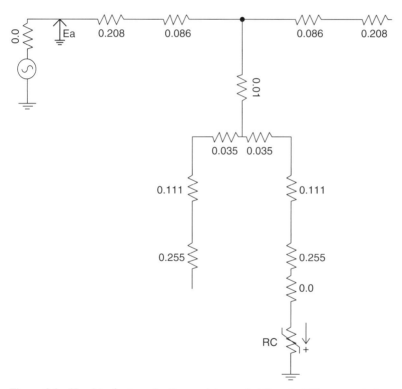

Figure 6.6   Hand-to-foot conduction, resistances in kilowats. RC represents the variable earth resistance. Note that the current flows through the torso. © IEEE (2009a).

of various body parts can be estimated for use in the equivalent circuit. Figure 6.6 shows an equivalent circuit based on the resistances for hand-to-foot conduction in Table 6.5. Figure 6.7 shows an equivalent circuit based on the resistances for foot-to-foot conduction in Table 6.6.

Calculated current values are shown in Table 6.8 for bare feet. Current values calculated for any type of footwear show no risk of shock or fibrillation. With bare

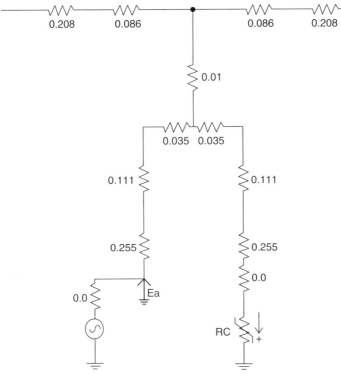

Figure 6.7 Foot-to-foot conduction, resistances in kilowatts. RC represents the variable earth resistance. Note that current flow does not go through the heart. © IEEE (2009a).

TABLE 6.8   Contact Resistance of a Single Bare Foot and Calculated Foot to Foot Current

Material	ρ (Ωm)	R (Ω)	5th Percentile (kΩ)	5th Percentile (mA)	50th Percentile (kΩ)	50th Percentile (mA)	95th Percentile (kΩ)	95th Percentile (mA)
Concrete, moist	30	94	0.7	**175**	0.9	**127**	2.1	**56**
Concrete, air dry	910	2.8×10^3	3.6	**33**	4.0	**30**	5.6	**22**
Gravel, dry	106	3.1×10^6	3.1×10^3	0.0	3.1×10^3	0.0	3.1×10^3	0.0
Gravel, wet (fresh)	8.5×10^3	2.7×10^4	27.2	4.4	27.4	4.4	28.6	4.2
Gravel, wet (salt)	24	75	0.7	**180**	0.9	**129**	2.1	**57**

Footprint radius assumed to be 8 cm. Two shoes with one on a conducting medium were assumed.

TABLE 6.9   Contact Resistance of a Single Bare Foot and Calculated Hand to Foot Current

Material	ρ (Ωm)	R (Ω)	5th Percentile (kΩ)		50th Percentile (kΩ)		95th Percentile (kΩ)	
Concrete, moist	30	94	1.5	**77.9**	1.7	**68.9**	2.6	**45.3**
Concrete, air dry	910	2.8×10^3	11.4	10.5	11.7	10.2	12.9	9.3
Gravel, dry	106	3.1×10^6	3.1×10^3	0.0	3.1×10^3	0.0	3.1×10^3	0.0
Gravel, wet (fresh)	8.5×10^3	2.7×10^4	28.0	4.3	28.2	4.3	29.1	4.1
Gravel, wet (salt)	24	75	1.5	**78.9**	1.7	**69.6**	2.6	**45.6**

Footprint radius assumed to be 8 cm. Two shoes with one on a conducting medium were assumed.

feet, risk of respiratory tetanus is present in the currents in bold, on concrete, and wet salt gravel. There will be some perception of a shock on wet fresh gravel as well. Table 6.9 shows the same calculations for hand-to-foot conduction, with the expected results of lower current levels. The lower current levels must be balanced against the fact that a hand-to-foot current path has greater chance of passing through the heart, and thus a smaller current could pose a greater risk.

6.6   HUMAN RESPONSE TO ELECTRICAL SHOCK VARIES WITH EXPOSURE CONDITIONS, CURRENT MAGNITUDE, AND DURATION

The electrocution equation (6.1) is a simplified time exposure factor for electrical shock. The IEC (IEC, 2005, p. 51) defines several zones, beginning with a reversibility zone of muscular contractions (AC-3, 0.5 mA line b to curve c1), followed by zones where the probability of ventricular fibrillation (PVF) is less than 5% (AC-4-1, between curves c1 and c2), between 5% and 50% (AC-4-2, between curves c2 and c3), and greater than 50% (AC-4-3, above curve c3). These are based on a current path of left hand to both feet. The IEC curves can be expressed mathematically as follows:

b: lower limit of reversibility zone of muscular contractions

$$t(I) = \frac{200}{\sqrt{I}} \quad \text{for} \quad I \geq 10\,\text{mA} \tag{6.5}$$

c1: lower limit of 5% PVF

$$t(I) = 0.2\sqrt{\frac{500 - I}{I - 40}} \tag{6.6}$$

c2: upper limit of 5% PVF; lower limit of 50% PVF

$$t(I) = 0.2\sqrt{\frac{1000 - I}{I - 50}} \tag{6.7}$$

c3: upper limit of 50% PVF

$$t(I) = 0.2\sqrt{\frac{1500 - I}{I - 80}} \tag{6.8}$$

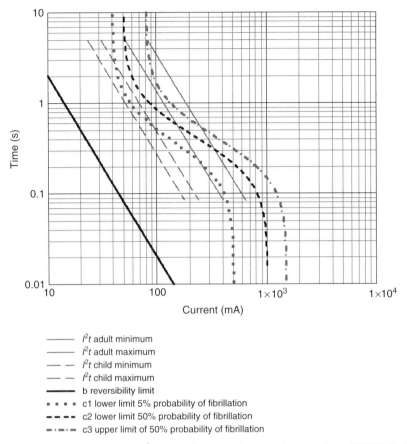

Figure 6.8   Comparison of I^2t curves from the electrocution equation with IEC fibrillation curves (b, c1, c2, and c3). © IEEE (2009a).

The comparison with the electrocution equation (6.1) is shown in Figure 6.8. The IEC curves provide a roughly similar standard.

The IEC curves can be adjusted by the heart factors to produce shifted curves for different exposure types as shown in Figures 6.9–6.12. These show that the fibrillation time thresholds are longer for other types of contact. Lower levels of currents also have time duration thresholds. For example, the magnitude of the sensation produced by an electric field can become painful if it is applied for 16 half-cycles, but is not felt at 4 half-cycles (Reilly, 1998, p. 264).

6.7   MEDICAL IMAGING AND SIMULATIONS

Further investigation of electrical shock effects would help to more precisely characterize risks and improve electrical safety. Bioelectromagnetism is an established area of research. Electrical impedance tomography (EIT) (Skipa, Sachse, and

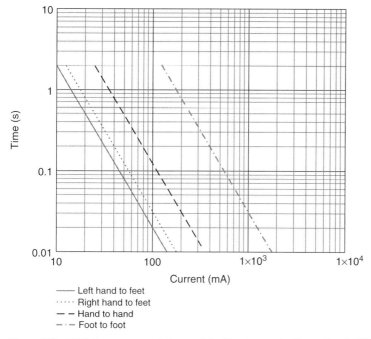

Figure 6.9   IEC damage curve b (reversible disturbances) adjusted by IEC heart-current factor for various current paths. © IEEE (2009a).

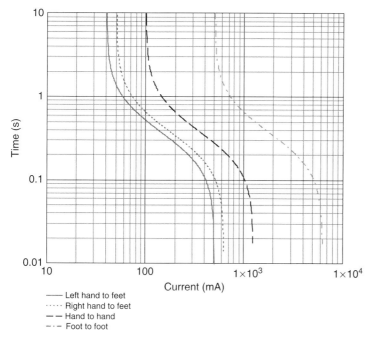

Figure 6.10   IEC fibrillation curve c1 (lower limit of 5% probability) adjusted by IEC heart-current factor for various current paths. © IEEE (2009a).

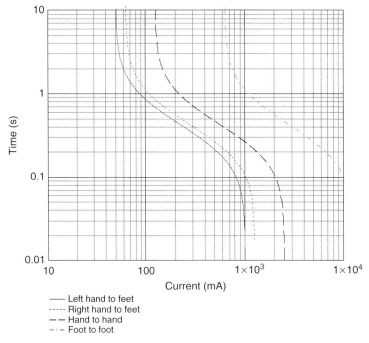

Figure 6.11 IEC fibrillation curve c2 (upper limit of 5% probability and lower limit of 50% probability) adjusted by IEC heart-current factor for various current paths. © IEEE (2009a).

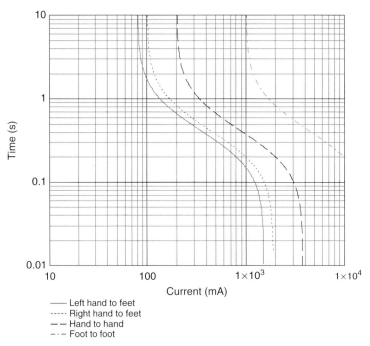

Figure 6.12 IEC fibrillation curve c3 (upper limit of 50% probability) adjusted by IEC heart-current factor for various current paths. © IEEE (2009a).

Dössel, 2000) is a medical imaging procedure where electrical currents are applied to the body and the voltage drop measured at other locations. Images are reconstructed using by solving inverse nonlinear field equations. Finite element method (FEM) models have been constructed for EIT, electrocardiography (ECG), and other bioelectric procedures (Pursula *et al.*, 2000). Magnetic resonance imaging (MRI) has also been proposed to visualize the flow of electrical currents within the body (Ozbek *et al.*, 2001). The flow of current within the body during an electrical shock has been simulated (Freschi *et al.*, 2013) for comparison with the current distributions in the IEC standards. These and similar models have be used for analyzing electrical injuries and the effectiveness of safety procedures and equipment. The FEM takes into account the volume impedance of each organ, volume currents, and the effects of electric and magnetic fields. Coupled with this are medical imaging techniques developed for cardiac monitoring that can measure and trace electrical currents within the body in three dimensions.

CHAPTER 7

FUNDAMENTALS OF GROUND GRID DESIGN

7.1   INTRODUCTION TO GROUND GRID DESIGN

The grounding or earthing system is a total set of measures used to connect the electrically conductive components of a power system to earth. The design of electrically safe workplaces has been developed at considerable depth in the field of grounding system design for electrical substations. The effects of potentials and currents on workers in substations has been carefully evaluated with regard to the electrical properties of the human body and the magnitudes of electrical current and voltage which are harmful, resulting in designs which are much safer than they otherwise would have been. This process has been greatly helped by industry standards which set forth design procedures and calculation techniques which will result in a safe design within the selected parameters.

The grounding system is an essential part of both high- and low-voltage electrical power networks, and has at least four important roles:

1. To protect against lightning by

    o providing an electrically and mechanically robust path for current to flow to ground;

    o limiting potential differences across electrical insulation on stricken towers;

    o reducing the number of flashovers that occur.

2. To minimize energy for correct operation of the power system by

    o providing unambiguous identification of faults, so that the correct protection systems operate;

    o providing low zero-sequence impedance for return of the unbalanced fraction of AC system currents.

3. To minimize energy to ensure electrical safety by

    o rapidly identifying system faults, leading to reduce fault duration;

    o limiting touch or step voltages to levels that restrict body currents to safe values.

Principles of Electrical Safety, First Edition, Peter E. Sutherland.
© 2015 John Wiley & Sons, Inc. Published 2015 by John Wiley & Sons, Inc.

4. To eliminate some hazards and reducing the energy of others to contribute to electromagnetic compatibility.

All of these functions are provided by a single grounding system. Some elements of this system may have specific electrical purposes, but all elements are normally bonded or coupled together, forming one system to be designed or analyzed.

7.2 SUMMARY OF GROUND GRID DESIGN PROCEDURES

Design of a ground grid is part of the overall design of a substation. The ground rods are driven in and the ground grid constructed before the surface layer of gravel is poured and the above-ground portions of the substation constructed. The design goals, as listed in IEEE Standard 80-2000 are

"To provide means to dissipate electric currents into the earth without exceeding any operating and equipment limits."

"To assure that a person in the vicinity of grounded facilities is not exposed to the danger of critical electrical shock."

The critical parameters in the design of the ground grid are listed in Table 7.1. The ground grid procedure is listed in Table 7.2.

7.2.1 Site Survey

The shape and area of the substation are determined and ground resistivity measurements are taken to determine ρ.

7.2.2 Conductor Sizing

The most important factor in terms of overcurrent phenomena is the sizing of the ground grid conductors.

The design procedure makes two assumptions:

1. Adiabatic heating of the conductor. The instantaneous power per unit length is

$$p(t) = i^2(t)\frac{\rho}{A} \ \ \mathrm{W/m} \tag{7.1}$$

TABLE 7.1   Critical Parameters in Ground Grid Design

Symbol	Name	Equation	Typical values	Units
I_G	Maximum grid current	$I_G = D_f \times I_g$	0.5–10	kA
t_f	Fault duration	—	0.25–1.0	s
t_s	Shock duration	$t_s = t_f$	0.25–1.0	s
ρ	Soil resistivity	Measured	10^1–10^4	$\Omega\,\mathrm{m}$
ρ_s	Resistivity of surface layer measured	—	21–6×10^6 (Wet) to 4000–10^9 (Dry)	$\Omega\,\mathrm{m}$

TABLE 7.2   Sample Procedure for Ground Grid Design

Step	Description	Results
1	Site survey, soil resistivity test	A, ρ
2	Conductor size, zero sequence current, fault-clearing time	$A_{mm}{}^2$, $3I_0$, t_c
3	Step and touch potentials	E_{step}, E_{touch}
4	Conductor loop design, conductor spacing, ground rod locations	Various dimensions
5	Estimated resistance of grounding system in uniform soil	R_G
6	Recalculate ground current and fault duration[a]	I_G, t_f
7	If GPR < tolerable touch voltage, go to step 12	$I_G R_G < E_{touch}$
8	If GPR > tolerable touch voltage, calculate mesh and step voltages	E_m, E_s, various K
9	If mesh voltage < tolerable touch voltage, go to step 10, otherwise go to step 11	$E_m < E_{touch}$
10	If step voltage < tolerable touch voltage, go to step 12, otherwise go to step 11	$E_s < E_{touch}$
11	If mesh or step voltage > tolerable touch voltage, revision of design is required	Decision
12	Add equipment ground conductors, additional grid conductors, ground rods as needed	Final design

[a] Based on current splits, worst-case fault and future expansion.

The joule heating is

$$Q = \int_0^{tc} p(t)dt = \int_0^{tc} i^2(t)\frac{\rho_r(1 + \alpha_r(T - T_a))}{A}\, dt = \frac{\rho_r \alpha_r}{A} I_F^2 t_c \ \text{J/m} \tag{7.2}$$

2. The thermal capacity per unit volume remains constant (this is usually true for short fault durations). The temperature rise from ambient (T_a) to the maximum conductor temperature (T_m) occurs in the fault clearing time t_c (Sverak, 1981).

$$Q = TCAP \times A \int_{T_a}^{T_m} \frac{dT}{K_0 + T} \ \text{J/m} \tag{7.3}$$

$$TCAP \times A = \frac{\rho_r}{A \ln\left(\dfrac{K_0 + T_m}{K_0 + T_a}\right)} I_F^2 t_c \tag{7.4}$$

$$A^2 = \frac{\alpha_r \rho_r t_c}{TCAP \ln\left(\dfrac{K_0 + T_m}{K_0 + T_a}\right)} I_F^2 \ \text{m}^2 \tag{7.5}$$

The conductor size can then be determined:

$$A_{m^2} = \frac{I_F}{\sqrt{\left(\dfrac{TCAP}{t_c \alpha_r \rho_r}\right) \ln\left(\dfrac{K_0 + T_m}{K_0 + T_a}\right)}} \ \text{m}^2 \tag{7.6}$$

For standard conductor sizes in mm^2,

$$A_{mm^2} = A_{m^2} \times 10^6 \ \text{mm}^2 \tag{7.7}$$

For asymmetrical faults,

$$I_F = D_f I_f \tag{7.8}$$

For English units,

$$A_{\text{kcmil}} = 197.4 A_{\text{mm}^2 \text{ kcmil}} \tag{7.9}$$

where

As^2 = conductor cross sectional area in m^2.

A_{mm}^2 = conductor cross sectional area in mm^2.

I_F = rms total fault current in kA.

TCAP is the material thermal capacity in $J/(m^3 \times °C)$. Typical values are 3.42×10^6 for soft-drawn copper, 3.85×10^6 for 40% conductivity copper-clad steel wire, and 3.28×10^6 for a steel conductor.

t_c = time duration of the current flow in s.

A_{kcmil} = conductor cross sectional area in kcmil.

α_r is the thermal coefficient of resistivity at 20°C reference in $°C^{-1}$. Typical values are 3.93×10^{-3} for soft-drawn copper, 3.78×10^{-3} for 40% conductivity copper-clad steel wire, and 1.60×10^{-3} for a steel conductor.

ρ_r is the resistivity of the ground conductor at 20 °C reference in Ω m. Typical values are 17.2×10^{-9} for soft-drawn copper, 44.0×10^{-9} for 40% conductivity copper-clad steel wire, and 159.0×10^{-9} for a steel conductor.

$$\alpha_r = \frac{1}{K_0 + T_a}$$

I_f is the rms symmetrical fault current in kA.

It is assumed that the symmetrical rms fault current $I_f \approx 3I_0$ from the ground fault calculations for the substation.

T_m is the material-fusing temperature in °C. Typical values are 1083 °C for soft-drawn copper, 1084 °C for 40% conductivity copper-clad steel wire, and 1510 °C for a steel conductor.

T_a is the ambient temperature, typically 40 °C.

$K_0 = 1/\alpha_r - T_a$ is the thermal constant for the material. Typical values are 234 for soft-drawn copper, 245 for 40% conductivity copper-clad steel wire, and 605 for a steel conductor.

D_f is the decrement factor. Where fault durations are less than 1 s or the X/R ratio is greater than 5, the asymmetry of fault current waveforms produces additional heating, which must be taken into account:

$$D_f = \sqrt{1 + \frac{\tau_a}{t_c}\left(1 - e^{\frac{-2t_c}{\tau_a}}\right)} \tag{7.10}$$

where

τ_a = is the time constant $(X/R)2\pi f$ in s

This is illustrated in Figure 7.1 for several X/R ratios.

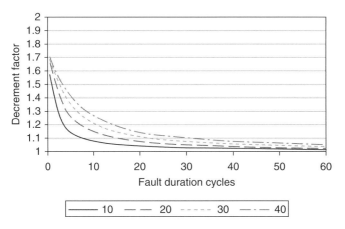

Figure 7.1 Decrement factor versus fault duration for four different X/R ratios.

TABLE 7.3 Fusing Currents in Symmetrical kA for Annealed Soft-Drawn 100% Conductivity Copper Conductors Versus X/R Ratio. All Clearing Times 0.5 s.

	Cross-sectional Area (mm^2)							
X/R	33.6[a]	35	67.4[b]	70	95	107.2[c]	120	150
0	13.4	14.0	26.9	27.9	37.9	42.7	47.8	59.8
10	13.7	14.3	27.6	28.6	38.9	43.8	49.1	61.4
20	14.1	14.7	28.3	29.3	39.8	44.9	50.3	62.9
30	14.4	15.0	28.9	30.0	40.8	46.0	51.5	64.4
40	14.7	15.4	29.6	30.7	41.7	47.0	52.7	65.8

[a]#2 AWG.

[b]#2/0 AWG.

[c]#4/0 AWG.

The conductors of a ground grid should be designed for a particular maximum fault current, X/R ratio, and clearing time. Examples are shown in Table 7.3. A safety factor is usually applied in the design to allow for future growth in fault current magnitudes.

7.2.3 Step and Touch Voltages

If a person standing on a surface whose potential has risen owing to the flow of ground current touches a grounded object, they experience a touch voltage (Figure 7.2). A Thévenin equivalent circuit of the person exposed to the touch voltage is also shown. Whether the touch voltage is hazardous can be determined by comparison with the calculated safe level of touch voltage for that substation. Similarly, if a person is standing on the surface, and the flow of ground current causes a dangerous voltage drop to occur between their feet, they are exposed to a step voltage (Figure 7.3). A Thévenin equivalent circuit of the person exposed to the step voltage is also shown.

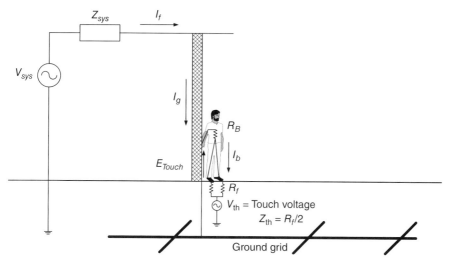

Figure 7.2   Touch voltage.

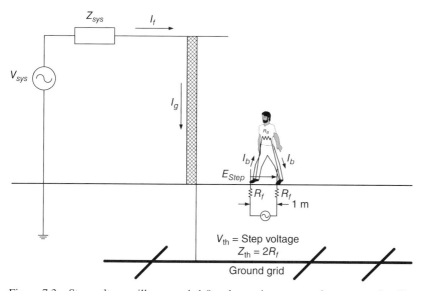

Figure 7.3   Step voltage will approach 1.0 as $h_s \to \rho$ increases and as $\rho_s \to \rho$. It will approach 0 as $\rho_s \to 0$.

The safe levels of step and touch potentials are defined on the basis of a person's body weight and the length of exposure. The usual standards used are for 50 and 70 kg (110 and 154 lb) body weights. The step and touch potentials are calculated using

$$E_{XW} = (1000 + m \cdot A \cdot C_s \cdot \rho_S) \frac{K_W}{\sqrt{t_s}} \tag{7.11}$$

where

$X=$ either "step" or "touch"

$W=$ either "50" or "70"

$1000 \ \Omega =$ the typical resistance of the human body

$m=$ 0 for metal-to-metal touch voltage, 1 otherwise

$A=$ 6 for "step" or 1.5 for "touch"

$\rho_s =$ the resistivity of the surface material in Ω m

$\rho_s = \rho =$ if there is no surface layer.

Normally $\rho_s > \rho$

$K_W =$ 0.116 for "50" or 0.157 for "70"

$t_s =$ the duration of the shock current, s

$C_s =$ the surface layer-derating factor

$C_s =$ 1.0 if there is no surface layer, otherwise an approximate formula (within 5% of computer models) may be used:

$$C_s = 1 - \frac{0.09\left(1 - \frac{\rho}{\rho_s}\right)}{2h_s + 0.09} \qquad (7.12)$$

where

$h_s =$ depth of the surface material in m

C_s will approach 1.0 as h_s increases and as $\rho_s \to \rho$. It will approach 0 as $\rho_s \to 0$.

7.2.4 Ground Grid Layout

Using the shape and area previously determined, a grid is laid out at a depth, h, with spacing D, and total length of buried conductor L_C. If ground rods, unequal spacing, or a shape other than square are used, other parameters will apply as well.

7.2.5 Ground Resistance Calculation

The resistance of the grounding grid, R_g, can be estimated using Sverak's equation:

$$R_g = \rho\left[\frac{1}{L_T} + \frac{1}{\sqrt{20A}}\left(1 + \frac{1}{1 + h\sqrt{20/A}}\right)\right] \qquad (7.13)$$

where

$R_g =$ resistance in Ω

$\rho Z =$ earth resistivity in Ω m

$L_T =$ total length of buried conductors in m

$A =$ total area of the grid in m^2.

7.2.6 Calculation of Maximum Grid Current

The maximum current that is used in ground potential rise calculations is not always the same as the maximum current used for conductor sizing. The maximum grid current is

$$I_G = D_f S_f 3I_0 \tag{7.14}$$

where

S_f is the current split factor determined from the detailed substation short-circuit calculations:

$$S_f = \left| \frac{Z_{eq}}{Z_{eq} + R_g} \right| \tag{7.15}$$

where

$$Z_{eq} = \frac{1}{\dfrac{1}{Z_{eq-l}} + \dfrac{1}{Z_{eq-f}}} \tag{7.16}$$

and by Endrenyi's method,

$$Z_{eq-l} = 0.5Z_{s-l} + \sqrt{R_{tg}Z_{s-l}} \tag{7.17}$$

$$Z_{eq-f} = 0.5Z_{s-f} + \sqrt{R_{dg}Z_{s-f}} \tag{7.18}$$

where

R_{tg} is the transmission line ground impedance in Ω/section
R_{dg} is the distribution line ground impedance in Ω/section
R_g is the station ground impedance in Ω
Z_{s-l} is the impedance of the overhead static wire for the transmission line in Ω/span
Z_{s-f} is the impedance of the overhead static wire for the distribution line in Ω/span

7.2.7 Calculation of Ground Potential Rise (GPR)

The ground potential rise is

$$E_{GPR} = R_g I_G \tag{7.19}$$

This is compared with the touch potentials $E_{\text{Touch }50}$ and $E_{\text{Touch }70}$. If it is larger, the mesh voltage is calculated.

7.2.8 Calculation of Mesh Voltage, E_m

The basis of the design procedure is to minimize the "mesh voltage," (E_m) which is the maximum touch voltage within the area of the ground grid, which is taken to mean at the center of the corner mesh, the usual point of maximum.

$$E_m = \frac{\rho K_m K_i I_G}{L_m} \tag{7.20}$$

where

ρ is the earth resistivity in Ω m

K_m is the geometrical factor, defined as:

$$K_m = \frac{1}{2\pi}\left[\ln\left[\frac{D^2}{16hd} + \frac{(D+2h)^2}{8Dd} - \frac{h}{4d}\right] + \frac{K_{ii}}{K_h}\ln\left[\frac{8}{\pi(2n-1)}\right]\right] \qquad (7.21)$$

K_i is the irregularity factor, defined as:

$$K_i = 0.644 + 0.148n$$

The variables in K_m and K_i are defined as follows:

D = spacing between parallel conductors, in m

h = depth of the grid, in m

d = diameter of the grid conductor, in m

K_{ii} = inner conductor factor, defined as

$K_{ii} = 1$ if there are ground rods along the perimeter or in the corners and along the perimeter and throughout the grid area.

If there are no ground rods or only a few ground rods, with none located on the corners or on the perimeter, K_{ii} is defined as

$$K_{ii} = \frac{1}{(2n)^{2/n}} \qquad (7.22)$$

K_h is the grid depth factor, defined as:

$$K_h = \sqrt{1 + \frac{h}{1.0\ m}} \qquad (7.23)$$

n is the effective number of parallel conductors in a grid:

$$n = n_a n_b n_c n_d \qquad (7.24)$$

where

$$n_a = \frac{2L_T}{L_P} \qquad (7.25)$$

$$n_b = \begin{cases} 1 & \text{for square grids} \\ \sqrt{\dfrac{L_p}{4\sqrt{A}}} \end{cases} \qquad (7.26)$$

$$n_c = \begin{cases} 1 & \text{for square and rectangular grids} \\ \left[\dfrac{L_x L_y}{A}\right]^{\frac{0.7A}{L_x L_y}} \end{cases} \qquad (7.27)$$

$$n_d = \begin{cases} 1 & \text{for square and rectangular and L-shaped grids} \\ \dfrac{D_m}{\sqrt{L_x^2 L_y^2}} \end{cases} \qquad (7.28)$$

and

L_C is the total length of conductor in the horizontal grid, in m
L_x is the maximum length of the grid in the x direction, in m
L_y is the maximum length of the grid in the y direction, in m
L_m is the effective buried length in m.

$$L_m = L_C + \left[1.55 + 1.22 \left(\frac{L_r}{\sqrt{L_x^2 + L_y^2}} \right) \right] L_R \qquad (7.29)$$

where

L_r = individual ground rod length, in m

L_R = total ground rod length, in m

D_m = maximum distance between any two points on the grid, in m

If the mesh voltage is higher than $E_{\text{Touch 50}}$ and $E_{\text{Touch 70}}$, the design must be modified. Once the mesh voltage is minimized, the step voltage is brought within limits as well.

7.2.9 Calculation of Step Voltage, E_s

The maximum step voltage is assumed to take place for a 1 m stride across the perimeter of the ground grid at its most extreme corner:

$$E_S = \frac{\rho K_S K_i I_G}{L_S} \qquad (7.30)$$

where
L_S is the effective length of buried conductor:

$$L_S = 0.75 L_C + 0.85 L_R \qquad (7.31)$$

K_S is the spacing factor for step voltage:

$$K_S = \frac{1}{\pi} \left[\frac{1}{2h} + \frac{1}{D+h} + \frac{1}{D} \left(1 - 0.5^{n-2} \right) \right]$$
$$\text{when} \quad 0.25 \text{ m} < h < 2.5 \text{ m} \qquad (7.32)$$

If the step voltage is higher than $E_{\text{Step 50}}$ and $E_{\text{Step 70}}$, the design must be modified.

7.2.10 Detailed Design

After the safe design has been obtained, detail such as equipment ground conductors, additional grid conductors, ground rods as needed for surge arresters, and other equipment can be added. A final design review is performed.

7.3   EXAMPLE DESIGN FROM IEEE STANDARD 80

This calculation is performed using improved and more uniform calculation procedures than those used in the referenced standard. Design data is shown in Table 7.4.

Step 1.   *Data collected from field measurements.*

$$A = 70 \text{ m} \times 70 \text{ m}$$

$$\rho = 400 \ \Omega \text{ m}$$

Step 2.   *Determination of conductor size.* The single-line diagram is shown in Figure 7.4. System analysis is performed on a 100 MVA base. The utility source impedances are converted by first finding the base impedance at 115 kV:

$$Z_{b115} = \frac{V_b^2(\text{kV})}{S(\text{MVA})} = \frac{115^2}{100} = 132.25$$

The source positive and zero sequence impedances are then converted to per unit. The negative sequence impedance is assumed to be equal to the

TABLE 7.4   Design Data for Ground Grid Example

Name	Symbol	Value	Units
Max. fault clearing time	t_f	0.5	s
Positive sequence impedance	Z_1	4.0+j10.0	Ω @ 115 kV
Zero sequence impedance	Z_0	10.0+j40.0	Ω @ 115 kV
Line-line voltage at fault location	V_{LL}	115	kV
Current division factor	S_f	0.6	NA
Soil resistivity	ρ	400	Ω m
Crushed rock resistivity (wet)	ρ_S	2500	Ω m
Crushed rock layer thickness	h_s	0.10	m
Grid burial depth	H	0.5	m
Area available for ground grid	A_a	73 $\times$ 84	m
Transformer positive and zero sequence impedances	Z_1, Z_0	Z=9% @ 15 MVA	Ω @ 13 kV

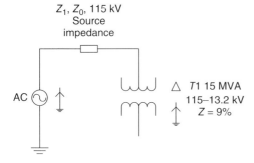

Z_1, Z_0, 115 kV
Source
impedance

AC

△ *T1* 15 MVA
115–13.2 kV
$Z = 9\%$

Figure 7.4   Single-line diagram for example.

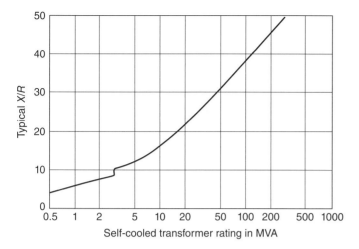

Figure 7.5   *X/R* ratio of transformers. (*Source*: © IEEE (1993))

positive sequence impedance, which is normal if the substation is remote from generator sources.

$$Z_{1pu} = \frac{Z_1}{Z_{b115}} = \frac{4 + j10}{132.25} = 0.0302 + j0.0756 \ pu$$

$$Z_{0pu} = \frac{Z_0}{Z_{b115}} = \frac{10 + j40}{132.25} = 0.0756 + j0.3025 \ pu$$

The transformer impedance will then be converted to the 100 kVA base. First, the X and R components must be found. For a 15 MVA transformer, the typical X/R ratio is 20 (Figure 7.5).

The impedance is then calculated in rectangular form:

$$\theta = \tan^{-1}\left(\frac{X}{R}\right) = \tan^{-1}(20) = 87.14°$$

$$Z_{T1} = Z(\cos\theta + j\sin\theta) = 0.0045 \quad + j0.090$$

The impedance is converted to the 100 MVA base:

$$Z_{T100} = Z_{T1}\frac{S_{new}}{S_{old}} = (0.0045 + j0.090)\frac{100}{15} = 0.029 + j0.60$$

The positive sequence equivalent circuit is shown in Figure 7.6.
The negative sequence equivalent circuit is shown in Figure 7.7.
The zero sequence equivalent circuit is shown in Figure 7.8.

For a single-line to ground fault on the transformer primary, the equivalent circuits are combined resulting in the equivalent impedance of $Z = 0.136 + j0.45 \ pu = 0.474\angle73.3°$. The zero sequence current is then

$$I_0 = \frac{E_a}{Z} = \frac{1.0\angle0°}{0.474\angle73.3°} = 2.11\angle - 73.3°$$

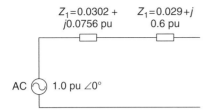

Figure 7.6   Positive sequence equivalent circuit.

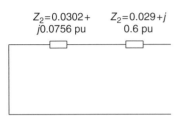

Figure 7.7   Negative sequence equivalent circuit.

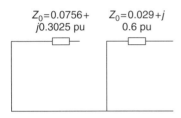

Figure 7.8   Zero sequence equivalent circuit.

The fault current can be calculated:

$$I_f = 3I_0 = 3(2.11\angle - 73.3°) = 6.33\angle - 73.3°$$

This process is illustrated in Figure 7.9.

The base current at 115 kV is

$$I_{b13.2} = \frac{S_b}{\sqrt{3}V_b} = \frac{100 \times 10^6}{\sqrt{3} \times 115 \times 10^3} = 0.502 \times 10^3 \text{ A}$$

then the single-line-to-ground fault current is

$$I_f = I_a I_{b13.2} = 6.33\angle - 73.3° \times 0.502 \times 10^3 \text{ A} = 3.18 \times 10^3 \text{ A}$$

at $X/R = \tan(73.3°) = 3.33$.

For a single-line to ground fault on the transformer secondary, the equivalent circuits are combined to the equivalent impedance of $Z = 0.117 + j1.876 \ pu = 1.879\angle 86.4°$. The zero sequence current is then

$$I_0 = \frac{E_a}{Z} = \frac{1.0\angle 0°}{1.879\angle 86.4°} = 0.532\angle - 86.4°$$

The fault current can be calculated:

$$I_f = 3I_0 = 3(0.532\angle - 86.4°) = 1.596\angle - 86.4°$$

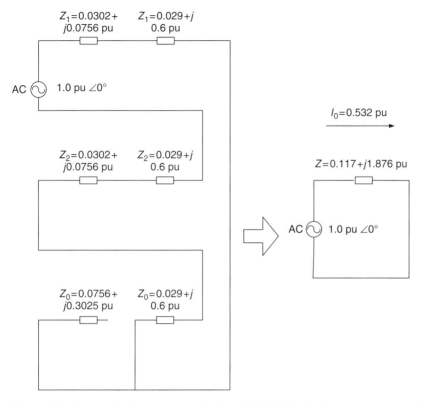

Figure 7.9   Symmetrical components solution of 115 kV single-line-to-ground ground fault.

This process is illustrated in Figure 7.10.

The base current at 13.2 kV is

$$I_{b13.2} = \frac{S_b}{\sqrt{3}V_b} = \frac{100 \times 10^6}{\sqrt{3} \times 13.2 \times 10^3} = 4.37 \times 10^3\,\text{A}$$

then the single-line-to-ground fault current is

$$I_f = I_a I_{b13.2} = 1.596\angle - 86.43° \times 4.37 \times 10^3\text{A} = 6.97 \times 10^3 \text{A}\angle - 86.43°.$$

The X/R ratio is calculated as $X/R = \tan(86.43°) = 16.0$.

Select the conductor size using equations (6-1)–(6-5), for a 40% copper-clad steel wire.

Given $t_c = 0.5$ s, $f = 60$ Hz and $\tau_a = (X/R)/(2\pi f) = 16/(2\pi \times 60) = 0.042$s, the decrement factor from equation (6-5) is

$$D_f = \sqrt{1 + \frac{\tau_a}{t_c}\left(1 - e^{\frac{-2t_c}{\tau_a}}\right)} = \sqrt{1 + \frac{0.042}{0.5}\left(1 - e^{\frac{-2 \times 0.5}{0.042}}\right)} = 1.0$$

Thus by equation (6-3),

$$I_F = D_f I_f = 1.0 \times 6.97 = 6.97 \ \text{kA}$$

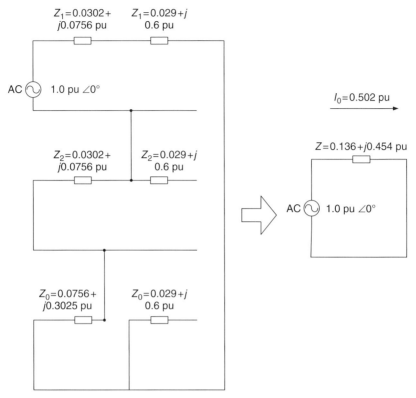

Figure 7.10 Symmetrical components solution of 13.2 kV single-line-to-ground ground fault.

Given $TCAP = 3.85 \times 10^{-6} \text{J}/(\text{m}^3 \, ^\circ\text{C})$, $\alpha_{20} = 3.78 \times 10^{-3} \, ^\circ\text{C}^{-1}$, $\rho_{20} = 44.0 \times 10^{-9} \, \Omega\text{m}$, $T_m = 1084 \, ^\circ\text{C}$, $T_a = 40 \, ^\circ\text{C}$, $K_0 = 245$, by equations (6-1), (6-2) the required conductor size is at least

$$A_{\text{mm}^2} = \frac{I_F \times 10^6}{\sqrt{\left(\dfrac{TCAP}{t_c \alpha_r \rho_r}\right) \ln \left(\dfrac{K_0 + T_m}{K_0 + T_a}\right)}}$$

$$= \frac{6.97 \times 10^3 \times 10^6}{\sqrt{\left(\dfrac{3.85}{0.5 \times 3.78 \times 10^{-3} \times 4.40}\right) \ln \left(\dfrac{245 + 1084}{245 + 40}\right)}} = 26.1 \ \text{mm}^2$$

This would result in the choice of a standard metric size of 35 mm², or the US size of #2 AWG (33.6 mm²).

However, the usual minimum size conductor for this application is #2/0 AWG (67.4 mm²), for considerations of mechanical strength and durability. The equivalent standard metric size is 70 mm².

The diameter of this conductor is

$$d = 2\sqrt{\frac{A}{\pi}} = 2\sqrt{\frac{70}{\pi}} = 9.44 \text{ mm.}$$

The recommended size for a 30% copper-clad steel wire, grounding conductor is 70 mm² or 2/0 AWG.

Step 3.   *Calculation of step and touch potentials.* First, calculate the surface layer derating factor:

$$C_S = 1 - \frac{0.09\left(1 - \dfrac{\rho}{\rho_s}\right)}{2h_s + 0.09} = 1 - \frac{0.09\left(1 - \dfrac{400}{2500}\right)}{2 \times 0.1 + 0.09} = 1 - 0.26 = 0.74$$

Calculation of step and touch potentials, assuming 70 kg weight,

$$E_{XW} = (1000 + m \cdot A \cdot C_S \cdot \rho_S)\frac{K_W}{\sqrt{t_s}}$$

$$E_{\text{Step 70}} = (1000 + 1 \times 6 \times 0.74 \times 2500)\frac{0.157}{\sqrt{0.5}}$$

$$E_{\text{Step 70}} = 2687 \text{ V}$$

$$E_{\text{Touch 70}} = (1000 + 1 \times 1.5 \times 0.74 \times 2500)\frac{0.157}{\sqrt{0.5}}$$

$$E_{\text{Touch 70}} = 838 \text{ V}$$

Step 4.   *Determine dimensions of ground grid.* Given the available area of 73 m × 84 m, the grid will be dimensioned as 70 m × 70 m. The distance between conductors is:

$$D = 7 \text{ m}$$

The total length of conductor required is

$$L_C = 2 \times 11 \times 70 = 1540 \text{ m}$$

Since there are no ground rods, the total length

$$L_T = L_C$$

The burial depth is

$$h = 0.5 \text{ m}$$

Step 5.   *Determination of ground grid resistance.*
From Step 1,

$$A = 70 \text{ m} \times 70 \text{ m} = 4900 \text{ m}^2.$$

$$\rho = 400 \text{ } \Omega \text{ m.}$$

From Step 4,

$$h = 0.5 \text{ m}$$

$$L_C = 1540 \text{ m}$$

$$R_g = \rho \left[\frac{1}{L_T} + \frac{1}{\sqrt{20A}} \left(1 + \frac{1}{1 + h\sqrt{20/A}} \right) \right]$$

$$= 400 \left[\frac{1}{1540} + \frac{1}{\sqrt{20 \times 4900}} \left(1 + \frac{1}{1 + 0.5\sqrt{20/4900}} \right) \right] = 2.78 \ \Omega$$

Step 6. *Recalculate ground current and fault duration.*
From Step 2,
$$D_f = 1.0$$

$3I_0 = 3180$ A (for 115 kV system)
By Endrenyi's method,

$$Z_{eq-l} = 0.5Z_{s-l} + \sqrt{R_{tg}Z_{s-l}}$$
$$= 0.5(8.246 + j3.658) + \sqrt{10(8.246 + j3.658)} = 13.415 + j3.797$$

$$Z_{eq-f} = 0.5Z_{s-f} + \sqrt{R_{dg}Z_{s-f}}$$
$$= 0.5(0.732 + j0.732) + \sqrt{25(0.732 + j0.732)} = 5.064 + j2.312$$

where

$$Z_{eq} = \frac{1}{\frac{1}{Z_{eq-l}} + \frac{1}{Z_{eq-f}}} = \frac{1}{\frac{1}{13.415 + j3.79} + \frac{1}{5.064 + j2.312}} = 3.696 + j1.497$$

S_f is the current split factor:
$$S_f = \left| \frac{Z_{eq}}{Z_{eq} + R_g} \right| = \left| \frac{3.696 + j1.497}{(3.696 + j1.497) + 2.78} \right| = 0.6$$

where

R_{tg} is the transmission line ground impedance = 10 Ω/section

R_{dg} is the distribution line ground impedance = 25 Ω/section

Z_{s-l} is the impedance of the overhead static wire for the transmission line = $8.246 + j3.658\Omega$/span

Z_{s-f} is the impedance of the overhead static wire for the distribution line = $0.732 + j0.732$ Ω/span

$$I_G = D_f S_f 3I_0$$
$$= 1.0 \times 0.6 \times 3180$$
$$= 1908 \ A$$

Step 7. *Calculate ground potential rise (GPR).*
$$E_{GPR} = R_g I_G = 1908 \times 2.78 = 5304 \ V$$
$$E_{GPR} = 5304 \ V >> 838 \ V$$

Step 8. *Calculate mesh voltage, E_m.*

From Step 1,
$$\rho = 400 \ \Omega \ \text{m}.$$

From Step 2,
$$d = 9.44 \times 10^{-3} \ \text{m}.$$

From Step 4,

$D = 7$ m

$h = 0.5$ m

$L_T = 1540$ m

$$n_a = \frac{2L_T}{L_P} = \frac{2 \times 1540}{280} = 11$$

$$n_b = \begin{cases} 1 \text{ for square grids} \\ \sqrt{\dfrac{L_p}{4\sqrt{A}}} \end{cases} = 1$$

$$n_c = \begin{cases} 1 \text{ for square and rectangular grids} \\ \left[\dfrac{L_x L_y}{A}\right]^{\frac{0.7A}{L_x L_y}} \end{cases} = 1$$

$$n_d = \begin{cases} 1 \text{ for square and rectangular and } L\text{-shaped grids} \\ \dfrac{D_m}{\sqrt{L_x^2 L_y^2}} \end{cases} = 1$$

n is the effective number of parallel conductors in a grid:
$$n = n_a n_b n_c n_d = 11 \times 1 \times 1 \times 1 = 11$$

Since there are no ground rods,
$$K_{ii} = \frac{1}{(2n)^{2/n}} = \frac{1}{(2 \times 11)^{2/11}} = 0.57$$

K_h is the grid depth factor, defined as
$$K_h = \sqrt{1 + \frac{h}{1.0 \ \text{m}}} = \sqrt{1 + \frac{0.5}{1.0}} = 1.23$$

K_m is the geometrical factor, defined as
$$K_m = \frac{1}{2\pi}\left[\ln\left[\frac{D^2}{16hd} + \frac{(D+2h)^2}{8Dd} - \frac{h}{4d}\right] + \frac{K_{ii}}{K_h}\ln\left[\frac{8}{\pi(2n-1)}\right]\right]$$

$$= \frac{1}{2\pi}\left[\ln\left[\frac{7^2}{16 \times 0.5 \times 0.00944} + \frac{(7+2\times0.5)^2}{8\times7\times0.00944} - \frac{0.5}{4\times0.00944}\right] + \frac{0.57}{1.23}\ln\left[\frac{8}{\pi(2\times11-1)}\right]\right]$$

$$= 0.92$$

K_i is the irregularity factor, defined as:

$$K_i = 0.644 + 0.148n = 0.644 + 0.148 \times 11 = 2.272$$

K_{ii} is the inner conductor factor, defined as:
L_m is the effective buried length in m.

$$L_m = L_T$$

with no ground rods.

$$E_m = \frac{\rho K_m K_i I_G}{L_m}$$
$$= \frac{400 \times 0.92 \times 2.27 \times 1908}{1540} = 1036 \ V$$
$$E_m = 1036 \ V > 838 \ V$$

Step 9.   *If mesh voltage < tolerable touch voltage, go to step 10, otherwise go to step 11. Ground rods with length $L_r = 7.5$ m, having a total length $L_R = 20 \times 7.5 = 150$ m, are added to the periphery and steps 4–8 repeated:*
The total length of conductor required is:

$$L_T = 2 \times 11 \times 70 + 20 \times 7.5 = 1540 + 150 = 1690 \ m$$

$$R_g = \rho \left[\frac{1}{L_T} + \frac{1}{\sqrt{20A}} \left(1 + \frac{1}{1 + h\sqrt{20/A}} \right) \right]$$

$$= 400 \left[\frac{1}{1690} + \frac{1}{\sqrt{20 \times 4900}} \left(1 + \frac{1}{1 + 0.5\sqrt{20/4900}} \right) \right] = 2.75 \ \Omega$$

$$I_G = 1908 \ A$$

$$E_{GPR} = 1908 \times 2.75 = 5247 \ V >> 838 \ V$$

$K_{ii} = 1$ if there are ground rods along the perimeter or in the corners and along the perimeter and throughout the grid area.
L_P is the length of the perimeter of the grid $= 4 \times 70 = 280$ m.
L_x is the maximum length of the grid in the x-direction $= 70$ m.
L_y is the maximum length of the grid in the y-direction $= 70$ m.
L_m is the effective buried length in m:

$$L_m = L_C + \left[1.55 + 1.22 \left(\frac{L_r}{\sqrt{L_x^2 + L_y^2}} \right) \right] L_R$$

$$= 1540 + \left[1.55 + 1.22 \left(\frac{7.5}{\sqrt{70^2 + 70^2}} \right) \right] \times 7.5 \times 20$$

$$= 1540 + 246 = 1786 \ m$$

$$K_m = \frac{1}{2\pi} \left[\ln \left[\frac{D^2}{16hd} + \frac{(D+2h)^2}{8Dd} - \frac{h}{4d} \right] + \frac{K_{ii}}{K_h} \ln \left[\frac{8}{\pi(2n-1)} \right] \right]$$

$$= \frac{1}{2\pi} \left[\ln \left[\frac{7^2}{16 \times 0.5 \times 0.00944} + \frac{(7+2\times 0.5)^2}{8 \times 7 \times 0.00944} - \frac{0.5}{4 \times 0.00944} \right] \right.$$

$$\left. + \frac{1}{1.23} \ln \left[\frac{8}{\pi(2 \times 11 - 1)} \right] \right]$$

$$= 0.82$$

$$E_m = \frac{\rho K_m K_i I_G}{L_m}$$

$$= \frac{400 \times 0.82 \times 2.27 \times 1908}{1786} = 796 \text{ V}$$

$$E_m = 796 \text{ V} < 838 \text{ V}$$

Step 10. *If step voltage < tolerable touch voltage, go to step 12, otherwise go to step 11.*

L_S is the effective length of buried conductor:

$$L_S = 0.75 L_C + 0.85 L_R$$
$$= 0.75 \times 1540 + 0.85 \times 150 = 1283 \text{ m}$$

K_S is the spacing factor for step voltage:

$$K_S = \frac{1}{\pi} \left[\frac{1}{2h} + \frac{1}{D+h} + \frac{1}{D} \left(1 - 0.5^{n-2} \right) \right]$$

$$= \frac{1}{\pi} \left[\frac{1}{2 \times 0.5} + \frac{1}{7+0.5} + \frac{1}{7} \left(1 - 0.5^{11-2} \right) \right] = 0.406$$

$$E_S = \frac{\rho K_S K_i I_G}{L_S}$$

$$= \frac{400 \times 0.406 \times 2.27 \times 1908}{1283} = 548 \text{ V}$$

$$E_s = 548 \text{ V} < 2687 \text{ V}$$

Step 11. *If mesh or step voltage > tolerable touch voltage, revision of design is required.*

Step 12. *Add equipment ground conductors, additional grid conductors, ground rods as needed.*

CHAPTER 8

SAFETY ASPECTS OF GROUND GRID OPERATION AND MAINTENANCE

8.1 INTRODUCTION

However well a grounding system is designed, inevitably changes in the power system, deterioration of materials, environmental changes, and equipment failure will decrease the inherent safety of the original system. It is important to perform regular inspections, maintenance, and testing of any system to ensure its continued safety. This is especially true of grounding systems because of their exposure to the elements, unwanted human and animal damage, and to electrical events such as lightning strikes, overloads, and short circuits. This is to emphasize that safe design is not the end of the safety process, but only the beginning. Equipment and installations must remain electrically safe throughout their entire useful life, until their final de-energization and demolition changes the nature of the hazards to other than electrical.

It is crucial for the user to be aware of ground-fault current magnitude at the service entrance to their premises. The protection that ground grids provide against step and touch potentials is only good up to the expected level and duration of ground fault currents, as originally communicated by the electric utility in the design phase. Higher current to ground can cause malfunctioning and damage of equipment as well as diminish system reliability. Unexpected high fault currents can also produce thermal and mechanical damages to a customer's ground grid and ground grid connections with consequential permanent loss of effectiveness of the same (Mitolo, Sutherland, and Natarajan, 2010a). In order to prevent these problems from occurring, a ground grid assessment utilizing updated data should be carried out on a regular basis. This chapter is adapted from EPRI (2006).

8.2 EFFECTS OF HIGH FAULT CURRENTS

There are many possible causes for increased ground potential rise (GPR) in substations (IEEE, 2000) which may be aggravated by high fault currents:

Principles of Electrical Safety, First Edition. Peter E. Sutherland.
© 2015 John Wiley & Sons, Inc. Published 2015 by John Wiley & Sons, Inc.

1. drying of the soil, increasing the ground resistance;
2. excessive voltage drops in the conductors and connectors due to high currents;
3. fusing, melting, and connector failures;
4. arcing, burning, and open circuits;
5. altered current flow paths, further increasing voltage drops;
6. corroded or otherwise damaged conductors and connectors (examples from a high voltage substation are shown in Figures 8.1–8.3);
7. thinning of the protective surface layer of crushed stone or gravel;
8. weeds and shrubs growing in surface layer (an example from a high voltage substation is shown in Figure 8.4);
9. mixing of the surface layer with soil and dust, decreasing its resistivity;
10. failure of static wire, ground, or neutral wire connections from transmission and distribution lines to the substation.
11. reduction in electrical safety causing increased step and touch potentials.

Figure 8.1   Deteriorated grounding conductor and cracked concrete foundation.

Figure 8.2 Deteriorated grounding conductor, broken concrete support, and insulator debris.

Figure 8.3 Deteriorated grounding conductor and broken insulator debris.

The mechanisms by which these and other factors can cause injury or death are the step and touch potentials. First, the ground resistance is calculated using Sverak's equation (IEEE, 2000, p. 65):

$$R_g = \rho \left[\frac{1}{L_T} + \frac{1}{\sqrt{20A}} \left(1 + \frac{1}{1 + h\sqrt{20/A}} \right) \right] \tag{8.1}$$

where:

R_g = resistance in Ω

ρ = earth resistivity in Ω m

L_T = total length of buried conductors in m

A = total area of the grid in m^2

Figure 8.4   Vegetation and deteriorated concrete.

The GPR is

$$E_{GPR} = R_g I_G \qquad (8.2)$$

where, when exposed to high fault currents, ρ can increase as a result of drying, L_T can decrease due to conductor damage, all increasing R_g.

Similarly, the mesh voltage

$$E_m = \frac{\rho K_m K_i I_G}{L_M} \qquad (8.3)$$

where (all as defined in (Villas and Portela, 2003))

K_m = geometrical factor

K_i = irregularity factor

L_M = effective buried length in m

The mesh voltage increases with I_G, ρ and decreasing L_M. The step voltage

$$E_s = \frac{\rho K_S K_i I_G}{L_S} \qquad (8.4)$$

where

L_S = effective length of buried conductor

K_S = spacing factor for step voltage.

The step voltage also increases with I_G, ρ and decreasing L_S. The constants K are geometrical and do not change with increasing fault current.

8.3 DAMAGE OR FAILURE OF GROUNDING EQUIPMENT

8.3.1 Thermal Damage to Conductors Due to Excessive Short-Circuit Currents

An increase in fault current will decrease the fusing time of the grid conductors:

$$t_c = \left(\frac{A_{\text{kcmil}}}{I \cdot D_f \cdot K_f} \right)^2 \tag{8.5}$$

where

t_c = fusing time of the conductor in s

A_{kcmil} = conductor cross sectional area in kcmil

I = rms symmetrical fault current in kA (It is assumed that the symmetrical rms fault current $I \approx 3I_0$ from the ground fault calculations for the substation).

K_f is the material-fusing constant. Typical values are 7.00 for soft-drawn copper, 10.45 for 40% conductivity copper-clad steel wire, and 15.95 for a steel conductor.

D_f is the decrement factor, used when fault durations are less than 1 s or the X/R ratio is greater than 5.

Examples of fusing current calculation results are shown in Table 8.1. It is recommended that the ground grid conductor thermal limit be plotted as a point on a time-current curve, Figure 8.5, and an $i^2 t$ line be extended upward from it to represent the ground grid thermal damage curve. If this curve is exceeded, the conductors

TABLE 8.1 Test Currents and Fusing Currents for Annealed Soft-Drawn 100% Conductivity Copper Conductors

			#4/0 AWG	500 kcmil
Test	Peak or rms Symmetrical	Time (s)	Current (kA)	Current (kA)
Electromagnetic force	Peak	0.2	57.6	136
Minimum fault	rms symmetrical	10.0	8.6	20.2
Fusing current	rms symmetrical	1.0	21.4	50.4
Fusing current	rms symmetrical	10.0	9.5	22.4

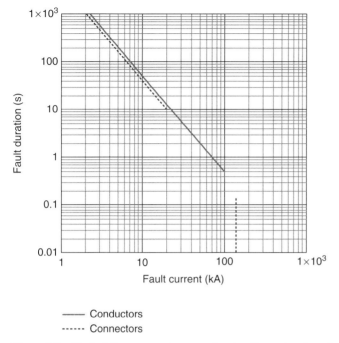

Figure 8.5   Typical time–current curves showing thermal and mechanical withstand for 500 kcmil copper ground grid conductors (assuming 0.5 s fault-clearing time) and connectors.

may be subject to fusing, melting, or other forms of thermal and mechanical damage. A vertical line may represent the mechanical damage curve.

8.3.2   Connector Damage Due to Excessive Short-Circuit Stresses

This primarily affects permanent connections. Grounding grid connections should withstand short-circuit electromagnetic forces up to the 1.0 s fusing current of the conductors to which they are attached, and heating from fault currents for up to 90% of the fusing current for 10 s.

The test specifications in IEEE Standard 837-2002 (IEEE, 2002a) call for the construction of a control conductor, as shown in Figure 8.6, of length L_{CC1} and resistance R_{CC1}. A test loop containing up to four connector assemblies under test is assembled. Each connector assembly consists of two conductor samples. If the conductors are stranded, then equalizers must be used at each end where a connector is not in place. The purpose of the equalizers is to establish an equipotential plane across the ends of the conductor strands.

The electromagnetic force test applies an asymmetrical waveform with the specifications listed in Table 8.2. The sample resistance R_{Total} after the first test is not

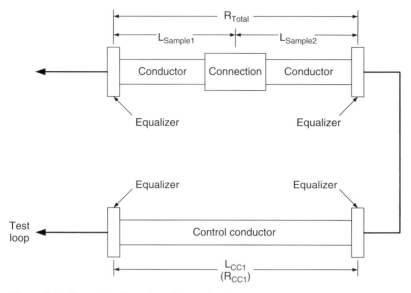

Figure 8.6 Short-circuit testing of ground grid connectors. The test loop contains one through four connector assemblies.

TABLE 8.2 Fault Current Tests for Connectors

Test	Force	Thermal
Std. 837-2002 Section	7.3	11.
Fault current duration	0.2 s	10.0 s
Fusing current duration	1.0 s	10.0 s
Fault current X/R	20	N/A
Fault current peak/rms	2.7	$\sqrt{2}$
Fault current rms/fusing current rms	1.0	0.9

to exceed 110% of R_{CC1} and 150% after three tests. The fusing test applies a symmetrical fault current, as listed in Table 8.2, repeated three times. The evaluation of the test consists of disassembly and dissection of the connection and inspection for signs of melting or other damage. Results of example calculations are shown in Table 8.1. As long as the fault magnitude and duration fall under the two points defined by these tests, as listed in Table 8.2, there should be no failures of connectors due to excessive fault currents.

8.3.3 Drying of the Soil Resulting in Increased Soil Resistivity

A current density of less than 200 A/m^2 for 1 s is recommended. Heating of soil whose temperature is above the freezing point has negligible effect on its resistivity. The effect of moisture content on soil resistivity is given in the IEEE Green Book

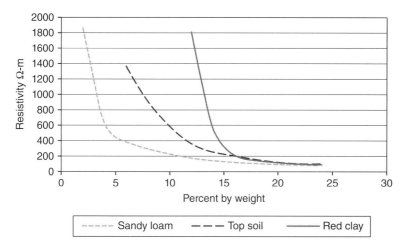

Figure 8.7   Effect of moisture content on soil resistivity.

(IEEE, 2007a) for several soil types. Resistivity is quite constant above 22% moisture content, but increases dramatically below that. Figure 8.7 shows this effect for three different types of soil. Experimental test results from Georgia Power show significant variation in ground grid resistance with the seasons, but little change when the integrated ground resistance including the transmission line shield wires was considered (EPRI, 1992).

   Without performing three-dimensional electromagnetic simulations, there is no easy way to calculate the drying of soil by the passage of electrical current and the consequent increase in resistivity.

8.4   RECOMMENDATIONS

In order to prevent these effects, the following regular recalculation of ground grid parameters should be made in addition to normal ground grid maintenance, to ensure that fault current levels are not exceeded:

1. ground resistance measurements;
2. short-circuit calculations, usually done with specialized software;
3. recalculation of step and touch potentials, usually done with specialized software.

   There are various measures that can be taken to reinforce a ground grid. These include

1. adding ground grid conductors, decreasing the spacing of conductors;
2. increasing the area of the grid;
3. adding parallel conductors around the perimeter of the ground grid;
4. adding ground rods, with closer spacing at the perimeter;

5. diverting fault currents to other paths;

6. limiting total fault current;

7. barring access to hazardous areas;

8. connecting overhead ground wires from transmission lines;

9. decreasing tower-footing resistances;

10. reinforcing or replacing connectors between above-ground components and the ground grid;

11. increasing thickness of upper layer of crushed stone;

12. using soil treatment to lower resistivity;

13. using deep ground wells, which may be useful when ground rods are not effective.

CHAPTER *9*

GROUNDING OF DISTRIBUTION SYSTEMS

9.1 STRAY CURRENTS IN DISTRIBUTION SYSTEMS

Electrical shock hazards can exist in many situations where there is no direct contact with any electrical conductors or equipment. This chapter will discuss some of the hazards which are produced by electrical utility distribution systems. As with other power systems, distribution systems are designed to be safe, and to be in accordance with codes and standards whose purpose is to enforce safe design practices. Nonetheless, as these systems cover wide areas and are subject to a nearly infinite variety of conditions, serving all manner of loads for many human purposes, it is inevitable that safety hazards will exist. There are a variety of distribution systems in the world, with different voltages, grounding configurations, numbers of conductors, and so forth. This chapter in particular addresses mostly the multigrounded neutral system prevalent in North America. Further information on other systems may be found in Mitolo (2009b).

Stray currents, sometimes called *objectionable currents* are part of the same phenomenon called *stray voltage* (Burke and Untiedt, 2009; Sutherland, 2011). Electric shocks to humans are documented in an EPRI technical brief on swimming pools (EPRI, 2000). A detailed description of stray voltages caused by the multigrounded neutral system and their effects on both humans and farm animals is found in the USDA handbook (USDA, 1991). These can cause injuries to humans and farm animals in a similar manner to step and touch potentials in substations, only in the home and farm environment. Accidents to humans typically result from shock hazards at swimming pools, bathtubs, basements, and other wet locations. These have led to the development of the ground fault current interrupter (GFCI). Accidents to farm animals are typically from cows receiving shocks in barns from contact with metal objects, especially during milking. Low levels of electrical current, not large enough to cause painful shocks, may cause reduced milk production. This chapter will begin with a discussion of the typical electric utility distribution system for homes and farms in the United States, and the domestic wiring system that it feeds. The flow of currents for both unbalanced loads and ground faults will be estimated, along with the voltages that are associated with the currents. The level of hazard for various configurations will be calculated. Finally, methods of remediation will be explored.

Principles of Electrical Safety, First Edition. Peter E. Sutherland.
© 2015 John Wiley & Sons, Inc. Published 2015 by John Wiley & Sons, Inc.

9.2 THREE-PHASE MULTIGROUNDED NEUTRAL DISTRIBUTION LINE

The utility multigrounded neutral system, widely used in the United States, is intended to share load current between earth and a conductor (Mitolo, 2010). A wide variety of other distribution systems and grounding methods are used throughout the world (Mitolo, Tartaglia, and Panetta, 2010). The phase conductors are mounted on insulators on the top of poles is shown in Figure 9.1. The three-phase version seen in Figure 9.2 is similar, except for having only one phase conductor on the poles. The neutral conductor is mounted on insulators on the side of the pole. At distances of approximately every 0.4 km (¼ mi), a ground is applied. The ground consists of a grounding rod with ground resistance of 25 Ω or less. This is connected to the

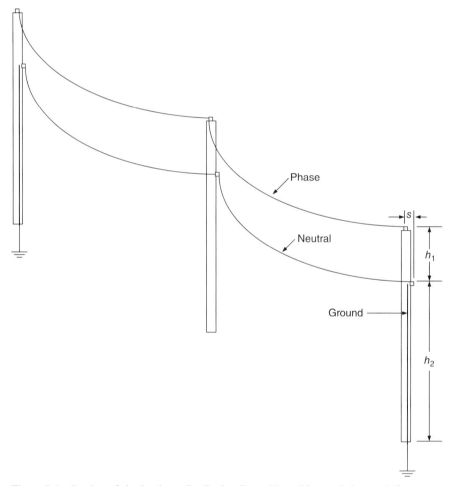

Figure 9.1   Portion of single-phase distribution line with multigrounded neutral (frequency of grounding exaggerated).

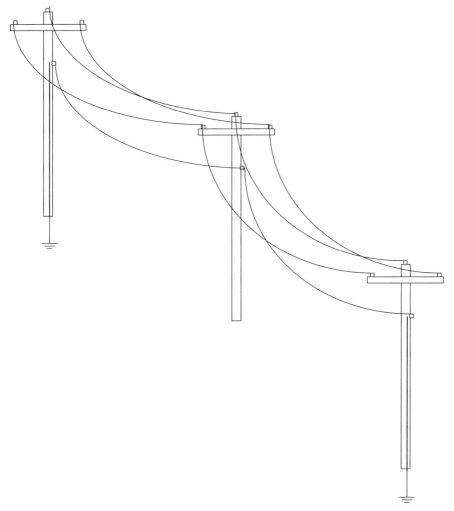

Figure 9.2   Portion of three-phase distribution line with multigrounded neutral (frequency of grounding exaggerated) (*Source*: © IEEE (2011)).

neutral conductor by means of a wire running up the pole. The neutral conductor is thus used also as the system-grounding conductor. System analysis methods will be developed for the three-phase system, as the single-phase system can be analyzed as a special case of the three-phase system. The analysis presented here is based on the method of Kersting (Kersting, 2003, 2007, 2009). The safety concerns with this system are

- the step and touch voltages which may develop as a result of excessive ground currents;
- overvoltages on the system neutral conductor that may pose a hazard to utility workers.

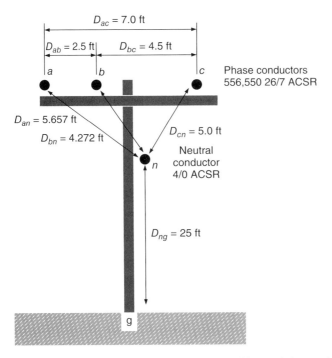

Figure 9.3   Line configuration: three-phase multigrounded neutral distribution line.

This section considers the medium-voltage distribution system. The 120/240 V secondary systems will be examined in the next section.

The three-phase, four-wire multigrounded neutral system can be modeled using the impedance matrix method, where the voltages are measured with reference to ground (Figure 9.3). The load voltage equation is written in the Z-matrix method:

$$V' = V - Z_p \cdot I \tag{9.1}$$

or in expanded form:

$$
\begin{bmatrix} V'_{ag} \\ V'_{bg} \\ V'_{cg} \\ V'_{ng} \end{bmatrix}
=
\begin{bmatrix} V_{ag} \\ V_{bg} \\ V_{cg} \\ V_{ng} \end{bmatrix}
-
\begin{bmatrix}
\hat{z}_{aa} & \hat{z}_{ab} & \hat{z}_{ac} & \hat{z}_{an} \\
\hat{z}_{ba} & \hat{z}_{bb} & \hat{z}_{bc} & \hat{z}_{bn} \\
\hat{z}_{ca} & \hat{z}_{cb} & \hat{z}_{cc} & \hat{z}_{cn} \\
\hat{z}_{na} & \hat{z}_{nb} & \hat{z}_{nc} & \hat{z}_{nn}
\end{bmatrix}
\cdot
\begin{bmatrix} I_a \\ I_b \\ I_c \\ I_n \end{bmatrix}
\tag{9.2}
$$

The impedance terms are calculated from Carson's equations, eliminating the need for a fifth row and column of the matrix for the ground impedances Z_{dd}:

$$\hat{z}_{ii} = r_i + r_d + j\frac{\omega\ \mu_0 \times 10^3}{2\ \pi} \left(\ln\left(\frac{1}{\mathrm{GMR}_i}\right) + D \right)\ \ \Omega/\mathrm{km} \tag{9.3}$$

$$\hat{z}_{ij} = r_d + j\frac{\omega\ \mu_0 \times 10^3}{2\ \pi} \left(\ln\left(\frac{1}{D_{ij}}\right) + D \right)\ \ \Omega/\mathrm{km} \tag{9.4}$$

where

$$\frac{\omega\mu_0 \times 10^3}{2\pi} = 0.0754 \ \Omega/\text{km} \quad \text{for a 60 Hz system}$$

$r_i =$ conductor resistance, in Ω/km

$r_d =$ equivalent earth resistance $= 0.05922 \ \Omega/\text{km}$

$\text{GMR}_i =$ conductor geometric mean radius (GMR)

$D_{ij} =$ distance between conductors i and j

$D =$ equivalent earth logarithmic term $= 7.93402$

The source voltage matrix is

$$V_{\text{Source}} = \begin{bmatrix} V_{an} \\ V_{bn} \\ V_{cn} \\ V_{ng} \end{bmatrix} \tag{9.5}$$

The load voltage matrix is

$$V_{\text{Load}} = \begin{bmatrix} V'_{an} \\ V'_{bn} \\ V'_{cn} \\ V'_{ng} \end{bmatrix} \tag{9.6}$$

The line current is

$$I_{\text{Line}} = \begin{bmatrix} I_a \\ I_b \\ I_c \\ I_n \end{bmatrix} \tag{9.7}$$

The loads are modeled as constant impedances to simplify the analysis. The load impedance matrix is

$$Z_{\text{Load}} = \begin{bmatrix} z_{an} & 0 & 0 & 0 \\ 0 & z_{bn} & 0 & 0 \\ 0 & 0 & z_{cn} & 0 \\ 0 & 0 & 0 & 0 \end{bmatrix} \tag{9.8}$$

The grounding impedance matrix is

$$Z_G = \begin{bmatrix} 0 & 0 & 0 & 0 \\ 0 & 0 & 0 & 0 \\ 0 & 0 & 0 & 0 \\ 0 & 0 & 0 & z_{ng} \end{bmatrix} \tag{9.9}$$

For one segment of a multigrounded distribution line, as shown in Figure 9.4, equation (9.1) becomes

$$V_{\text{Load}} = V_{\text{Source}} - (Z_p \cdot l_s) \ I_{\text{Line}} \tag{9.10}$$

Adding in the load voltage drop,

$$0 = V_{\text{Source}} - (Z_p \cdot l_s + Z_{\text{Load}} + Z_G) \ I_{\text{Line}} \tag{9.11}$$

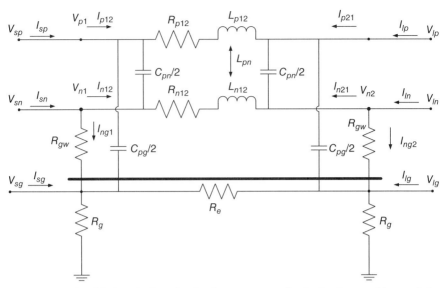

Figure 9.4   Simplified equivalent circuit of one segment of a single-phase multigrounded neutral distribution line.

The earth currents are the sum of the currents at each node:

$$I_d = I_a + I_b + I_c + I_n \tag{9.12}$$

The procedure for solving one segment of a multigrounded line is summarized as follows:

1. Solving for line current,

$$I_{\text{Line}} = (Z_p \cdot l_s + Z_{\text{Load}} + Z_G)^{-1} V$$

2. Solving for the segment end voltage,

$$V' = V - (Z_p \cdot l_s) I_{\text{Line}}$$

Solving for earth currents,

$$I_d = I_a + I_b + I_c + I_n$$

Example:

Using the example line from Kersting, with grounding resistance of 25 Ω, and a line length of 1.6 km = 1 mi, as shown in

$$R_i = 0.1859 \ \Omega/\text{mi}, \quad D_{ii} = 0.927 \ \text{ft}, \quad \text{GMR}_i = 0.0313 \ \text{ft}$$
$$R_n = 0.592 \ \Omega/\text{mi}, \quad D_{nn} = 0.563 \ \text{ft}, \quad \text{GMR}_n = 0.00814 \ \text{ft}$$
$$D_{ab} = 2.5 \ \text{ft}, \quad D_{bc} = 4.5 \ \text{ft}, \quad D_{ac} = 7.0 \ \text{ft}$$
$$D_{an} = 5.657 \ \text{ft}, \quad D_{bn} = 4.272 \ \text{ft}, \quad D_{cn} = 5.0 \ \text{ft}$$

$$\hat{z}_{ii} = r_i + 0.05922 + j0.0754 \left(\ln \left(\frac{1}{GMR_i} \right) + 7.93402 \right) \; \Omega/km$$

$$\hat{z}_{nn} = r_n + 0.05922 + j0.0754 \left(\ln \left(\frac{1}{GMR_n} \right) + 7.93402 \right) \; \Omega/km$$

$$\hat{z}_{ij} = 0.05922 + j0.0754 \left(\ln \left(\frac{1}{D_{ij}} \right) + 7.93402 \right) \; \Omega/km$$

Then,

$$Z_p = \begin{bmatrix} 0.175 + j0.859 & 0.059 + j0.529 & 0.059 + j0.451 & 0.059 + j0.468 \\ 0.059 + j0.529 & 0.175 + j0.859 & 0.059 + j0.485 & 0.059 + j0.477 \\ 0.059 + j0.451 & 0.059 + j0.485 & 0.175 + j0.859 & 0.059 + j0.489 \\ 0.059 + j0.468 & 0.059 + j0.477 & 0.059 + j0.489 & 0.427 + j0.961 \end{bmatrix} \Omega/km$$

$$Z_{Load} = \begin{bmatrix} 30.409 + j16.413 & 0.000 & 0.000 & 0.000 \\ 0.000 & 49.242 + j16.185 & 0.000 & 0.000 \\ 0.000 & 0.000 & 20.733 + j15.550 & 0.000 \\ 0.000 & 0.000 & 0.000 & 0.000 \end{bmatrix} \Omega$$

$$Z_G = \begin{bmatrix} 0.000 & 0.000 & 0.000 & 0.000 \\ 0.000 & 0.000 & 0.000 & 0.000 \\ 0.000 & 0.000 & 0.000 & 0.000 \\ 0.000 & 0.000 & 0.000 & 25.0 \end{bmatrix} \Omega$$

$$V_{Source} = \begin{bmatrix} 7199.6\angle 0.0 \\ 7199.6\angle - 120.0 \\ 7199.6\angle 120.0 \\ 0.0\angle 0.0 \end{bmatrix}$$

and

$$V_{Load} = \begin{bmatrix} 7128.1\angle - 1.4 \\ 7224.3\angle - 119.9 \\ 6965.9\angle 119.2 \\ 99.7\angle - 64.2 \end{bmatrix}$$

$$I_{Line} = \begin{bmatrix} 206.3\angle - 29.7 \\ 139.4\angle - 138.1 \\ 268.8\angle 82.3 \\ 4.0\angle - 64.2 \end{bmatrix}$$

With these values,

$$I_d = \sum I_{Line} = 131.5\angle - 149.2$$

A person, having 1000 Ω of resistance with their feet grounded and touching the neutral wire, will conduct 99.7 mA.

If the ground resistance is reduced to 1.0 Ω,

$$V_{ng} = 45.1\angle - 102$$

$$I_n = 45.1\angle - 102$$

$$I_d = \sum I_{Line} = 78.6\angle - 162.0$$

A person with 1000 Ω of resistance will conduct 45.1 mA.
Even if the ground resistance is reduced to zero, and $V_{ng} = 0$,

$$I_n = 61.8\angle -125.3$$
$$I_d = \sum I_{\text{Line}} = 78.6\angle -162.0$$

However, a 1000 Ω person will conduct 1 mA or more as long as the grounding resistance is 0.016 Ω or greater. Another factor that can be used to reduce shock hazard is to reduce the distance between grounds. With 25 Ω grounding, $l_s = 0.4$ km results in 26.9 mA. To achieve 1 mA, $l_s = 0.015$ km.

In conclusion, the safe levels of stray current can be determined on the basis of the phase and neutral currents, the grounding resistance, and the spacing of grounds on the poles.

If the neutral is ungrounded, the unbalanced phase current will result in a hazardous voltage on the neutral conductor. Assuming the grounding resistance is now 1000 Ω, $l_s = 1.6$ km results in 102.7 V and $l_s = 0.4$ km results in 27.1 V. To achieve 1.0 V, $l_s = 0.015$ km.

9.3 SECONDARY SYSTEMS: 120/240 V SINGLE PHASE

The majority of domestic services are fed with the familiar center-tapped 120/240 V single-phase transformer (Kersting, 2009; Mitolo, Liu, and Qiu, 2013). Electrical safety in the home revolves around two major issues, electrical shock from contact with energized conductors and electrical fires due to short circuits (Mitolo, 2009b). Here, we will examine the electrical shock hazards. The voltage from a conductor, whether it is hot or neutral to ground, will produce a current flow through the human body, which can then cause the sensation of shock, injury, or death. In order to minimize hazards in the home, limitations are placed on voltage magnitude, current magnitude, and duration of exposure. The voltage magnitude is limited by means of grounding, the current magnitude by the series impedance of the system and the human body resistance, and the duration by the time to trip of the protective device, normally a ground fault circuit interrupter (GFCI). By means of circuit analysis, we can determine the voltage and current magnitudes, and thus the extent of the risks involved.

The simplified equivalent circuit, Figure 9.5, consists of the following elements:

1. Utility system single-phase Thèvenin equivalent source.
2. Utility grounding conductor and ground rod, modeled as a resistance.
3. Utility transformer: single-phase 120/240 V secondary.
4. Service entrance conductors, triplex cable with bare neutral and insulated phase conductors. These are relatively heavy cables, the neutral often being used for structural support. They are modeled as series and mutual inductances and resistances.

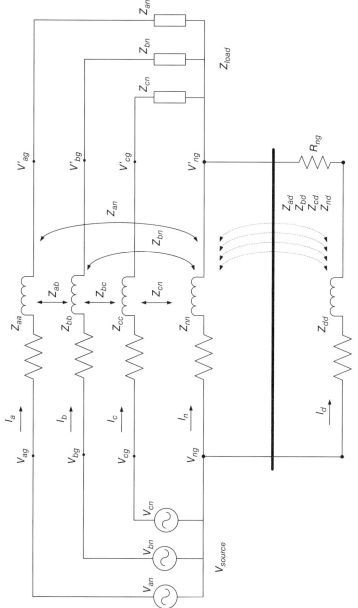

Figure 9.5 Equivalent circuit of one segment of a three-phase multigrounded neutral distribution line, showing mutual impedances (*Source*: © IEEE (2011)).

155

5. Service entrance, location of revenue metering equipment and service entrance panel.

6. Premises ground, connected to service entrance panel, modeled as a resistance.

7. Premises wiring, consisting of insulated phase and neutral conductors and uninsulated grounding conductor (not shown). These are small-gauge conductors, and may be modeled as resistances.

8. Loads, which may be 120 or 240 V.

The detailed transformer model is shown in Figure 9.6. The transformer is characterized by the following data:

1. Transformer volt-ampere rating, S in kVA
2. Primary voltage, V_0
3. Secondary voltage, V_1, $V_1 + V_2$
4. Transformation ratio: $n_t = V_0/V_1$
5. Impedance $Z_T = R_T + jX_T$ as a percentage on the kVA base.

The transformer impedance is divided into three impedances, Z_0, Z_1, and Z_2:

$$Z_0 = 0.5 \times R_T + j0.8 \times X_T$$
$$Z_1 = R_T + j0.4 \times X_T \qquad (9.13)$$
$$Z_2 = R_T + j0.4 \times X_T$$

The transformer impedance must be converted from percentage to ohms, using the per-unit system:

Determine the primary and secondary full load currents: $I_0 = S/V_0$, $I_1 = I_2 = S/(2 \times V_1)$.

Calculate the base impedances for the primary and secondary: $Z_{0\,base} = V_0/I_1$, $Z_{1\,base} = Z_{2\,base} = V_1/I_1$.

Convert percentage impedance to ohms: $Z_X = (Z_X/100) \times Z_{X\,base}$, where $X = 0, 1, 2$.

Convert primary ohms to secondary base: $Z_{0s} = Z_0/n_t^2$.

The service entrance cable is modeled as in equations (9.3) and (9.4):

$$\hat{z}_{ii} = \left[0.05922 + r_d + j0.0754 \left(\ln \left(\frac{1}{GMR_i} \right) + 7.93402 \right) \right] \ \Omega/km \quad (9.14)$$

$$\hat{z}_{ij} = \left[r_d + j0.0754 \left(\ln \left(\frac{1}{D_{ij}} \right) + 7.93402 \right) \right] \ \Omega/km \quad (9.15)$$

The loads are modeled as constant impedance to simplify the analysis. The source impedance is assumed to be much less than the transformer impedance and is neglected. The premises wiring impedance is assumed to be much less than the load impedance and is neglected. The final equivalent circuit is shown in Figure 9.7.

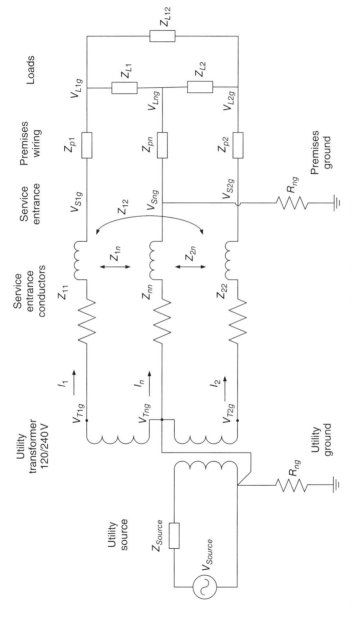

Figure 9.6   Equivalent circuit of domestic service with premises wiring and load (*Source*: © IEEE (2011)).

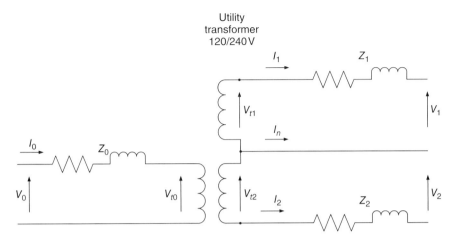

Figure 9.7   Detailed model of distribution transformer (*Source*: © IEEE (2011)).

9.3.1   Example of Stray Currents — Touching a Grounded Conductor

Calculate the current through a person touching a grounded object.

Source voltage = 7200 V, equivalent circuit source voltage = 120 V.

Transformer specifications: $S = 50$ kVA, 7200/120/240 V, $Z = 1.1 + j1.8\%$.

Transformer impedance conversion:

$$Z_{1\text{ base}} = (120\text{ V})^2/50 \times 10^3 \text{ VA} = 0.288 \ \Omega$$

$$Z_0 = 0.5 \times R_T + j0.8 \times X_T = 0.0055 + j0.0144 \text{ pu} = 0.0191 + j0.05 \ \Omega \ @ \ 120 \text{ V}$$

$$Z_1 = Z_2 = R_T + j0.4 \times X_T = 0.011 + j0.72 \text{ pu} = 0.0382 + j \ 0.025 \ \Omega \ @ \ 120 \text{ V}$$

$$Z_0 + Z_1 = Z_0 + Z_2 = 0.0573 + j0.075 \ \Omega \ @ \ 120 \text{ V}$$

Service entrance cable: phase conductors #2/0 AA, neutral conductor #2/0 ACSR, length = 100 ft (30.5 m).

For aluminum conductors, $r = 0.755 = 0.469 \ \Omega/\text{km}$, GMR = 0.0135 ft = 0.00411 m, $D = 0.447$ in. = 0.0114 m.   Insulation   thickness = 60 mils = 0.00152 m.

$$\hat{z}_{11} = \hat{z}_{22} = \left[0.05922 + r_d + j0.0754 \left(\ln\left(\frac{1}{GMR_i}\right) + 7.93402\right)\right] \frac{\Omega}{\text{km}} \times \ell_s$$

$$= \left[0.05922 + 0.469 + j0.0754 \left(\ln\left(\frac{1}{0.00411}\right) + 7.93402\right) \frac{\Omega}{\text{km}}\right] \times 0.0305 \text{ km}$$

$$= 0.0161 + j0.0126 \ \Omega$$

$$\hat{z}_{12} = \hat{z}_{21} = \left[0.05922 + j0.0754 \left(\ln\left(\frac{1}{2 \times 0.00152}\right) + 7.93402\right) \frac{\Omega}{\text{km}}\right]$$

$$\times \ 0.0305 \text{ km}$$

$$= 0.00181 + j0.0315 \ \Omega$$

For ACSR conductors, $r = 0.895$ Ω/mi $= 0.556$ Ω/km, GMR $= 0.00510$ ft $= 0.00155$ m, $D = 0.477$ in. $= 0.0121$ m.

$$\hat{z}_{nn} = \left[0.05922 + r_i + j0.0754\left(\ln\left(\frac{1}{GMR_i}\right) + 7.93402\right) \frac{\Omega}{km}\right] \times \ell_s$$

$$= \left[0.05922 + 0.556 + j0.0754\left(\ln\left(\frac{1}{0.00510}\right) + 7.93402\right) \frac{\Omega}{km}\right] \times 0.0305 \ km$$

$$= 0.0187 + j0.0304 \ \Omega$$

$$\hat{z}_{1n} = \hat{z}_{2n} = \hat{z}_{n1} = \hat{z}_{n2} = \left[0.05922 + j0.0754\left(\ln\left(\frac{1}{0.00152}\right) + 7.93402\right) \frac{\Omega}{km}\right]$$

$$\times 0.0305 \ km$$

$$= 0.00181 + j0.0332 \ \Omega$$

Loads: $Z_{L1} = 2.88 \ \Omega \ \angle 18.2° = 2.74 + j0.90$, $Z_{L2} = 1.44 \ \Omega \ \angle 25.8° = 1.30 + j0.63$.

Writing the matrix equations,

$$[V] = [Z][I]$$

$$[I] = [Z]^{-1}[V]$$

$$\begin{bmatrix} I_1 \\ I_2 \\ I_n + I_g \end{bmatrix} = \begin{bmatrix} \hat{z}_0 + \hat{z}_1 + \hat{z}_{11} + \hat{z}_{L1} & \hat{z}_{12} & \hat{z}_{1n} \\ \hat{z}_{21} & \hat{z}_0 + \hat{z}_2 + \hat{z}_{22} + \hat{z}_{L2} & \hat{z}_{2n} \\ \hat{z}_{n1} & \hat{z}_{n2} & \hat{z}_{nn} \| 2R_{ng} \end{bmatrix}^{-1} \begin{bmatrix} V_{S1} \\ V_{S2} \\ 0 \end{bmatrix}$$

For this example, by varying R_g,

$R_g = 25\Omega$, $I_g = 19.8$ mA, and a 1000 Ω human in parallel with R_g sees 0.71 mA.

$R_g = 1.0\Omega$, $I_g = 20.4$ mA, and a 1000 Ω human in parallel with R_g sees 0.72 mA.

$R_g = 0.1\Omega$, $I_g = 26.1$ mA, and a 1000 Ω human in parallel with R_g sees 0.84 mA.

Showing that for this condition of imbalance, the shock sensation is below the 1.0 mA threshold of perception.

9.3.2 Example of Stray Currents — With One Conductor Shorted to Neutral

Calculate the current through a person touching a grounded object when Load 1 is shorted phase to neutral. For this example, by varying R_g,

$R_g = 25 \ \Omega$, $I_g = 1721$ mA, and a 1000 Ω human sees 61.4 mA,

$R_g = 1.0 \ \Omega$, $I_g = 1747$ mA, and a 1000 Ω human sees 61.8 mA,

$R_g = 0.1 \ \Omega$, $I_g = 1965$ A, and a 1000 Ω human sees 63.5 mA,

showing that for this fault condition, the shock is above the 30 mA threshold for paralysis, and approaches the 75 mA threshold for fibrillation.

9.4 REMEDIATION OF STRAY-CURRENT PROBLEMS

The shock hazard is caused by the ground loop between the ground at the pole and the ground at the service entrance. By touching the grounded neutral, a person places themselves in parallel with the grounding resistance through which current may be flowing, and a voltage drop present. This results in stray current flowing through the human body.

Removing the ground loop can be done by two methods:

1. Removing the ground at the secondary of the utility transformer, by removing the bond between the secondary neutral and primary ground. The equivalent circuit is shown in Figure 9.8. The simplified equivalent circuit used in the calculations above is modified as shown in Figure 9.9. In either example considered, the unbalanced circuit or the line to neutral fault, no current will flow from neutral to ground, and no stray current would flow through a human touching the neutral.

2. Removing the premises ground and leaving all grounding in the utility system. The equivalent circuit is shown in Figure 9.10. The simplified equivalent circuit used in the calculations above is modified as shown in Figure 9.11. In either example considered, the unbalanced circuit or the line to neutral fault, no current will flow from neutral to ground, and no stray current would flow through a human touching the neutral.

The first method is preferable because it requires making the fewest changes to the system, and retains the premises ground, which is considered to provide an extra

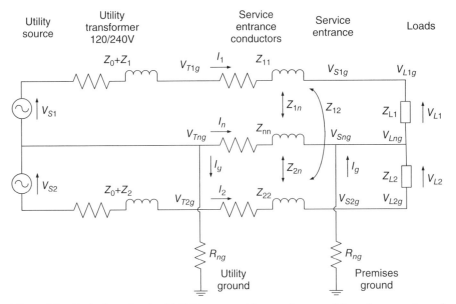

Figure 9.8 Equivalent circuit of 120 V system with ground removed from the secondary of the utility transformer for analysis (*Source*: © IEEE (2011)).

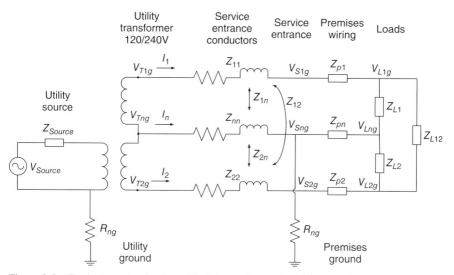

Figure 9.9   Equivalent circuit of modified domestic service with ground removed from the secondary of the utility transformer (*Source*: © IEEE (2011)).

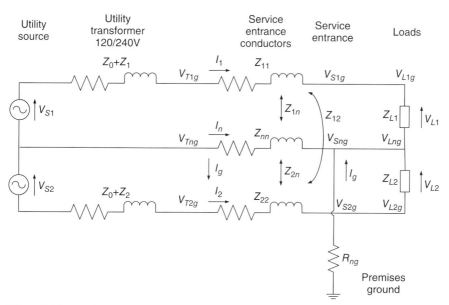

Figure 9.10   Equivalent circuit of modified 120 V system with ground removed from secondary of transformer for analysis (*Source*: © IEEE (2011)).

margin of safety, and cannot be removed easily in existing systems. The household wiring system becomes similar to an industrial power system, with a separately derived ground at the secondary of the distribution transformer. The grounding system is constructed in two isolated zones, one for the medium-voltage distribution and one for the low-voltage distribution. This prevents, for example, current from a

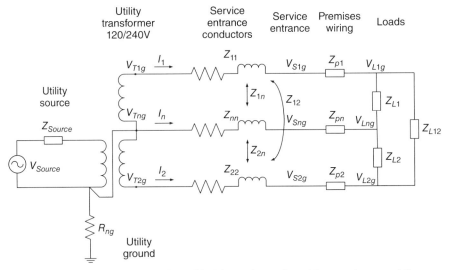

Figure 9.11 Equivalent circuit of modified domestic service with ground removed from secondary of transformer (*Source*: © IEEE (2011)).

medium-voltage ground fault flowing into a low-voltage system. It prevents a ground fault in the low-voltage system from tripping medium-voltage protection systems, improving protective device coordination (Figure 9.12).

In some cases, coupling devices have been used in place of the neutral–ground bond at the transformer, which only allow high fault currents to pass, but block low

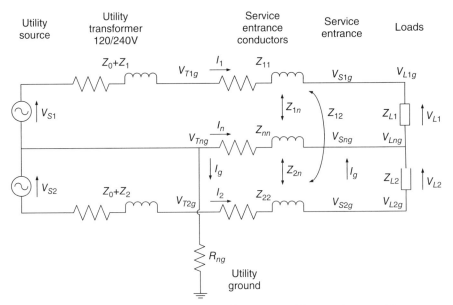

Figure 9.12 Equivalent circuit of modified 120 V system with ground removed from secondary of transformer for analysis (*Source*: © IEEE (2011)).

currents due to unbalances. This method keeps the system unsafe when faults occur, and is not recommended.

These methods do not eliminate the ground loops inherent in the multigrounded utility distribution system. However, they do make the system much safer in the user's premises by eliminating the most important ground loop, which has the greatest impact on the people using electric power. What is important regarding the multi-grounded system involving grounding the neutral on utility poles is to avoid making contact with the vertical ground wires on the poles. The best practice in this area is to insulate the ground wire, preventing human contact with the conductor.

9.5 GROUNDING AND OVERVOLTAGES IN DISTRIBUTION SYSTEMS

Distribution primary circuits typically have one of the following wire configurations:

- Three-wire ungrounded.
- Three-wire unigrounded.
- Four-wire unigrounded neutral system.
- Four-wire multigrounded neutral system.

The ungrounded circuit historically has been selected for those systems where service continuity is of primary concern. One of the issues with an ungrounded system is transient overvoltages due to restriking or intermittent ground faults. These type of faults can and do develop substantial overvoltages on ungrounded electrical systems with respect to ground. There have been many documented cases of measured line-to-ground voltages of 3 per unit or higher resulting in equipment damage. In all instances, the cause has been traced to a low-level intermittent arcing ground fault on an ungrounded system. One possible solution is to convert the ungrounded system to a high-resistance grounded system by deriving a neutral point through grounding transformers (Blackburn, 1998, pp. 210–233; Roberts *et al.*, 2003).

An ungrounded feeder in a power system might look like Figure 9.13. The source would be a distribution substation transformer with an ungrounded delta or wye secondary. Three-phase loads would also be ungrounded delta or wye connected. Single-phase loads would be connected phase to phase. The transformers and feeder conductors will contain distributed capacitances, which form the unintentional grounding system (Figure 9.14). In the transformer, these are mostly from windings to the grounded core. In cables, these are mostly from phase conductors to grounded shields. In overhead lines, these are both phase to phase and phase to ground.

The sequence network derived from this feeder configuration results in the distributed capacitances being shunted across the positive-, negative-, and zero-sequence networks, as shown in Figure 9.15. The capacitances have little effect on the positive- and negative-sequence networks, as they are shunted by the source and transformer inductance. However, capacitance is the dominant component of the zero-sequence network, as the zero-sequence capacitive reactance is usually much larger than the conductor impedance.

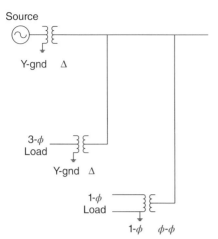

Figure 9.13 Ungrounded three-phase feeder.

Lines or cables

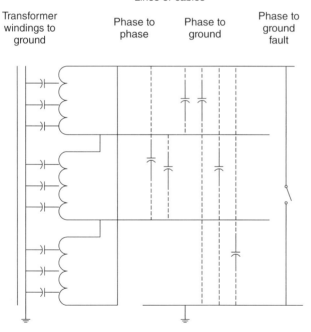

Figure 9.14 Distributed capacitances in ungrounded three-phase feeder.

When a phase-A-to-ground fault occurs as shown in Figure 9.14, this results in a series interconnection of the sequence networks. The network of Figure 9.15 reduces to the source voltage across the zero-sequence capacitance in Figure 9.16. By the use of symmetrical component analysis, the fault current in phase A can be shown to be

$$I_a = \frac{3V_S}{X_{0C}}$$

where $X_{0C} = X_{0S} + X_{0L}$ (9.16)

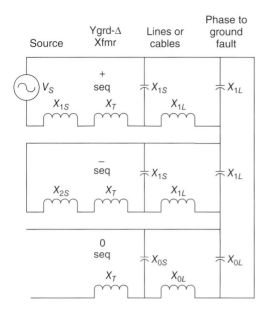

Figure 9.15   Interconnected sequence networks of ungrounded feeder.

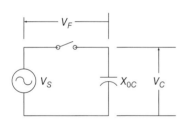

Figure 9.16   Simplified sequence network for phase-to-ground fault.

$$\text{and } X_{0C} \gg Z_{0L}$$

Similarly, the voltages in the unfaulted phases are

$$V_b = (a^2 + 1) \ V_S \tag{9.17}$$

$$V_c = (a + 1) \ V_S \tag{9.18}$$

where $a = 1\angle 120°$. Thus, the voltage from the faulted phase A to ground will be zero, while the other two phases will have their voltage to ground increased by $\sqrt{3}$. This is the magnitude of overvoltages that were reported in the 4.8 kV systems. In order to prevent failures due to overvoltages, all equipment in ungrounded systems should be rated for full line-to-line voltage, and not line-to-neutral.

In ungrounded systems where there is very little damping from phase to ground connected relaying, metering, or fault detectors, multiple restrikes from arcing ground faults can lead to voltage buildups to several times peak line to neutral voltage. An example of the type of overvoltages that can occur is shown in the simulation results of Figure 9.17. Here, the arcing fault clears itself at the current zero which occurs during the first voltage peak in the graph. System capacitance holds the voltage across the gap to peak phase to ground voltage. The arc restrikes half a cycle later during the next voltage peak. This adds two times peak voltage to the capacitor. A second

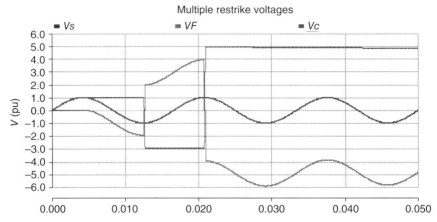

Figure 9.17   Simulation of overvoltages due to multiple restrike. Breaker opens and arc extinguishes at 0.25 cycles, restrikes occur at 0.75 cycles and again at 1.25 cycles, resulting in overvoltages of 3 and 5 per unit.

restrike half a cycle later repeats the cycle, resulting in five times the peak voltage. By this time, insulation failure is likely to result in a permanent fault.

The presence of a resistance in parallel with the distributed capacitance to ground will reduce these overvoltages. The ground detection units, which consist of meters and zero-sequence overvoltage relays, provide such a resistance fed from grounded-wye delta banks of voltage transformers. Additionally, the voltage transformers supply the $3V_0$ polarization selector for the ground directional relays on the feeders.

Transient overvoltages caused by multiple restrikes can be mitigated by rapid detection and clearing of ground faults, the use of higher insulation levels, or by converting to a grounded system.

Ferroresonance is a phenomenon where the nonlinear magnetizing reactance of a (usually) ungrounded and unloaded or lightly loaded transformer resonates with the distributed capacitances in lines or cables feeding the transformer. The resulting voltage oscillation is a distorted, often chaotic, waveform, with peaks as high as five per unit. This will most likely occur during single-phase switching of transformers in systems with high distributed phase to ground capacitances. Ferroresonance can occur in power systems with either grounded or ungrounded sources. However, the primary of the transformer that is involved in ferroresonance is usually ungrounded.

Ferroresonance results in distinctive oscillations that can be detected and recorded by a transient recorder or power quality monitoring system. Monitoring systems may be put in place to verify whether ferroresonance is occurring. These should include oscillographic recordings of voltage transients and the timing of switching events.

9.6   HIGH-RESISTANCE GROUNDING OF DISTRIBUTION SYSTEMS

The high-resistance grounding bank consists of three single-phase transformers, such as distribution transformers, with a wye-grounded primary and broken delta secondary across which a resistor is placed (Figure 9.18). The function of the grounding bank is to add a resistance across the distributed zero-sequence capacitance (Figure 9.19). The value of the grounding resistance is usually made equal to the system charging reactance. This will result in a single phase to ground fault current of 1.4 times the system charging current. Typical fault currents in high-resistance grounded systems are in the range $1-10$ A.

The grounding resistor will limit system overvoltages due to multiple restrike. For example, Figure 9.20 shows the restrike simulation of the previous example with high-resistance grounding equal to the system capacitive reactance. The rationale for choosing the resistance to equal the charging capacitance is shown in Figures 9.21 and 9.22. These are plotted for a 5.04 kV (2.9 kV phase to ground) system with 3.48 mF charging capacitance and $R = 762$ Ω. Lowering the resistance below this value will result in a rapid increase in fault current from 5 to 10 A, while doubling it will increase the transient overvoltages from 2 to 2.5 per unit.

The grounding resistor will not limit overvoltages due to single-line-to-ground faults. The equipment insulation level must continue to be line to line as in ungrounded systems.

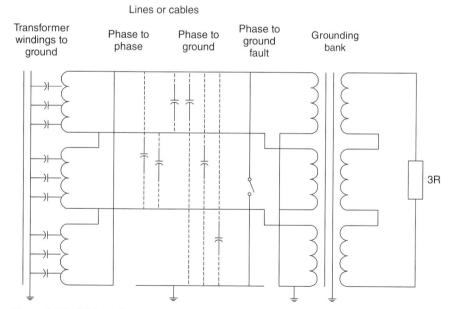

Figure 9.18   High-resistance wye-broken delta grounding bank added to the system of Figure 9.13.

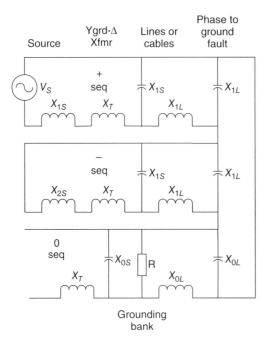

Figure 9.19 Interconnected sequence networks with high-resistance grounding bank.

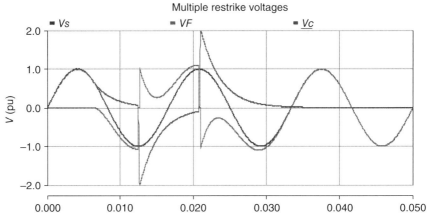

Figure 9.20 Restrike overvoltages of Figure 9.17 limited by high-resistance grounding with $R = XC$ to less than 2.5 per unit.

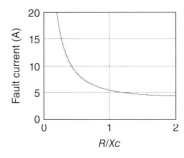

Figure 9.21 Fault current versus resistance in a typical high-resistance grounded system.

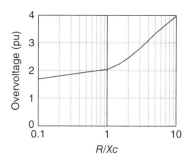

Figure 9.22   Overvoltage due to restrike versus resistance in a typical high-resistance grounded system.

The grounding resistor will limit system overvoltages due to ferroresonance as long as the grounding resistance is less than or equal to XC (Mason, 1956, p. 209). In sizing high-resistance grounding systems, resistances in the range $0.5-1.0$ times the system charging reactance are acceptable.

9.6.1   Methods of Determining Charging Current

It is recommended that charging current measurements be performed on one or more banks in order to verify calculated results. The charging current is normally calculated by summing the capacitances to ground ($^-$F/phase) of all connected equipment, such as

- Cables
- Overhead lines
- Power transformers
- Instrument transformers (voltage)
- Motors
- Generators
- Surge capacitors.

Any power factor correction capacitors should be connected phase to phase and not phase to ground. Since cables are usually the largest contributors to the total capacitance, the total length of cables and the capacitance to ground per phase per unit length are the key pieces of data required.

In low-voltage systems, measurement of ground fault current on an intentionally grounded phase is feasible (IPC, undated). This method has also been used on medium-voltage systems, but lack of a solid ground to measure to may cause inaccuracy in the results. If a zero-sequence window Current Transformer (CT) is installed on a circuit, charging current can be measured directly.

Charging current can be measured during grounding bank installation.

When the grounding bank is being installed, connect the three secondary windings in delta, without the resistor (Figure 9.23) and measure the delta current. In the absence of a fault, this should be close to zero. Apply three times rated secondary voltage to the broken delta and measure the current. The charging current

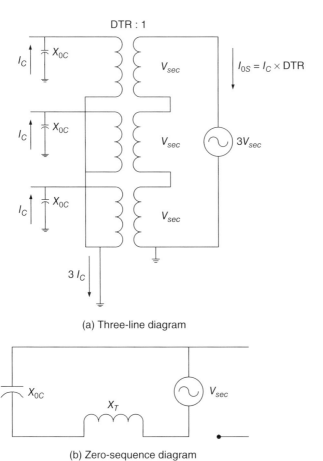

(a) Three-line diagram

(b) Zero-sequence diagram

Figure 9.23   Measurement of charging current using voltage source and grounding bank.
(a) Three-line diagram and (b) zero-sequence diagram.

I_C will be the measured delta current I_{0S} divided by the ratio of the distribution transformer, DTR:

$$I_C = \frac{I_{0S}}{DTR} \tag{9.19}$$

With a resistor in the broken delta (Figure 9.24), the resistance and capacitance are in parallel across the voltage source. The magnitude of the measured current is increased due to the added resistance:

$$I_{OR} = I_{0S}\frac{\sqrt{R^2 + X_{0C}^2}}{R} \tag{9.20}$$

If $R = X_{0C}$, then

$$I_{OR} = \sqrt{2} \cdot I_{0S} \tag{9.21}$$

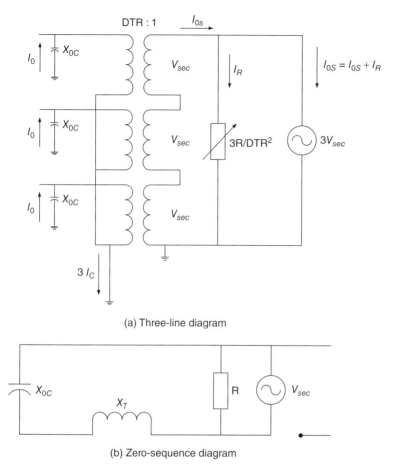

(a) Three-line diagram

(b) Zero-sequence diagram

Figure 9.24   Field tuning of grounding resistor. (a) Three-line diagram and (b) zero-sequence diagram.

Break the delta and install the resistor. Adjust the taps on the resistor until the current is at its closest to 1.414 times the current previously measured. Remove the voltage source.

The application of high-resistance grounding banks to ungrounded three-phase distribution systems has been shown to be useful in decreasing transient overvoltages due to multiple restrikes and ferroresonance. Simulation and research results show that a resistance in the range 0.5–1.0 times the zero sequence capacitive reactance will be most effective in reducing these overvoltages.

It is difficult to measure zero-sequence charging currents in an ungrounded system due to the lack of a ground to measure against. A new method was proposed for measuring charging currents and tuning high-resistance grounding banks. This method only requires the use of an AC voltage source and an ammeter.

CHAPTER *10*

ARC FLASH HAZARD ANALYSIS

10.1   INTRODUCTION TO ARC FLASH HAZARDS

For many years, it had been assumed that electrical shock was the greatest hazard presented by electrical work. However, more recently, it has been recognized that electric arc flash and blast effects produce greater severity of injuries and more hospital admissions than electrical shock (Lee, 1982). The problem is more severe than that indicated by the data because fatalities are not reflected in hospital admissions data. An arc flash hazard is a dangerous condition associated with the release of energy caused by an electric arc. Generally, an electric arc is formed when two conductor materials (e.g., copper) separated by a gap are shorted together, whether from inadvertent contact or insulation failure, causing passage of substantial electric currents through the gap between the conductor materials. The arc that develops is not only composed of current flowing in air, but also in vaporized conductor material, usually copper or aluminum. The temperature of the arc can reach 35,000 K, which is several times hotter than the surface of the sun. The only higher temperatures that can be produced on earth are from extremely large lasers. The heat produced by the arc can cause severe, often fatal, burns over a large portion of the body and ignition of clothing and other materials. The blast caused by the rapid expansion of copper from the heat can cause hearing loss, equipment doors to blow off their hinges, and people to be blown many feet across a room. The shrapnel associated with arc blast, consisting of small copper particles, can also cause severe injuries and death.

Since these very high temperatures vaporize the conductor materials, extremely explosive pressures are created as a result of metal vaporization and rapid heating of the air in the gap by electric current passing through it. The vaporization of metal and heating of the surrounding air results in a very rapid blast due to the buildup of high pressures. The blast causes molten metal and equipment parts to violently spew from the point of the blast. People near arc flash hazards may be subject to severe burns, ruptured eardrums, collapsed lungs, and forces that can violently knock them back.

There are several approaches to remediating arc flash hazards. One approach is to use engineering methods to reduce the amount of short-circuit current associated with the electric arc, as well as to reduce the duration of the arc. Reducing the amount of energy associated with the electric arc and the duration of the arc results in fewer injuries. Another approach to remediating arc flash hazards is to require people working on or near exposed energized conductors to wear personal protective equipment

Principles of Electrical Safety, First Edition. Peter E. Sutherland.
© 2015 John Wiley & Sons, Inc. Published 2015 by John Wiley & Sons, Inc.

(PPE) (e.g., heavy jackets, protective hoods, face shields, and gloves) designed specifically for arc flash hazards. A third approach to remediating arc flash hazards involves using procedural or administrative methods to warn people of the potential for the occurrence of arc flash hazards if one is going to work on or be near exposed energized conductors that could experience an electric arc. Procedural or administrative methods may include applying warning labels on or near exposed energized conductors. For example, a typical label could state that if one is going to work on equipment having exposed energized conductors, then workers have to wear authorized PPE and maintain a predetermined distance from the conductor materials that could be subject to an arc flash hazard.

The procedures and policies set forth in ANSI/AIHA Z10-2012 Occupational Health and Safety Management Systems (AIHA, 2012) can be used as part of an electrical safety program to mitigate the hazards of arc flash and electrical shock. The actions described in the previous paragraph are part of the "Hierarchy of Control" set forth in Z10-2012:

"**A.** Elimination;

B. Substitution of less hazardous materials, processes, operations or equipment;

C. Engineering;

D. Warnings;

E. Administrative controls, and

F. Personal protective equipment."

The above-described approaches to remediating arc flash hazards are not always effective for arc flashes caused by electric arcs which produce an incident energy that exceeds 40 cal/cm^2. Some design techniques which reduce the possibility of such high levels of incident energy are discussed in this chapter. Design for safety is critical because engineering methods that reduce the amount of energy associated with the electric arc and the duration of the arc may not be available for use with levels greater than 40 cal/cm^2. Furthermore, currently available PPE may not provide protection against blasts that can arise from energy levels that are greater than 40 cal/cm^2. Similarly, the hazards associated with blasts that can arise from energy levels that are greater than 40 cal/cm^2 essentially make the use of warning labels ineffectual for their intended purposes. Because the above-described approaches to remediating arc flash hazards are not suitable for arc flashes caused by electric arcs having energy that exceeds 40 cal/cm^2, workers need to shut down the equipment and work on it while de-energized. However, shutting down equipment and working on it while de-energized is not always an optimal solution, especially in facilitates where it is desirable to keep equipment operating 24 h a day, 7 days a week.

For these reasons, arc flash hazard analysis has developed as a field within electrical engineering. There are three major aspects to this:

1. *Safety by prevention.* Systems may be designed to minimize arc flash hazards. Protective device selection and adjustments are made to minimize exposure time and incident energy.

2. *Operational safety.* Design of work practices and procedures in order to reduce or eliminate work on energized electrical conductors. This includes methods such as arc flash labels, energized work permits, lock-out-tag-out, and application of safety grounds.

3. *Safety by protection.* The use of PPE such as clothing, face protection, eye protection, and hearing protection. These steps are used to protect the worker after the arc flash hazard has been minimized as much as possible by other methods.

Once arc flash hazard analysis became a requirement for many industrial power system studies, the reduction of arc flash hazards became an important concern (Sutherland *et al.*, 2009). Inevitably, the factors that lead to reductions in arc flash hazards do not always lead to an improvement in other areas, in particular, protective device coordination. These examples cover several particular cases where an arc flash hazard analysis of an older industrial plant yielded several cases of buses where the incident energy was above 40 cal/cm^2, making an "Extreme Danger" label necessary. By changing protective device settings and, if necessary, the devices themselves, the incident energy can be reduced. Methods to maintain coordination, if possible, are discussed.

An arc flash hazard analysis usually begins with a study of a facility based on the procedures and methodology of the National Fire Protection Association (NFPA) Standard 70E (NFPA, 2012), and IEEE 1584–2002 (IEEE, 2002b). However, an arc flash hazard analysis is normally performed in conjunction with a suite of other power systems studies, such as

1. Load flow

2. Short circuit

3. Protective device coordination

The results of an arc flash hazard analysis are labels to be placed on equipment, which give, among other things,

1. The flash protection boundary, which is the distance within which a person must wear PPE. This is the distance at which exposed skin will be at risk of a just-curable second-degree burn. It is not the distance at which there is no risk of injury.

   This boundary must be reasonable, such that people who are not working on the equipment can still perform their functions. For example, it should not extend beyond the fence of an outdoor substation or prohibit opening the door of an indoor substation.

2. The working distance for which the incident energy is calculated and the PPE is specified. Persons or body parts closer than this distance will require additional protection.

3. The incident energy, in cal/cm^2, to a person at the working distance specified. If the incident energy is above 40 cal/cm^2, no approach is possible without de-energization of the equipment.

When no arc flash study is performed, hazard/risk categories are determined with the help of Table 130.7[C] (16) in NFPA 70E-2012 (NFPA, 2012). This table should be used with caution, because it was prepared using many assumptions, which are detailed in the footnotes. The effect of upstream current limiting fuses (CLF) is not taken into account. The use of CLFs with a clearing time of ½ cycle (8.3 ms in a 60 Hz system, 10 ms in a 50 Hz system) or less can significantly reduce the incident energy. Specific fault currents, trip times, working distances, and arc gaps were used for the different voltage ranges covered.

Arc flash hazard levels depend on fault current magnitude and duration, I^2t. The power going into the arc resistance is a function of I^2R, where R is the arc resistance. This is multiplied by time, which is the arc duration. Since arcs, currents, and resistances are in constant variation, this is really a time-varying integral whose units are those of energy (joules or calories). When this energy is radiated to a person's skin, they will be exposed to incident energy (cal/cm^2), which is what cases the burn.

Mitigation techniques for reducing incident energy include (Hodder *et al.*, 2006; Brown and Shapiro, 2009). The factors that can be controlled include the following:

1. Selection of suitable time-current characteristics for protective devices.
2. Modification of circuit breaker and relay settings.
3. *Pickup*. This is the minimum current at which a device actuates. A lower pickup provides arc fault protection for a greater range of fault currents.
4. *Time delay*. A shorter time delay reduces the time to trip and lowers I^2t.
5. *Instantaneous pickup*. The operating time is typically the minimum possible for the circuit breaker being used. Lower instantaneous pickup settings reduce arc flash hazard.
6. Zone-selective interlocking.
7. Current-limiting low-voltage circuit breakers.
8. Fused circuit breakers.
9. Arc blast containment systems.
10. Optical sensors for arcing fault detection.

Protective devices in a power system are coordinated in time and in current pickup in order to provide for an orderly shutdown in case of a fault and to prevent blackouts. Changes in protective device settings solely to reduce arc flash hazards will inevitably result in a loss of coordination, resulting in failure to operate or delayed operation during faults, and unnecessary blackouts. Examples of changing a slow fuse for a fast fuse can be found in (Doan and Sweigart, 2003) and (Sutherland, 2009a).

Resolution of combined arc flash and coordination problems requires evaluation of multiple options. The solution of one problem may cause the other problem to reappear in another place. The presence of old equipment with unreliable characteristics complicates the assessment. Published time-current curves and breaker-opening times should not always be relied on. Proposed solutions may not always be implemented, and the engineer should be prepared for a long drawn out process before all problems are resolved.

A more extensive list of "design changes" and "overcurrent upgrades" is given in (Hodder *et al.*, 2006).

This is slightly different from the problem of the design of limited arc energy distribution systems (Das, 2005) although many of the same principles can be used.

10.2 FACTORS AFFECTING THE SEVERITY OF ARC FLASH HAZARDS

There are many formulae and procedures for calculating arc flash hazard results. These have been evaluated, and it has been found that the IEEE Standard 1584-2002 (IEEE, 2002b) method for up to 15 kV and the Lee method (Lee, 1971) for voltages above 15 kV give the most accurate results (Ammerman, Sen, and Nelson, 2009). The empirical method has been used to find the best curve fits for the variables on the basis of laboratory tests. Theory-based approaches have not been shown to be as accurate. The following variables are utilized:

I_{bf} = Bolted fault current, kA. This is the current for a three-phase zero-impedance short circuit with no arcing present, and thus a zero-voltage fault also. Bolted fault current for typical electrical systems are limited by the impedance of the generators or utility system (sources), and by the impedance of the lines, cables, and transformers that bring the current to the load (distribution system). Additional fault current may be contributed by synchronous and induction motors, which produce rapidly decaying fault currents due to magnetic field transients and load and motor inertia. For the IEEE Standard 1584-2002 (IEEE, 2002b) model, the available bolted fault currents ranged from 0.70 to 106 kA. Typical secondary low-voltage systems have bolted fault currents ranging from 1 to 50 kA. Bolted fault current is the primary factor in determining both arcing current and incident energy.

V = Voltage, kV. 0.208 kV < V < 1.5 kV. The system voltage is the nominal voltage based on system transformation ratios. The actual voltage during a short circuit may be considerably less. Tests for IEEE Standard 1584-2002 were conducted for the voltage range of 208–15 kV. All tests were performed for three-phase, alternating current (AC) systems. The arcs at 208 V were not always self-sustaining, so the caveat was added that the model does not apply for voltages of 240 V and below, fed by a single transformer rated less than 125 kVA. System voltage is a small factor in determining arcing current.

G = Electrode gap, mm. 13 mm < G < 152mm. The spacing between electrodes is dependent on the type of equipment and the system voltage. Typical values are 25 mm for low-voltage panelboards and MCCs, 32 mm for low-voltage switchgear, 104 mm for 5 kV switchgear, and 152 mm for 15 kV switchgear. For the tests in IEEE-1584-2002, vertical bus sections were used, both in open air and enclosed in a box. This is representative of switchgear, motor control center (MCC), and panelboard construction. The electrode gap was a small factor in determining both arcing current and incident energy.

Box gap. The spacing between the electrodes and the back of the box used in testing were representative of actual distances used in equipment. The box gap was not a factor in determining arcing current or incident energy. The presence or absence (in open conductor configurations) of a box gap was a small factor in determining the arcing current. See factor K_1 described below.

Grounding type. Power distribution systems may be operated with the system neutral connected to earth ground by a low-resistance cable, in which case the system is said to be solidly grounded. Solid grounding is the norm for electric utility transmission and distribution and for industrial and commercial low-voltage distribution in the United States. Grounding the neutral through a small resistor, called low-resistance grounding, is used in many industrial power systems at medium voltage in order to limit equipment damage due to ground faults. Low-resistance grounding typically limits ground fault current to the range of 100–400 A, and requires special ground fault protection relays to operate. For systems where service continuity is paramount, high-resistance grounding may be used. In this system, ground fault current is limited to 5–10 A, and an alarm is produced when a ground fault is detected. This may be investigated and removed at a time which is convenient, with the system still operating. However, if a second ground fault should occur, the resulting phase-to-phase fault may shut the system down.

The effectiveness of a grounding system, whether solid grounding, depending on ground rods and ground grids, or impedance grounded system, may be measured using the zero-sequence impedance ratios (Central Station Engineers, 1964, p. 464). A system may be said to be "effectively grounded" when

$$X_0/X_1 < 3.0 \quad \text{and} \quad R_0/X_1 < 1.0 \tag{10.1}$$

where

X_0 = zero-sequence reactance for a given point of the system

X_1 = positive-sequence reactance for a given point of the system

R_0 = zero-sequence resistance for a given point of the system.

For hand calculation, the user may simply examine the single-line diagram and determine whether a system is solidly grounded or resistance grounded. For computer calculation, the zero-sequence impedance method or any other which gives equivalent results should be utilized. The grounding type was a small factor in determining the arcing current. See factor K_2 described below.

Fault type. In three-phase systems, most faults will quickly escalate into three-phase faults; therefore, this type was used in the tests for IEEE 1584-2002. However, because single-phase systems do exist and may experience arcing faults, further testing to extend the model is underway.

t = Arcing time, in s. This is the time from arc inception to arc interruption. Arc interruption is usually performed by a fuse or circuit breaker. The mode of operation of these devices is to create a controlled arc within a safe enclosure, which is drawn out and extinguished, thus halting the flow of current. Arcing time may range from 0.0 s up to several seconds. Typically, arc flash calculations will limit the maximum arcing time calculated to 2 s. After this time, all persons are assumed to have been moved away from the vicinity of the arc. Arcing time is a linear factor in the calculation of incident energy.

Frequency. Tests for both 50 and 60 Hz systems were made for IEEE 1584-2002. No significant difference was found between the two. Tests for direct current (DC) and for other frequencies of AC, such as 400 Hz used in aircraft electrical systems, will be performed at a later date.

Electrode Materials tested for IEEE 1584-2002 included copper and aluminum, the most common conductor materials. No significant effect of electrode materials was found.

D = Distance from arc location to worker, in mm. The incident energy is inversely proportional to the distance D squared in open air, and raised to the distance exponent, x, in an enclosure. The working distance is either determined in the field, or standardized for different types of equipment at different voltages. Typically, an arm's length, 455 mm (18 in.), is used for low-voltage MCCs and panelboards, while longer distances are used for low-voltage switchgear, 610 mm (24 in.), and medium-voltage switchgear, 910 mm (36 in.).

The following factors were used in the calculations:

x = distance exponent, unitless. The incident energy is inversely proportional to the distance D squared in open air, and raised to the distance exponent, x, in an enclosure.

C_f = calculation factor, unitless. This is a statistical factor used in IEEE 1584-2002 to bring the curve-fitting results into a 95% confidence level. For system voltages below 1 kV, $C_f = 1.5$, while for system voltages above 1 kV, $C_f = 1.0$.

85% factor. For systems where the nominal voltage is less than 1 kV, the calculations for incident energy may not give the worst-case results using the arcing times determined with the calculated arcing current. For these systems, a second calculation is taken at 85% of calculated arcing current, with correspondingly longer arcing times (dependent on the protective device characteristics). The worst-case incident energy from these two calculations is used in the final result.

K for enclosure type in arcing current calculation:

$K = -0.153$ for open conductors (no enclosure).

$K = -0.097$ for box enclosure.

K_1 for enclosure type in incident energy calculation.

$K_1 = -0.792$ for open conductors (no enclosure).

$K_1 = -0.555$ for box enclosure.

K_2 for grounding type in incident energy calculation.

$K_2 = 0$ for ungrounded and high-resistance grounded systems.

$K_2 = -0.113$ for grounded systems.

E_B = energy to produce a just-curable second-degree burn, = 5.0 J/cm². This can be converted to calories per square centimeter, which is a commonly used set of units in the field, by dividing by 4.184. Thus $E_B = 1.2$ cal/cm².

The following are calculated values:

I_a = arcing current, kA, is the current in the electrical arcing fault:

$$I_a = 10^{K + 0.662 \cdot \log_{10}(I_{bf}) + 0.966\,V + 0.000526\,G + 0.5588\,V\log_{10}(I_{bf}) - 0.00304\,G\log_{10}(I_{bf})}$$

for $V < 1$ kV

$$I_a = 10^{0.00402+0.983\log_{10}(I_{bf})}$$

for $V \geq 1$ kV $\qquad$ (10.2)

E = Incident energy, J/cm^2. This is the radiant energy from the arc to which a person is exposed to at distance D for time t. The IEEE 1584-2002 procedure has two steps. The first is to compute normalized incident energy, E_n, at $t = 0.2$ s, and $D = 610$ mm.

$$E_n = 10^{K_1+K_2+1.081\cdot\log_{10}(I_a)+0.0011G} \qquad (10.3)$$

This normalized value is converted in the second step to the actual time and distance.

$$E = 4.184 \cdot C_f E_n \left(\frac{t}{0.2}\right)\left(\frac{610^x}{D^x}\right) \qquad (10.4)$$

Lee Method for incident energy (Lee, 1982). The IEEE 1584-2002 model extends only to 15 kV systems and for conductor gaps 13 mm $< G <$ 152 mm. The more conservative Lee model may be used outside this range.

$$E = 2.142 \div 10^6 \cdot VI_{bf}\left(\frac{t}{D^2}\right) \qquad (10.5)$$

D_B = Flash-protection boundary, in mm. The IEEE 1584-2002 model is

$$D_B = \left[4.184 \cdot C_f E_n \left(\frac{t}{0.2}\right)\left(\frac{610^x}{E_B}\right)\right]^{\frac{1}{x}} \qquad (10.6)$$

The Lee model is

$$E = \sqrt{2.142 \times 10^6 \cdot VI_{bf}\left(\frac{t}{E_B}\right)} \qquad (10.7)$$

10.3   EXAMPLE ARC FLASH CALCULATIONS

Given 480 V switchgear, in a solidly grounded system, with available bolted fault current of 40 kA, bus gap is 32 mm, and working distance is 610 mm. The interrupting time is 0.3 s.

$$I_a = 10^{K + 0.662\cdot\log_{10}(I_{bf}) + 0.966V + 0.000526\,G + 0.5588V\log_{10}(I_{bf})-0.00304G\log_{10}(I_{bf})}$$

$$= 10^{-0.097+0.662\cdot\log_{10}(40)+0.966\times0.48+0.000526\times32+0.5588\times0.48\log_{10}(40)-0.00304\times32\log_{10}(40)}$$

$$= 20.0 \text{ kA}$$

Next, compute normalized incident energy at $t = 0.2$ s, and $D = 610$ mm.

$$E_n = 10^{K_1 + K_2 + 1.081\cdot\log_{10}(I_a) + 0.0011G}$$

$$= 10^{-0.555 + -0.113 + 1.081\cdot\log_{10}(20) + 0.0011 \times 32}$$

$$= 5.92 \text{ J/cm}^2$$

This normalized value is converted in the second step to the actual time and distance.

$$E = 4.184 \cdot C_f E_n \left(\frac{t}{0.2}\right) \left(\frac{610^x}{D^x}\right)$$

$$= 4.184 \cdot 1.5 \cdot 5.92 \left(\frac{0.3}{0.2}\right) \left(\frac{610^{1.473}}{610^{1.473}}\right)$$

$$= 13.3 \ \text{J/cm}^2$$

The flash-protection boundary

$$D_B = \left[4.184 \cdot C_f E_n \left(\frac{t}{0.2}\right) \left(\frac{610^x}{E_B}\right)\right]^{\frac{1}{x}}$$

$$= \left[4.184 \cdot 1.5 \cdot 5.92 \left(\frac{0.3}{0.2}\right) \left(\frac{610^{1.473}}{5.0}\right)\right]^{\frac{1}{1.473}}$$

$$= 123.4 \ \text{mm}$$

10.4   REMEDIATION OF ARC FLASH HAZARDS

10.4.1   Example: Correcting an Arc Flash Problem When a Coordination Problem Requires Replacing Trip Units

In this older substation (Figure 10.1), the low-voltage circuit breakers use elec-tromechanical dashpot-type trip (Figure 10.2) units with wide characteristics

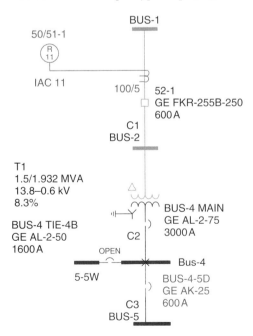

Figure 10.1   Single line diagram of older substation whose low-voltage circuit breakers use electromechanical dashpot-type trip units. (*Source*: © IEEE (2009a)).

Figure 10.2   Typical electromechanical trip unit for low-voltage circuit breaker.

(Figure 10.3). The main breaker has an instantaneous only characteristic (vertical bar on the time-current curve). The transformer protection is from an older inverse-time overcurrent relay, with nearly definite time characteristics. The relay and low-voltage circuit breakers do not coordinate in this situation, such that the only practical solution is an upgrade to modern electronic equipment. The 600 V transformer secondary and bus, both have high levels of incident energy, exceeding 40 cal/cm^2, as shown in Figure 10.4. Arc flash hazard risks can be reduced to less than 25 cal/cm^2 on the transformer secondary and 600 V bus by reducing the time dial of the relay from 1.5 to 0.5, as shown in Figures 10.5, 10.6, and Table 10.1. This does not significantly change the coordination, or lack thereof, because the actual tie breaker characteristics, defined by a dashpot-type trip unit, are probably not very close to the published curve. Dashpots of these old-type circuit breakers tend to suffer from drying out of oil and lack of the correct dashpot oil for maintenance.

The instantaneous unit on the main breaker is set high, at the same value as the instantaneous unit of the tie breaker. However, the main breaker has no short-time or long-time delay units. Therefore, no changes of primary relay setting or relay curve shape will alleviate the coordination problem.

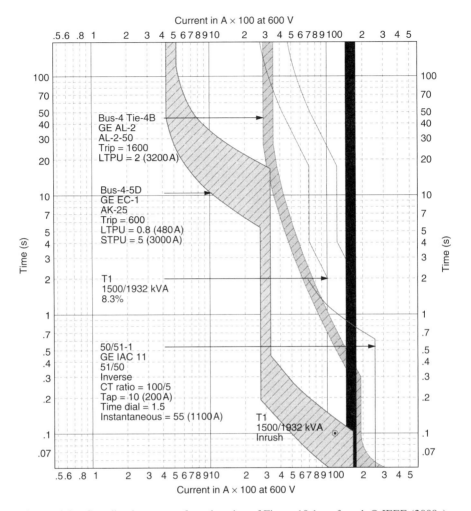

Figure 10.3 Coordination curves for substation of Figure 10.1, as found. © IEEE (2009a).

The only practical solution is to update the low voltage circuit breakers with solid-state trip units. However, the arc flash problem needs to be corrected now, while the circuit breakers might not be updated or replaced for some time. The arc flash hazard can be reduced, while the coordination problem should be addressed separately. When this occurs, the fact that the relay setting has been lowered to reduce arc flash hazard should not be forgotten.

10.4.2 Example: Correcting a Coordination Problem Without Introducing an Arc Flash Problem

In the existing configuration of this feeder, Figure 10.7, there is initially a severe arc flash problem, as shown in line 1 of Table 10.2, as well as a coordination problem,

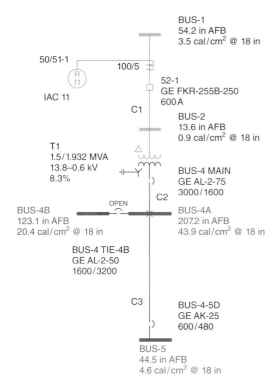

BUS-1
54.2 in AFB
3.5 cal/cm^2 @ 18 in

50/51-1

100/5

IAC 11

52-1
GE FKR-255B-250
600 A

C1

BUS-2
13.6 in AFB
0.9 cal/cm^2 @ 18 in

T1
1.5/1.932 MVA
13.8–0.6 kV
8.3%

BUS-4 MAIN
GE AL-2-75
3000/1600

C2

OPEN

BUS-4B
123.1 in AFB
20.4 cal/cm^2 @ 18 in

BUS-4A
207.2 in AFB
43.9 cal/cm^2 @ 18 in

BUS-4 TIE-4B
GE AL-2-50
1600/3200

C3

BUS-4-5D
GE AK-25
600/480

BUS-5
44.5 in AFB
4.6 cal/cm^2 @ 18 in

Figure 10.4   Arc flash results for substation of Figure 10.1, as found. © IEEE (2009a).

TABLE 10.1   Arc Flash Results at BUS-4 Versus Time Dial Setting for Relay 50/51-1 for Figure 10.1

Relay Time Dial Setting	Arc Flash Boundary (mm)	Incident Energy @ 455 mm (cal/cm^2)
1.5	5263	43.9
0.5	3297	22.0

Source: © IEEE (2009a).

Figure 10.8. This relay does not protect the 500 kVA transformer because the pickup is too high. It also does not coordinate with the main breakers on the 1500 kVA substation.

It was decided to upgrade the feeder with a relay having a very inverse time current curve for better coordination characteristics (Figure 10.9). If the relay is set to protect the 500 kVA transformer (Figure 10.10 and line 2 of Table 10.2), there is a coordination problem, in that the relay set to does not coordinate with the protection for the 1500 kVΛ transformer (Figure 10.11).

The other alternative of setting the relay for the 1500 kVA (Figures 10.12, 10.13, and line 3 of Table 10.2) transformer leaves the 500 kVA unit unprotected (Figure 10.14) and raises the incident energy to unacceptable levels.

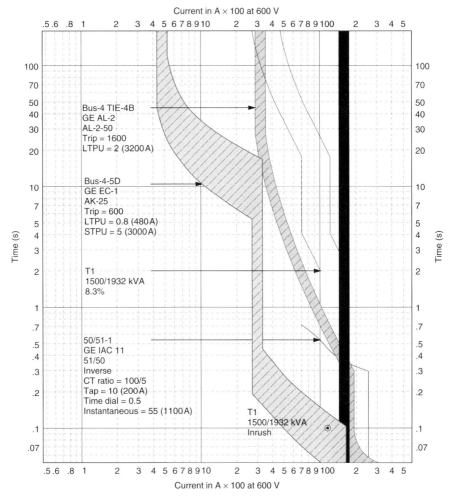

Figure 10.5   Coordination curves for substation of Figure 10.1, showing recommended changes © IEEE (2009a).

One solution to this dilemma is the use of a transformer primary fuse on the 500 kVA transformer, such that the incident energy is within acceptable limits on the 600 V bus. This will allow the primary relay R1 to be set to protect the 1500 kVA transformer.

It is important to select the proper fuse when remediating arc flash hazards. Two different fuses of the same rating and manufacturer, even if both are current-limiting fuses, may have completely different arc flash limitation characteristics. This is shown in Figure 10.15 and Table 10.3, where, for this situation the fuse in line 1 has an incident energy of 6.6 cal/cm^2, while that in Figure 10.16 and in Table 10.3, line 2 has 44.9 cal/cm^2, which is in the extreme danger category.

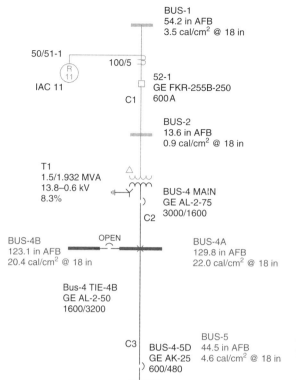

Figure 10.6   Arc flash results for substation of Figure 10.1, showing recommended changes. © IEEE (2009a).

Looking at the time-current curves (Figure 10.17), the Fuse 1 curve sweeps down and to the left, providing short clearing times across the range of high available fault currents. However, the Fuse 2 curve is steep, providing long clearing times at all but the highest fault currents. An example of changing a slow fuse for a fast fuse can also be found in (Doan and Sweigart, 2003).

10.5 COORDINATION OF LOW-VOLTAGE BREAKER INSTANTANEOUS TRIPS FOR ARC FLASH HAZARD REDUCTION

The practice of performing coordination studies for industrial, commercial, and institutional power systems has been regulated in many ways by the *National Electrical Code* (NEC) (NFPA, 2014) and *NFPA 70E, Standard for Electrical Safety in the Workplace* (NFPA 70E) (NFPA, 2012). In recent years, an apparent difficulty has arisen owing to requirements in the NEC for coordination of protective devices which are difficult to implement without increasing the incident energy.

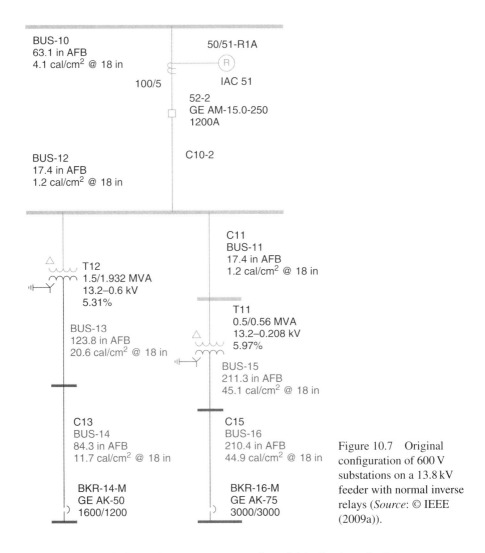

Figure 10.7 Original configuration of 600 V substations on a 13.8 kV feeder with normal inverse relays (*Source*: © IEEE (2009a)).

TABLE 10.2 Arc Flash Results at BUS-16 Versus Time Dial Setting for Relay R1

Line (See Text)	Relay	Tap Setting	Time Dial Setting	Arc Flash Boundary (mm)	Incident energy @ 455 mm (cal/cm^2)
1	Inverse	8.0	0.5	5344	44.9
2	Very inverse	1.5	1.5	2860	17.9
3	Very inverse	6.0	0.5	5344	44.9

Source: © IEEE (2009a).

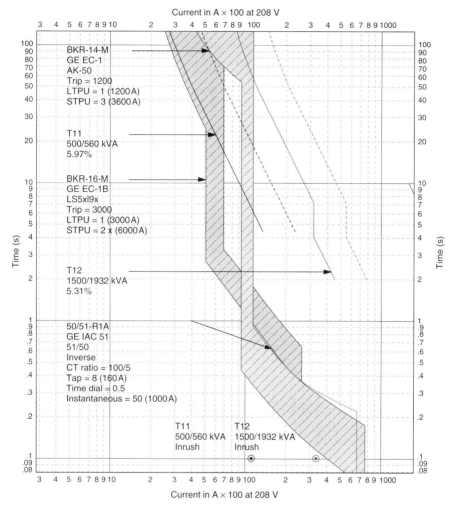

Figure 10.8   Original coordination of 600 V substations on a 13.8 kV feeder (*Source*: © IEEE (2009a)).

An emergency system (ES) is defined as a system which is legally required, and which automatically supplies power and light to designated areas, or to areas where power and light are essential for human health and safety. Clause 700.28 of the NEC, added in 2005, requires the protective devices of an ES to be selectively coordinated with all upstream devices (Valdes *et al.*, 2009; Valdes, Hansen, and Sutherland, 2012).

Clause 701.27 adds the same requirement for "legally required standby systems" (LRSS). An LRSS is defined as a standby system which is legally required, and which automatically supplies power to selected loads, which are not classified as ESs. Both types of systems are normally seen in hospitals, nursing homes, and other health-care facilities.

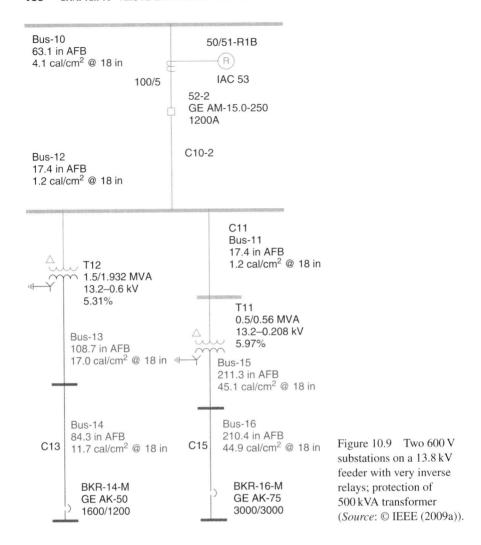

Figure 10.9 Two 600 V substations on a 13.8 kV feeder with very inverse relays; protection of 500 kVA transformer (*Source*: © IEEE (2009a)).

Another type of system where coordination is required is the "critical operations power system" (COPS). This is for facilities such as "public safety, emergency management, national security, or business continuity." Clause 708.54 provides similar requirements for coordination as the ones mentioned previously.

When the load consists of hoists or people movers, in particular, "elevators, dumbwaiters, escalators, moving walks, platform, and stairway chairlifts," NEC Clause 620.62 requires coordination when more than one machine is supplied by a single feeder. The purpose of this is to prevent the failure of one elevator (for example) to result in the shutdown of other elevators.

When an arc flash hazard analysis is performed for a hospital, it is often found that in order to lower incident energy levels, instantaneous trip devices must be used

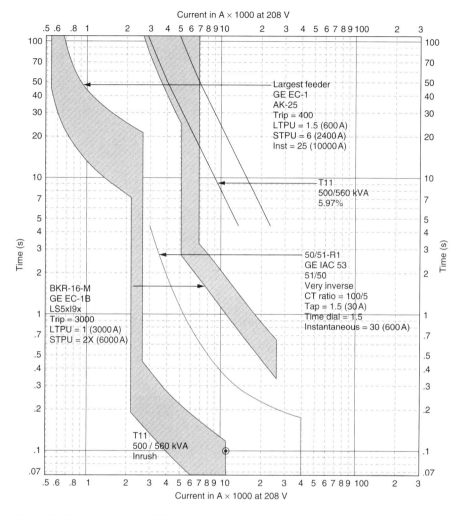

Figure 10.10   Protection of 500 kVA transformer (*Source*: IEEE (2009a)).

and set lower than is normal in a protective device coordination study alone (Hodder *et al.*, 2006; Brown and Shapiro, 2009). Recently, the "energy boundary curve" (EBC) method has been developed to aid in this analysis (Parsons *et al.*, 2008; Valdes, Crabtree, and Papallo, 2010b). In the following examples, the EBC technique will be applied to practical solutions to mitigate coordination in ES and LMSS.

10.5.1   Hospital #1 — Time–Current Curve Examples

This example concerns the emergency switchgear, designated "EMSW," in a typical hospital. The one-line diagram is shown in Figure 10.18, and the TCC in Figure 10.19. This diagram depicts the coordination between the main circuit breaker of the bus

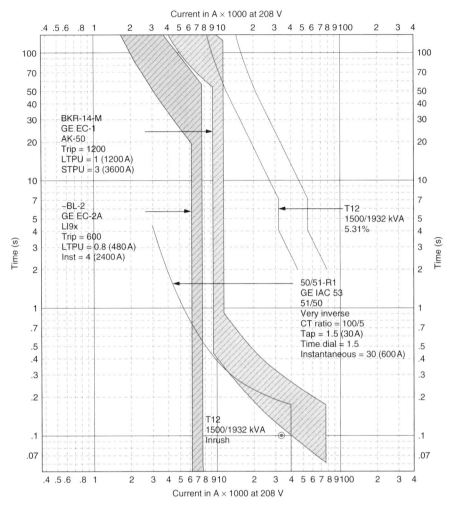

Figure 10.11   Protection of 1500 kVA transformer with 500 kVA settings (*Source*: © IEEE (2009a)).

EMSW, the largest feeder breaker of the switchgear, and the largest downstream breaker on the bus ES. With the settings as shown in the TCC, both busses EMSW and ES have incident energy at the working distance of greater than 40 cal/cm^2. This level of incident energy is greater than the recommended limit for arc flash PPE. The incident energy levels on both busses can be reduced to less than 0.4 cal/cm^2 by adjusting breaker settings as shown in Figure 10.20. Because the incident energy is less than 1.2 cal/cm^2, this places the working distance outside the arc flash boundary. By Clause 130.5 (B)(1) of NFPA 70E-2012 (NFPA, 2012), the worker is not required to wear arc-rated clothing or other PPE. However, the worker should still wear appropriate protective clothing, such as specified for Hazard Risk Category 0 in NFPA 70E Table 130.7(C)(16).

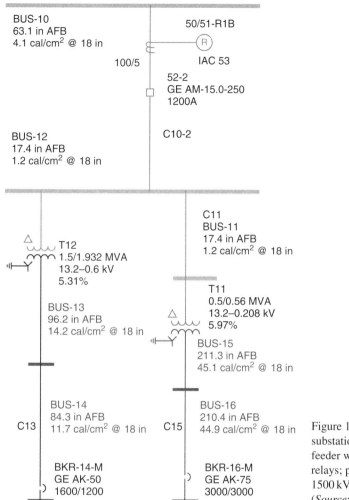

Figure 10.12   Two 600 V substations on a 13.8 kV feeder with very inverse relays; protection of 1500 kVA transformer (*Source*: © IEEE (2009a)).

The circuit breaker DP-CB-ES is a molded-case circuit breaker with a thermal magnetic trip unit. The frame size is 600 A and the trip unit pickup is set at 500 A. The instantaneous trip setting is reduced from 3250 to 3020 A.

The circuit breaker "ES" is a molded-case circuit breaker with an electronic trip unit. The frame size is 1200 A and the sensor rating 1200 A, with a trip unit having a 1000 A rating plug. The instantaneous trip setting is reduced from 7880 to 3020 A.

The 1600 A generator circuit breaker "GEN" is a power circuit breaker with an electronic trip unit. The frame size is 1600 A, and the tapped 1600 A current sensor is set at 1600 A. The instantaneous trip setting is reduced from 12,800 to 3200 A.

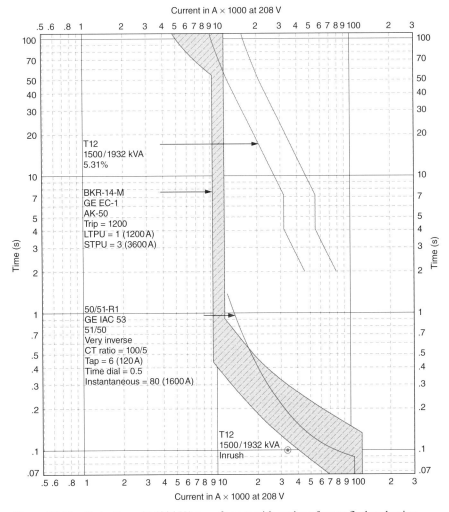

Figure 10.13 Protection of 1500 kVA transformer with settings for arc flash reduction (*Source*: © IEEE (2009a)).

However, when this is done, there is no coordination at the instantaneous level. We can determine which devices can be selective from the manufacturer's selectivity tables (General Electric Company, 2008a, b) (Mitolo, Cline, Hansen, Papallo, 2010). This selectivity will only apply if there are current-limiting circuit breakers or fuses in the circuit.

The arc flash hazard levels must also be checked with the ATS in the normal position. In this case, the results are similar, because the fault current is limited by the 300 kVA, 208 V transformer.

In this example, the generator breaker is not current limiting. The next combination circuit breaker "ES" with an electronic trip unit is, followed by the thermal

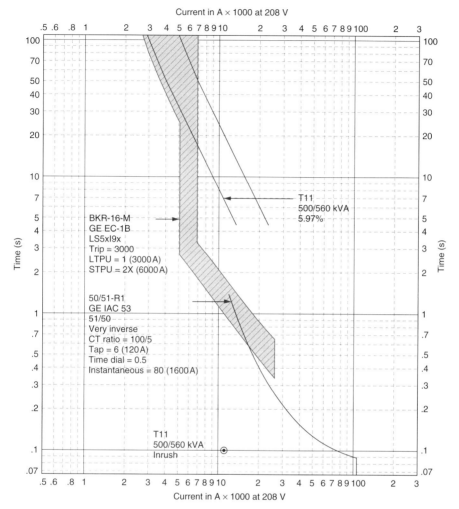

Figure 10.14   Protection of 500 kVA transformer with 1500 kVA settings (*Source*: © IEEE (2009a)).

magnetic MCCB "DP-CB-ES." This particular combination can only be coordinated by the TCC and not by instantaneous let-through current (General Electric Company, 2008b, p. 27). The next pair is the 600 A breaker "DP-CB-ES" and the 208 V, 1200 A circuit breaker, "DP-CB." Both are thermal magnetic MCCBs. This pair cannot be coordinated instantaneously and do not appear in the manufacturers tables. The TCC for this pair shows only normal coordination of curves and no special instantaneous trip modifications.

When the feeder is supplied by normal power (Figures 10.21 and 10.22), the available fault current increases from 11 to 16 kA on the 208 V bus. The coordination is now with a stored-energy circuit breaker, which does not have an instantaneous

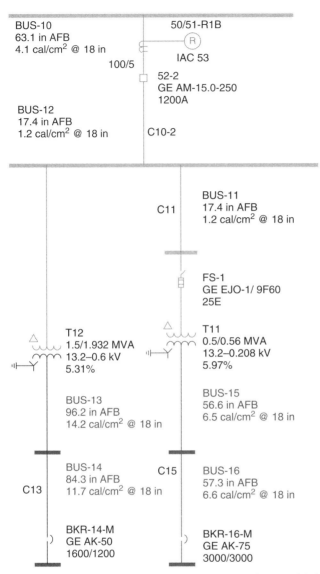

Figure 10.15   Two 600 V substations on a 13.8 kV feeder, with fast fuses on the smaller substation (*Source*: © IEEE (2009a)).

trip. In order to keep arc flash energies at the switchgear bus low, the coordination with the downstream 208 V breaker has been omitted.

10.5.2   Hospital #2 — Time – Current Curve Examples

This example concerns the coordination between the main and largest feeder breaker of the emergency bus in another typical hospital, down to the uninterruptible power

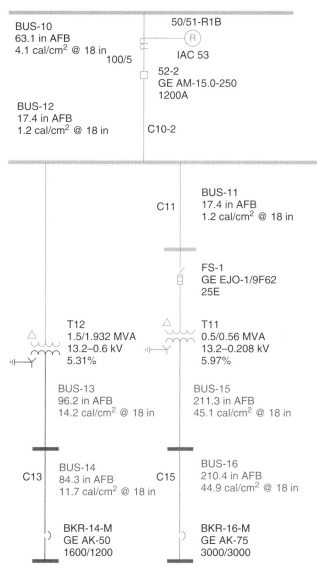

Figure 10.16   Two 600 V substations on a 13.8 kV feeder, with slow fuses on the smaller substation (*Source*: © IEEE (2009a)).

TABLE 10.3   Arc Flash Results at BUS-16 Versus Fuse Type (R1 at 6.0AT, 0.5TD)

Fuse (See Text)	Type	Fuse Size	Arc Flash Boundary (mm)	Incident energy @ 455 mm (cal/cm^2)
1	FAST	25E	1455	6.6
2	SLOW	25E	5344	44.9

(*Source*: © IEEE (2009a)).

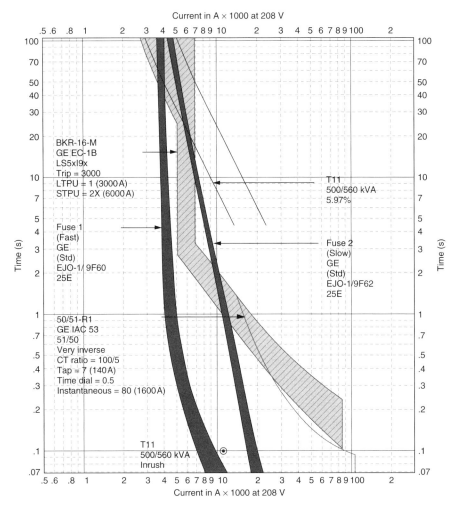

Figure 10.17   Comparison of fuse-time-current curves (*Source*: © IEEE (2009a)).

supply (UPS) in the ES. The one-line diagram is shown in Figure. The time-current curve in Figure 10.24 depicts coordination between main and largest feeder breaker of bus BEDP1, the largest feeder breaker of bus B1-EMRG, and the main breaker of bus BG IT UPS-MAIN.

The 800 A main circuit breaker MCB #B-68 of the emergency bus BEDP1 and the 600 A feeder breaker ATS-B1/B2 B70 are current limiting MCCBs with electronic trip units, which are "not instantaneously selective" (General Electric Company, 2008b, p. 37). The reason for this is that the ratio of rating plug sizes is too small. If the upstream breaker were 1000 A, or the downstream 500 A, they would be selective to a certain current level.

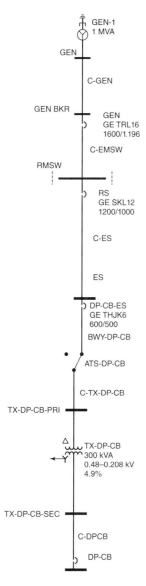

Figure 10.18   One-line diagram EMSW, Hospital #1.

The fault current level is 8.1 kA, which is below the defined current-limiting region of the circuit breakers (Figure 10.25). The available fault current is below where the 240 V the current-limiting curve leaves the square-root-of-two line (Valdes, Crabtree, and Papallo, 2010). The start of current limiting for the thermal magnetic circuit breaker BG-IT-UPS B61 can be seen clearly in the let-through energy curves (Figure 10.26). This is typical of systems with generators and low available fault current.

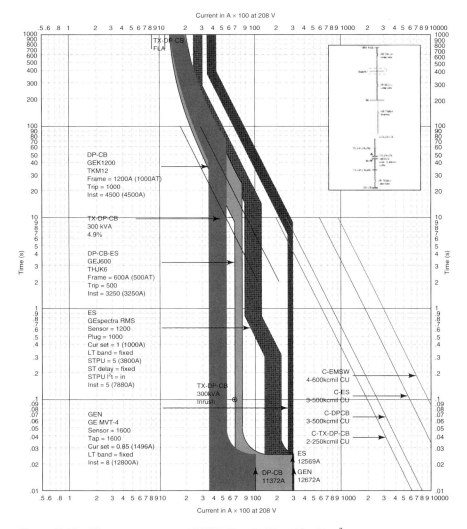

Figure 10.19   Time-current curves EMSW, Hospital#1, >40 cal/cm².

The second pair of circuit breakers to be evaluated is the 300 A thermal magnetic circuit breaker BG-IT-UPS B61 and the 125 A thermal magnetic circuit breaker BG-IT-UPS.. This pair cannot be coordinated instantaneously, and do not appear in the manufacturers tables. The resulting TCC (Figure 10.24) shows only normal coordination of curves, and no special instantaneous trip modifications. With these settings, the following arc flash incident energy levels have been calculated:

Bus BEDP1: 18.9 cal/cm²

Bus B1-EMRG: 14.8 cal/cm²

Bus BG IT UPS: 12.3 cal/cm²

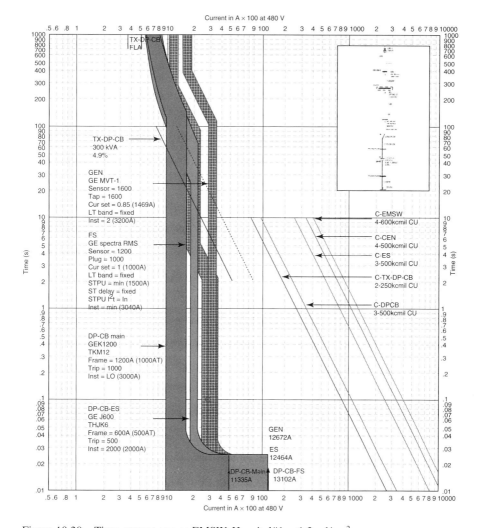

Figure 10.20   Time-current curves EMSW, Hospital#1, <1.2 cal/cm^2.

The arc flash hazard levels must also be checked with the ATS in the normal position.

When the feeder is supplied by normal power, as shown in the one-line diagram of Figure 10.27 and the time-current curve in Figure 10.28, the available fault current increases from 3.7 to 8.3 kA on the 208 V bus. The coordination is now between the thermal magnetic circuit breaker B1-EMG-SWL B9, which does not have instantaneous coordination, and the 300 A thermal magnetic circuit breaker BG-IT-UPS B61. In order to keep arc flash energies at the switchgear bus low, the coordination with the downstream 208 V breaker has been omitted. In this case, the results are better, with the bus B1-EMRG below 1.2 cal/cm^2 because of the lower

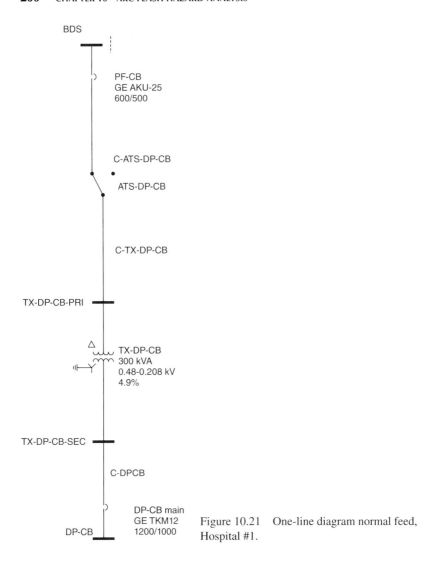

Figure 10.21 One-line diagram normal feed, Hospital #1.

setting on the upstream breaker. The lower-current emergency case becomes the higher arc flash hazard situation.

10.5.3 Hospital #3 — Time–Current Curve Examples

The one-line diagram of Figure 10.29 shows an 800 kVA emergency generator which can feed the emergency load via the main breaker GEN MAIN on the bus GENERATOR GEAR then through feeder breakers FDR G3 and FDR G4. Figure 10.30 depicts coordination between the main breaker and the two largest feeder breakers of the bus GENERATOR GEAR.

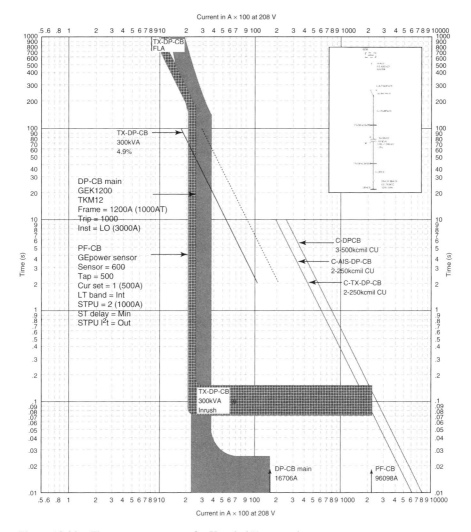

Figure 10.22   Time–current curves for Hospital#1, normal.

The first two breakers to be coordinated are electronic trip current-limiting MCCBs, which are the 1200 A main breaker followed by the largest feeder, which is rated at 400 A. This pair is selective up to an available fault current of 10.8 kA (General Electric Company, 2008b, p. 21). Because the available fault current is calculated to be 10.3 kA, the two breakers are selective.

When the coordination is examined using let-through current tables, it is found that the fault current level of 10.3 kA is below the defined current-limiting region of the circuit breakers GEN MAIN and FDR G3 (Figure 10.31). The available fault current is below where the 480 V current-limiting curve leaves the square-root-of-two line (Valdes, Crabtree, and Papallo, 2010). The start of current limiting can also be

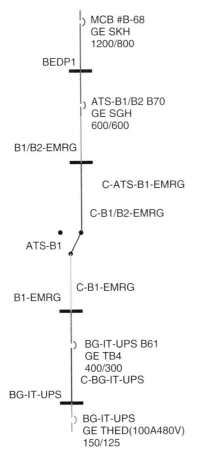

Figure 10.23   One-line diagram emergency feed, Hospital #2.

seen clearly in the let-through energy curves. This is typical of systems with genera-tors and low available fault current.

The second pair of circuit breakers to be examined are the 1200 A main breaker GEN MAIN followed by the 200 A feeder FDR G4. This pair is also selective up to an available fault current of 10.8 kA (General Electric Company, 2008b, p. 21). Because the available fault current is calculated to be 10.3 kA, the two breakers are selective.

With these settings, the following arc flash incident energy levels have been calculated:

Bus GENERATOR GEAR: 32.6 cal/cm^2

Feeder G3: 0.3 cal/cm^2

Feeder G4: 0.3 cal/cm^2

Thus, by using selectivity tables, breaker coordination and arc flash reduction can be achieved.

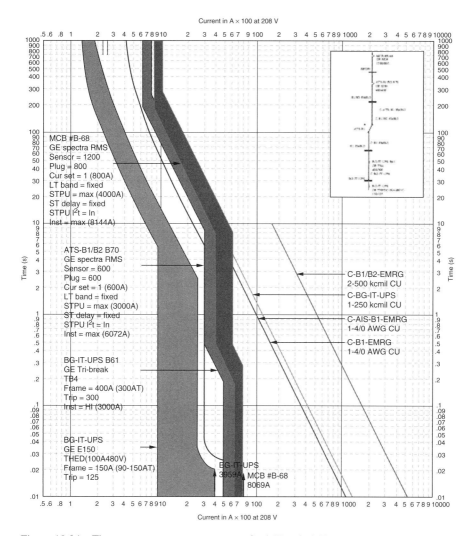

Figure 10.24   Time–current curves emergency feed, Hospital #2.

The arc flash hazard levels must also be checked with the ATS in the normal position. In this case, the results are similar, with the buses ATS3 and ATS4 at very low levels of incident energy.

The one-line diagram of Figure 10.32 shows a utility service which can feed the emergency load via the 3000 A insulated case circuit breaker (ICCB) and main breaker SERVICE MCB on the bus HV-MDB then through current-limiting MCCB feeder breakers HV-MDB-3 (200 A) and HV-MDB-5 (400 A). Figure 10.33 depicts coordination between the main and the two largest feeder breakers of the bus HV-MDB.

According to the selectivity tables, instantaneous coordination occurs up to 46 kA (General Electric Company, 2008b, p. 14). The rms let-through current of the

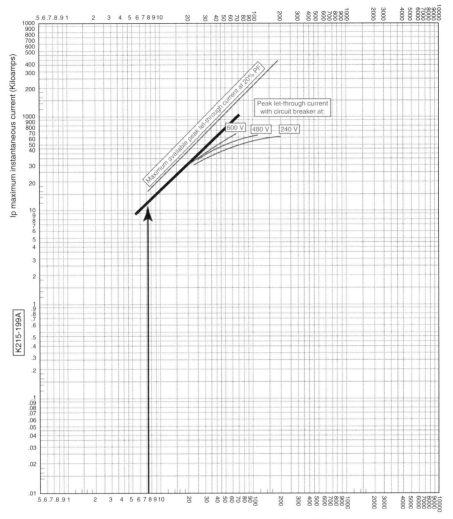

Figure 10.25 Let-through current curve for 600 A breaker in Hospital #2. Available fault current is 8096 A. The diagonal line is square root of two line.

MCCBs by the up-over-down method for 76 kA (Figure 10.34) is 38 kA. Because the 6000 A instantaneous pickup setting of the ICCB is below the let-through current, instantaneous coordination exists.

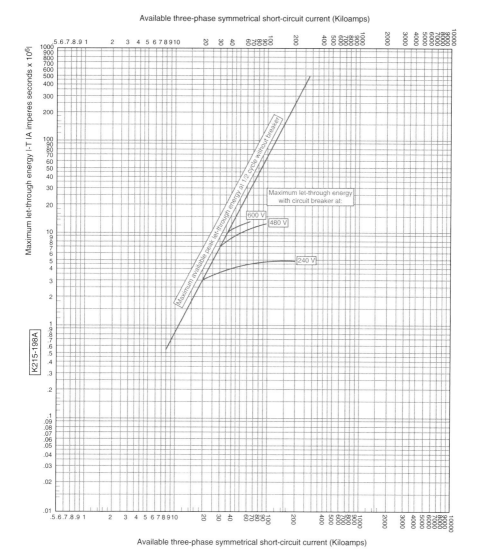

Figure 10.26   Let-through energy curves for 400 A thermal magnetic circuit breaker in Hospital #2.

10.6   LOW-VOLTAGE TRANSFORMER SECONDARY ARC FLASH PROTECTION USING FUSES

The secondary of a low-voltage transformer, typically 480 V, is a particularly hazardous area in regards to arc flash (Figure 10.35). As discussed previously (e.g.,

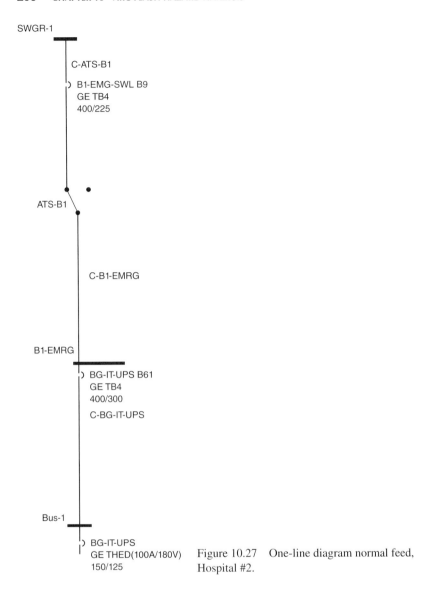

SWGR-1

C-ATS-B1

B1-EMG-SWL B9
GE TB4
400/225

ATS-B1

C-B1-EMRG

B1-EMRG

BG-IT-UPS B61
GE TB4
400/300

C-BG-IT-UPS

Bus-1

BG-IT-UPS
GE THED(100A/180V)
150/125

Figure 10.27   One-line diagram normal feed,
Hospital #2.

Figure 10.12), the primary protection is designed to protect the transformer from damage, and does not usually provide arc flash protection on the secondary side. The zone of arc flash hazard extends from the secondary terminals of the transformer, through the secondary transition compartment, to the line side terminals of the sub-station main breaker (Figure 10.36).

An arc flash hazard reduction system for use in a power substation can be designed (Gradwell, 2014) using current-limiting protection devices such as current-limiting protection fuses (e.g., a high-speed and high-current fuse) which

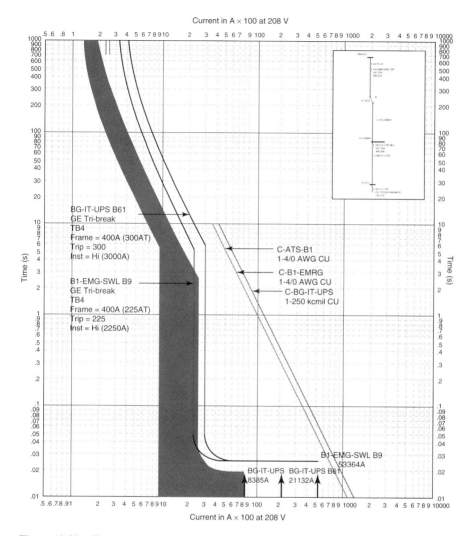

Figure 10.28 Time–current curves normal feed, Hospital #2.

are placed between a transformer and switchgear having a main circuit breaker, such that the current-limiting protection devices are directly connected to the secondary windings of the transformer (Figure 10.37). Placing the current-limiting protection devices between the secondary windings of the transformer and the switchgear provides arc flash hazard reduction capability in a region of a power substation that is designated as an "extreme danger" area because it can have incident energy levels that are higher than about $40\,\text{cal/cm}^2$. Benefits of this scheme include providing arc flash hazard reduction in regions of a power substation that can have incident energy levels that are greater than about $40\,\text{cal/cm}^2$. This enables workers to perform maintenance on such areas without having to shut down and de-energize equipment.

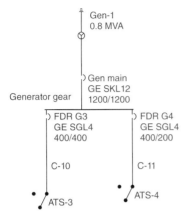

Figure 10.29   Hospital #3, emergency generator system one-line diagram.

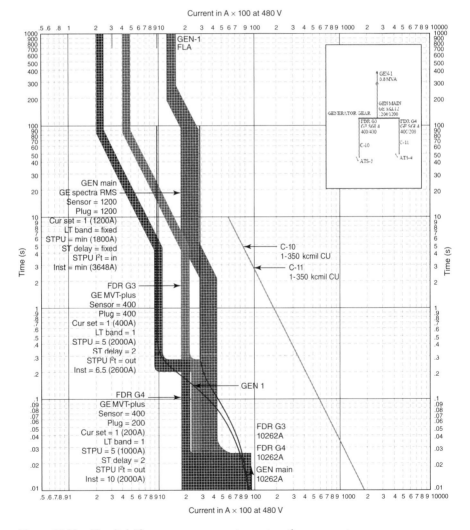

Figure 10.30   Hospital #3, emergency generator system time–current curves.

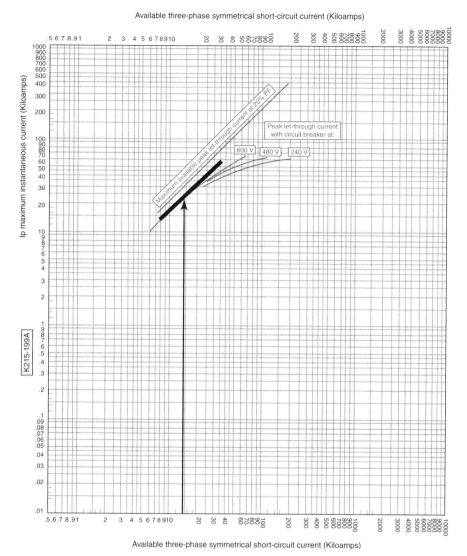

Figure 10.31   Let-through current curve for 400 A circuit breaker in Hospital #3 emergency system.

Performing maintenance on "extreme danger" regions without having to shut down and de-energize equipment results in increased productivity of the power substation.

The type of fuse which must be used for this application is not the typical current-limiting fuse used in electrical power distribution. It must rather be a high-speed device, such as is used for the protection of semiconductor devices, but yet has high enough ratings that it may be used in this application. In particular, it must have a continuous current rating higher than the maximum current output of the transformer, plus a safety margin, an interrupting rating higher than the maximum

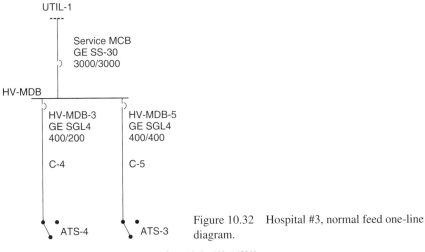

Figure 10.32 Hospital #3, normal feed one-line diagram.

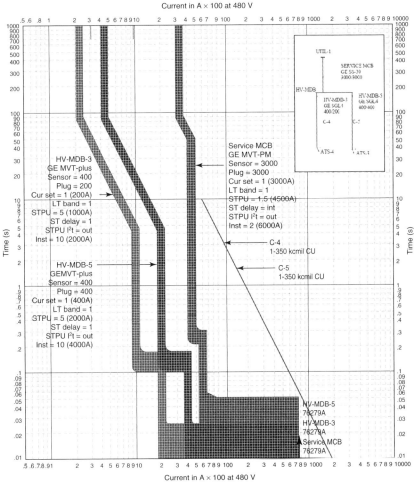

Figure 10.33 Hospital #3, normal system time–current curves.

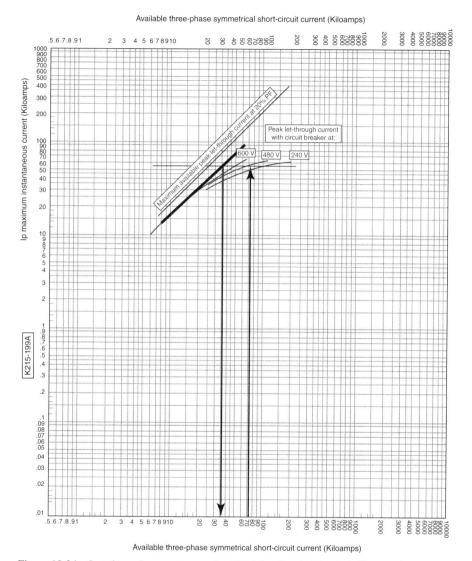

Available three-phase symmetrical short-circuit current (Kiloamps)

Figure 10.34 Let-through current curve for 400 A breaker in Hospital #3 normal system. The available fault current is 76 kA.

bolted fault current at the transformer secondary, again with a safety margin, and finally, a clearing time which is sufficient such that for an arcing fault with reduced current, the total incident energy is less than 40 cal/cm^2.

Square body fuses having characteristics such as: 550 V, 2000 A, clearing I^2t : 6.35×10^6 A^2 s, are one example of commercially available high-speed, high-current fuses that may be used. There are other high-speed, high-current fuses of comparable size, rating (e.g., 800, 1200, 1600, and 2400 A) and clearing time that may be used as a current-limiting protection device for limitation of arc flash hazards. The appearance (Figure 10.38), time-current curve (Figure 10.39), and peak let-through curve

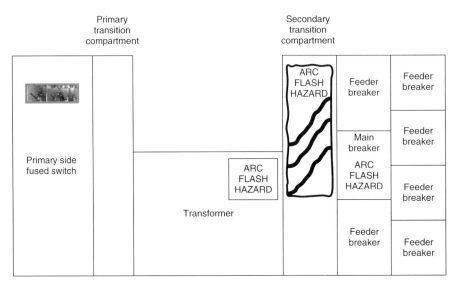

Figure 10.35 Elevation view of low-voltage substation showing arc flash hazard areas.

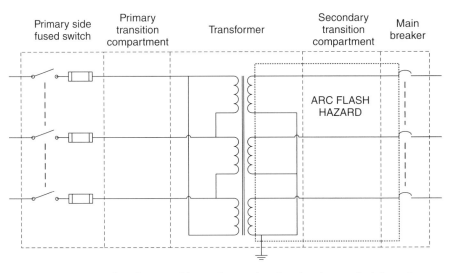

Figure 10.36 Three-line diagram of low-voltage substation showing arc flash hazard areas.

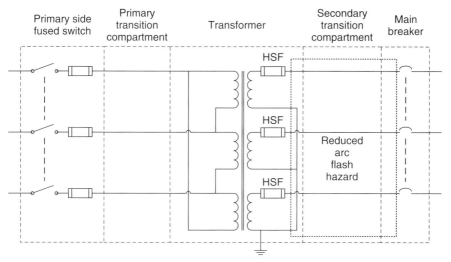

Figure 10.37 Three-line diagram of low-voltage substation showing arc flash reduction using high-speed fuses (HSF).

Figure 10.38 Appearance of typical high-speed, high-current, low-voltage fuse (Cooper Bussman).

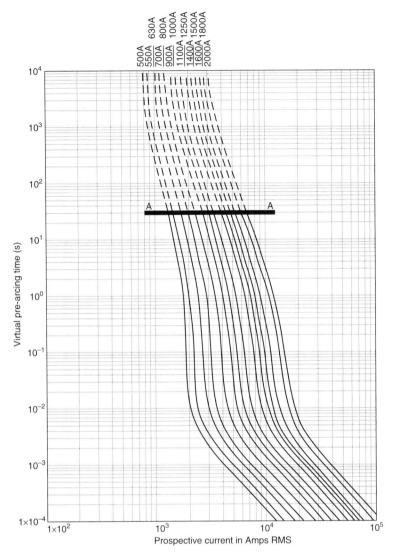

Figure 10.39   Typical time–current curve for high-speed, high-current, low-voltage fuse (Cooper Bussman).

(Figure 10.40) of a typical fuse which could be used in this application are shown here. The dashed line area in the time-current curve should not be used, and this area should be coordinated with the primary protection. A full arc flash hazard analysis should be performed for every application.

Other current-limiting protection devices may be suitable for use with this scheme. For example, current-limiting circuit breakers, "Is limiters" or "CLIP" devices may be used, as long as they can provide the necessary degree of current limitation.

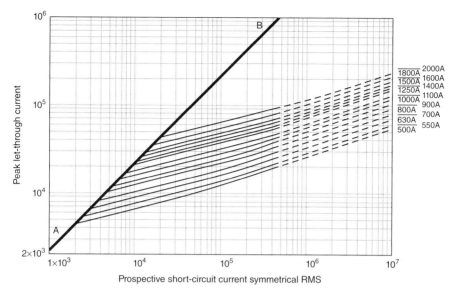

Figure 10.40 Typical peak let-through curve for high-speed, high-current, low-voltage fuse (Cooper Bussman).

CHAPTER **11**

EFFECT OF HIGH FAULT CURRENTS ON PROTECTION AND METERING

11.1   INTRODUCTION

The design of safe electrical systems relies not just on reducing hazards present from energized conductors in terms of shock and arc flash, but in rapidly de-energizing conductors whose operation presents a safety hazard. For example, if an overhead line conductor or insulator is broken and a short circuit to the earth occurs, presenting a severe safety hazard to any person in the vicinity, a fuse or circuit breaker must be present to interrupt and clear the fault as soon as possible (usually milliseconds) thus greatly reducing the hazards presented. These protection systems, which both detect and clear faults, must themselves be able to operate reliably and effectively, under conditions of high electrical stress, not only the elevated currents caused by short circuits but also the overvoltage transients which may be present as well. This chapter will deal with the effects of high fault currents, while voltage transients will be covered in a later chapter.

The literature on effects of high fault currents on protection and metering equipment will be reviewed. The capabilities and limitations of existing short-circuit protection devices will be included in the literature search. High fault currents are well known to cause saturation of iron core current transformers (CTs). This can adversely affect the performance of system protection devices. High fault currents can also exceed the range of operation of current-operated protective devices and produce high voltages in the CT secondary circuits. This chapter is adapted from EPRI (2006).

The National Electrical Code (NEC) (NFPA, 2014), Article 110.9, requires that

"Equipment intended to interrupt current at fault levels shall have an interrupting rating at nominal circuit voltage sufficient for the current that is available at the line terminals of the equipment."

No exception exists for contingencies such as when the fault current level only exists for a short time owing to switching or other operations. For example, a double-ended lineup of 15 kV switchgear, with a normally open tie circuit breaker with a closed transition has to be rated for the higher short-circuit duty which

Principles of Electrical Safety, First Edition. Peter E. Sutherland.
© 2015 John Wiley & Sons, Inc. Published 2015 by John Wiley & Sons, Inc.

occurs when the tie breaker is closed. This approximately doubles the fault level and increases CT saturation, both AC and DC.

IEEE Standard C37.110-2007, *IEEE Guide For The Application of Current Transformers Used for Protective Relaying Purposes* (IEEE, 2007b) provides recommendations for calculations for AC and DC CT saturation.

The performance of protective relaying systems in the presence of CT saturation has been discussed in, among others, IEEE standards (IEEE, 2007b), textbooks on relaying (Blackburn, 1998; Elmore, 1994), and IEEE committee reports (Linders *et al.*, 1995) (Power Systems Relaying Committee, 1976). Despite the great amount of previous work, this topic still causes great controversy and universally accepted application guidelines do not yet exist.

11.2 CURRENT TRANSFORMER SATURATION

CTs are connected in series with the circuit whose current is to be measured as shown in Figure 11.1, (EPRI, 2004b).

CTs are intended to deliver a secondary current that is directly proportional to the primary current with as little distortion as possible. In most cases, the secondary output current is usually reduced to less than 5 A. Although there are CTs with 1 A or 10 A secondaries, the most common rating in the United States is 5 A.

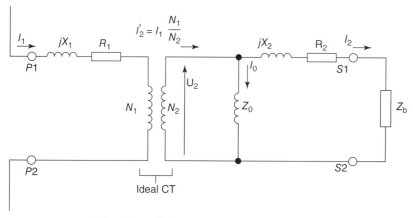

X_1 = Primary leakage reactance
R_1 = Primary winding resistance
X_2 = Secondary leakage reactance
Z_0 = Magnetizing impedence
R_2 = Secondary winding resistance
Z_b = Secondary load

Note: Normally the leakage fluxes X_1 and X_2 can be negelected

Figure 11.1 Current transformer equivalent circuit (*Source*: EPRI, 2006).

CTs are rated for a certain turns-ratio of operation. For example, a CT with a turns-ratio of 500 : 5 reduces 500 A on the primary to 5 A on the secondary. A properly designed CT circuit yields a secondary current of 5 A or less at rated primary current. Although the CT secondary and the relay are not intended for continuous operation at higher than 5 A, they are designed to withstand greater values of current for short periods. For example, short-circuit currents may be 20 or more times the normal current in a power system.

During normal operation, the CT secondary winding induces a magnetic flux that opposes and nearly cancels the primary induced flux. As a result, the flux density is very low and the resulting voltage at the secondary terminals is also very low. Relays, meters, or other connected devices are constructed with only a few turns of relatively large wire—this low impedance effectively functions as a short circuit across the CT secondary. The secondary voltage of a CT remains at a low value as long as the secondary circuit remains closed. An open circuit on the secondary side of a CT that still has current flow on the primary side can result in a dangerously high secondary voltage. Opening the secondary removes the opposing secondary flux, thus allowing the primary flux to generate a very high voltage at the secondary terminals. Electrical arcing caused by an open-circuited CT can injure personnel and damage equipment. When the primary side is carrying any current, great care must be taken to ensure that the secondary circuit remains closed at all times.

The CT's ability to produce a secondary current proportional to its primary current is limited by the highest secondary voltage that it can produce without saturation. Beyond a certain level of excitation (actual values are readily available from the manufacturer), the CT is said to *enter saturation*. When saturated, most of the primary current maintains the core flux, and the shape of both the exciting and secondary currents departs from the normal sine wave. The secondary voltage and current then collapse to zero, where they remain, until the next primary current zero is reached. The process is repeated each half-cycle and results in a distorted secondary waveform.

Saturation of a CT can prevent a protective relay from operating properly. For this reason, a CT must be carefully sized so that it will perform properly for the maximum expected fault current. Low-ratio CTs (e.g., 50 : 5 or 75 : 5) are particularly susceptible to saturation during fault conditions.

The following problems may exist if a CT is allowed to saturate:

False tripping. Differential relays used for transformer protection may respond to a through-fault condition. Numerical relays, however, pose far lower burden (0.5 VA) on CTs. Numerical relays are capable of detecting CT saturation and blocking the relay from tripping, minimizing the effect of false tripping.

Delayed tripping. A distorted secondary reproduction of the primary current can delay relay time response. This delay in tripping may result in de-energizing a larger portion of the system due to loss of relay coordination caused by the CT saturation.

Failure to trip. Failure to trip may occur if the CT secondary current is very low or extremely distorted. Backup relays must then respond to clear the fault. Digital relays, however, have the added advantage that they can detect these conditions and still deliver a trip because the relay is not dependent on the power that must be supplied by the CT to trip.

11.3   SATURATION OF LOW-RATIO CTs

High levels of fault current, especially when DC offset is present, cause the secondary current of a CT to be significantly distorted and diminished in magnitude, even with a very small burden. This is most significant with low-ratio CTs. CT saturation may cause overcurrent relays to misoperate or fail to operate, resulting in a failure of the protection system.

11.3.1   AC Saturation

AC saturation is a gradual process, whereas the rms value of the current increases and the ratio accuracy of the CT decreases. AC saturation begins to affect protective relay performance if the rms excitation voltage begins to exceed the "saturation voltage," where the rate of increase of the excitation current with respect to excitation voltage greatly increases. According to the definition given in the standard, the "saturation voltage," V_X, is the point of intersection of lines extended from the straight portions of the saturation curve. This is usually slightly higher than the "knee-point voltage," which is defined in the standard as the point on the saturation curve where a tangent drawn to the curve has an angle of 45° to the horizontal axis. See the illustration of the knee-point concept in Figure 11.2.

Section 4.5.3 of IEEE Standard C37.110-2007 (IEEE, 2007b) suggests that the effective CT ratio error will be significant if the calculated secondary voltage that the CT must support exceeds the saturation voltage, V_X, that is,

$$V_X > V_S, \text{ where}$$
$$V_S = I_S \times Z_S$$
$$= \frac{I_P}{N} \times Z_S \qquad (11.1)$$

where

V_S = CT secondary voltage

I_S = CT secondary current

I_P = CT primary current

N = CT turns ratio, and

Z_S = total secondary burden of the CT.

After an example calculation of this type, where V_S approaches V_X, one author states, "Although this is near the knee of the saturation curve, the small excitation

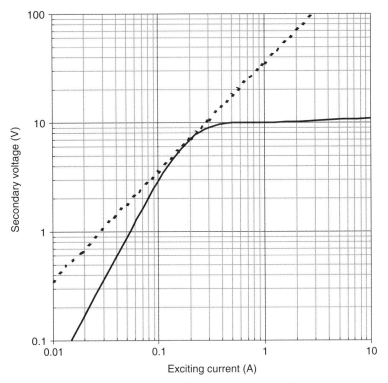

Figure 11.2   The "knee point" of a CT saturation curve is the point where a tangent to the curve forms a 45° angle with the horizontal axis (*Source*: EPRI, 2006).

current does not significantly decrease the fault current to the relays." (Blackburn, 1998, p. 146).

Another commonly used criterion is from Section 5.10 of the standard, "A rule of thumb frequently used in relaying to minimize the CT saturation effects is to select a CT with a C voltage rating at least twice that required for the maximum steady-state symmetrical fault current." It is furthermore stated in (Linders *et al.*, 1995) that "One basic rule-of-thumb has applied in the application of CT's, namely, The knee-point voltage of the CT as defined by the CT excitation curve should not be less than twice the voltage required to drive the maximum secondary symmetrical current through the combined burden of the relay, connecting wiring and CT." The "C" rating is often assumed to be approximately equal to the knee-point voltage, in fact, "the knee-point voltage may be 50–75% of the standard accuracy class voltage rating of the CT (e.g., C400)." (IEEE, 2007b, Section 4.5.3).

It is important to carefully consider the implications of AC saturation before applying these criteria. As stated above, the effect of AC saturation is to cause an error in the effective ratio of the CT such that the secondary current available to relays and other measuring devices will be less than what would be expected knowing the primary fault current and the nominal CT ratio. The fact that the calculated voltage exceeds the saturation voltage, V_X, does not mean that the CT collapses entirely—it is just that the ratio error increases significantly.

There is more to saturation that an increase in the effective ratio error of the CT. The presence of significant saturation also causes the waveform of the secondary current to depart from the normal sinusoidal pattern. Hence, when significant saturation is present, it is no longer possible to think in terms of conventional sinusoidal response.

Practically, an accurate measurement of current is only important in the case of time overcurrent relays. With instantaneous relays, it is only important to know that the available current will be greater than the setting of the relay. It is not important to know by how much the available secondary current will exceed the instantaneous relay setting or whether the waveform of that current will be a respectable sinusoid.

Therefore, the pertinent question is, "Given the performance characteristics of the CT and the associated CT burden, will the CT be 'reasonably linear' for all fault magnitudes for which the time overcurrent relay is expected to operate?" On the basis of this perspective, the following conditions should be checked:

For feeder relays having both time and instantaneous elements, will the calculated secondary voltage be less than the CT saturation voltage, V_X, for the maximum fault at which the time overcurrent relay is expected to operate, namely, the current level at which the instantaneous element is calibrated to pick up?

For main and tie relays with no instantaneous elements, will the calculated secondary voltage be less than the CT saturation voltage, V_X, for the maximum fault at which the time overcurrent relay is expected to operate, namely, the expected (calculated) maximum current through the circuit breaker associated with the CT in question?

11.3.2 DC Saturation

The standard (IEEE, 2007b) suggests two criteria regarding DC saturation. The first (Section 4.5.3) is that the effective CT ratio error will be significant when the following condition is met:

$$V_X > \frac{I_S \times Z_S \left(1 + \frac{X}{R} \times \frac{R_S + R_B}{Z_S} \right)}{1 - \text{per unit remanence}} \qquad (11.2)$$

where

$R_S + R_B$ = resistive component of the CT burden

$\quad X$ = primary side system reactance, and

$\quad R$ = primary side system resistance.

Alternatively, one author (Elmore, 1994. p. 80) suggests that if

$$V_X \geq I_S(R_S + R_B)\frac{X}{R}, \qquad (11.3)$$

DC saturation will not occur. If remanence is ignored, these two equations may be simplified as representing

$$V_X \geq Z_S \left(1 + \frac{X}{R}\right) \tag{11.4}$$

for equation (11.2) and

$$V_X \geq Z_S \frac{X}{R} \tag{11.5}$$

for equation (11.3).

These equations may be seen in some CT application calculations and are valid as long as there is no significant inductance in the CT secondary circuit or significant remanence in the CT itself. Dropping the factor of 1 in equation (11.4) simply means that X/R is assumed to be high.

Remanence is the tendency of the iron core of a CT to retain magnetic flux on the basis of prior history. Remanence flux levels of up to 80% of saturation level have been observed (IEEE, 2007b, par. 4.6.1). The worst case for remanence comes about when a DC continuity test is used to verify CT circuit continuity (Seveik and DoCarmo, 2000), but this remnant flux will dissipate if the CT is demagnetized following the test.

A more common issue comes up when the circuit breaker interrupts an offset fault current. Interruption of the DC component of this current leaves one or more CTs partially magnetized. This remnant flux will also dissipate with a few seconds of loading, if 60% of the saturation voltage is exceeded, but it can be a concern if the breaker subsequently recloses into a fault before flux dissipation can occur.

Remnant flux may either aid or oppose the magnetization imposed by DC transients. Practically, remnant flux is not a real concern in industrial applications where there is no automatic reclosing.

DC saturation does not occur instantaneously, but rather builds up with time. Hence, the second criterion to be considered in evaluating DC saturation is the time to saturate (in fundamental frequency cycles):

$$T_S = -\frac{X}{\omega R} \ln\left(1 - \frac{K_S - 1}{X/R}\right) \tag{11.6}$$

The saturation factor K_S, is defined as

$$K_S = \frac{V_X}{I_S (R_S + R_B)} \tag{11.7}$$

This formula does not include the effect of remanence, which will decrease the time to saturation.

The effect of DC saturation is to interfere with operation of instantaneous relays. Given the performance characteristics of the CT and the associated CT burden, will DC saturation occur quickly enough, and with sufficient severity, to interfere with the operation of these instantaneous relays? On the basis of this perspective, the following tests should be performed:

For feeder relays having both time and instantaneous elements, will the calculated secondary DC saturation voltage be less than the CT saturation voltage, V_X, for the maximum fault at which the instantaneous relay is expected to operate, namely, the maximum available fault current and will the "time to saturate" be shorter than the time required for the measuring algorithm in the instantaneous relay to respond to the CT secondary current prior to the point at which significant DC saturation appears (typically, one-half cycle)?

For main and tie relays with no instantaneous elements, DC saturation is not an issue.

If the effect of remanence flux is taken into account, all relay-CT combinations may have time to saturation of less than one cycle, effectively preventing relay operation. This situation is typical of most applications of overcurrent relays with CTs in medium-voltage switchgear. It is a generally accepted compromise that non-operation of overcurrent relays may occur when the remanence flux is large. For this reason, backup overcurrent protection and bus differential protection, which are not susceptible to CT saturation, are utilized in power systems where DC saturation may be a problem, such as in generating stations. IEEE standard C37.110-2007 (IEEE, 2007b, Section 4.5.3) states, "These requirements generally result in impracticably large CTs, and hence compensating steps must be taken to minimize saturation effects on the relay protection plan. Some high-speed instantaneous relays can operate before saturation has time to occur."

In many existing relaying systems, the time-overcurrent relay functions serve to back up the instantaneous relay functions. It is the instantaneous functions that may be susceptible to DC saturation. Because the formulae are based on certain assumptions, there is an error of $-0 + 0.5$ cycles in the time to saturation. The actual response of the relay to the saturated current waveform, Figure 11.3, is subject to many imponderables. An answer to the question of whether a particular instantaneous relay will respond within an acceptable time delay can only be answered by test.

The instantaneous relay may operate immediately if the time to saturation is long enough to allow it to operate. If the time to saturation is too short, the relay

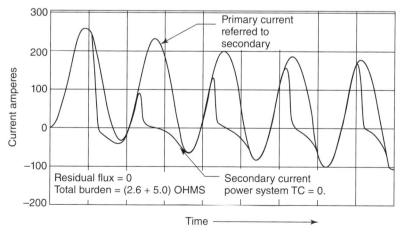

Figure 11.3 Typical waveforms of CT primary and secondary current with DC saturation (Power Systems Relaying Committee, 1976) (c) 1976 IEEE.

may operate several cycles later when the CT emerges from saturation. The maximum delay for instantaneous tripping is determined by the coordination time intervals (CTIs) defined for the protective system.

11.4 TESTING OF CURRENT TRANSFORMER SATURATION

When testing a specific CT for application in a particular situation, the tests should model the CT secondary circuits with the identical lead sizes and lengths. The tests should be conducted in a three-phase high-current short-circuit laboratory using system primary current and X/R ratios as determined from the short-circuit study. Tests should be made with all relays attached to the CT using maximum existing system settings. Tests should be made both with and without the overvoltage protective devices (OCPs) (surge suppressors) used on the CT secondaries. A typical test setup is shown in Figures 11.4 and 11.5. If the time to close the instantaneous contacts is less than the defined maximum time, the CT-relay combination will be deemed to have passed the test. If a relay fails the test with standard accuracy CTs, the test shall be repeated with high accuracy CTs. The CTs to be used if further tests fail are the next higher ratio standard and high accuracy units. The first CT-relay combination which passes the test should be utilized in the system.

Some protective relays contain an "overreach filter" which removes the DC component from an asymmetrical current wave. This will slow the response time by 10-15 ms (0.6-0.9 cycles). DC saturation occurs in the core of the CT and affects the current waveform that is input to the relay. The use of an overreach filter will not prevent DC saturation but will reduce the probability of tripping by inserting an

Figure 11.4   View of a test setup located in test cell.

Figure 11.5 Relay shown connected to a test setup. OCP modules shown immediately to the right.

additional time delay in the relay response. The use of the overreach filter on the relays is not recommended in this application. Actual tests should be performed with and without overreach filters (if present).

The paper "Relay Performance Considerations with Low Ratio CT's and High Fault Currents" (Power Systems Relaying Committee, 1976) stated that in order to eliminate CT saturation, the secondary circuit voltage should be less than $\frac{1}{2}$ the knee voltage. Testing would be performed to prove which of the CTs under evaluation would be required as a minimum to operate the relay.

The effects of DC saturation should also to be evaluated during testing.

The CTs to be used were to be short-circuit tested with maximum three-phase fault with maximum offset. A test is considered successful if one of the two relay operations occurred: either the instantaneous or the overcurrent function operated. Three successful sequential tests have to occur for a CT to be considered acceptable.

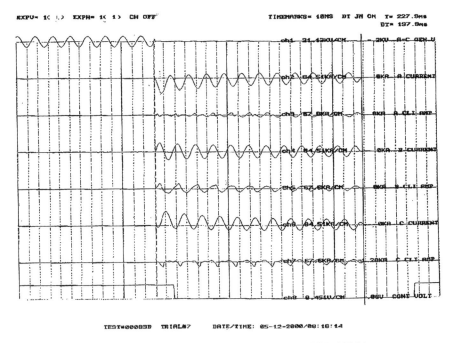

Figure 11.6   Example test report for CT saturation (*Source*: EPRI (2006)).

Also unknown are effects of OCPs. If any test failed, the OCP was to be removed and the test performed again. If the test was successful after the OCP removal, then all OCPs should be removed.

Three-phase testing is performed at a specified X/R ratio to verify that the instantaneous function of all relays will operate within required limits when using standard accuracy class CTs. Line-to-line testing could be used to prove that each relay's instantaneous function would not have operated with standard accuracy class CTs.

A test cell typically provides fault current directly to the test setup from a generator at a specified level of fault current, in kiloamperes, and X/R ratio. A sample of typical test current is shown in Figure 11.6.

According to the equation for the test current (Blackburn, 1998):

$$I(t) = I_m[\sin(\omega t + \alpha - \delta) - \sin(\alpha - \delta)e^{-t/\tau}] \tag{11.8}$$

where

I_m = peak AC, A

Δ = fault incidence angle, rad

τ = time constant of DC component, s

$\alpha = \tan^{-1}(\omega t)$, rad

$\omega = 2\pi f$, rad/s

$f =$ frequency, Hz

$t =$ time, s

The DC time constant, τ, affects the rate of decay of the DC component of the fault current. This time constant is controlled by the time constants of the generator and the series reactor.

It is difficult to reach firm conclusions from this data. However, the following tentative conclusions may be reached:

1. The minimum criterion to avoid AC saturation is that the maximum CT secondary voltage should be less than the so-called "knee-point" voltage or approximately the accuracy classification ("C" rating) voltage.

2. There is a conflict between the recommendations of the standard (IEEE, 2007b) and paper (Linders *et al.*, 1995) regarding whether 50% or 100% of the knee-point voltage should be compared with the maximum CT secondary voltage.

3. The criteria to avoid DC saturation of having the maximum CT secondary voltage times the X/R ratio of the faulted circuit be less than the knee-point voltage is unrealistic in most instances. Owing to the lack of a better guideline, it is recommended that disputed CT-relay combinations be tested.

4. Because of the issue of unequal DC offset of various phases raising doubts about the validity of testing of three-phase relays, phase-phase fault testing is recommended instead of three phase fault.

5. The issue of whether remnant flux is significant in industrial applications should be investigated further.

6. In the saturation calculations shown above, by far the greatest burden problem is the CT. The use of solid-state relays reduces the burden problem to only a small extent compared to the burden of the CT. Further efforts should be made to provide new CT designs, which reduce this burden.

7. Be very careful of source limitations when testing, that is, higher X/R may not necessarily be better as imbalance effects may be more prominent. It is better to match the testing X/R requirements to what you actually need. Be sure and get in advance the characteristics of the source from the testing facility (if you can) well before testing begins.

8. Since testing facilities are expensive, have all your calculations prepared and checked.

9. Be familiar with the programming of the relays prior to the testing as this saves time.

10. Use relays with waveform capture and event log capture. This is invaluable for analysis.

11. Bring a digital camera. It helps in keeping track of the testing setup.

12. Get as many results as possible on paper before to leaving the test facility for analysis purposes as it can take some time to obtain these results officially.

11.5 EFFECT OF HIGH FAULT CURRENTS ON COORDINATION

The protective device coordination in a radial feeder or for any distribution feeder is important. The coordination of protection devices is necessary to maintain selectivity (i.e., to remove the portion of the power system that is experiencing the fault). The inverse-time-current curve provides coordination at the same time that it offers the speed and accuracy needed to clear and the correct fault quickly. By the nature of the design, the inverse-time curves utilize the magnitude of the current to decide the tripping time, unlike the instantaneous relays with definite time.

The name of the inverse-time curve is self-explanatory. The higher the magnitude of the fault or the load current, the less time is required for the relay to operate. This provides quick removal of the faulted portion from the power system. Under normal load condition or for a low-magnitude fault current, the relay allows enough time for other downstream devices to clear the fault. Because the microprocessor relay has to process the current mathematically to decide the time to trip, the relay can be equipped with different types of inverse curves.

The user at the time of the study decides the curve to be used. The relay simply actuates that curve equation and processes accordingly. IEEE has created standard inverse curves (IEEE, 1997). The manufacturer of the microprocessor relay must comply with these curves. Depending on the time to trip for higher current or for lower current, different curves are available.

The following list includes a few of the many popular curves:

1. Moderately inverse
2. Inverse
3. Normally inverse
4. Very inverse
5. Extremely inverse

The curves for the moderately inverse, very inverse, and extremely inverse curves are set forth in (IEEE, 1997):

Moderately inverse:

$$i(t) = \frac{0.0515}{\left(\frac{I}{I_{pu}}\right)^{0.02} - 1} + 0.1140 \tag{11.9}$$

Very inverse:

$$i(t) = \frac{19.61}{\left(\frac{I}{I_{pu}}\right)^{2} - 1} + 0.491 \tag{11.10}$$

Extremely inverse:

$$i(t) = \frac{28.2}{2 - 1} + 0.1217 \tag{11.11}$$

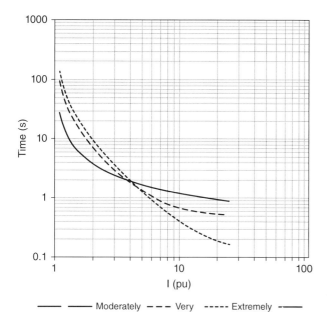

Figure 11.7   IEEE inverse-time overcurrent curves; time dial set at the midpoint.

Figure 11.7 depicts the three of these curves plotted according to the equations set forth by IEEE.

An example of miscoordination of a feeder circuit with a fault current increase is shown in Figure 11.8, where Fuse 1 clears fault F1 before the relay operates. The

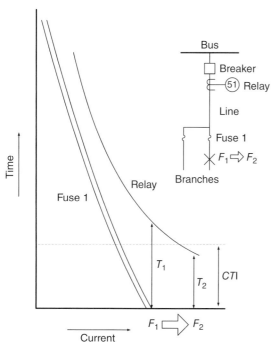

Figure 11.8   Coordination of a distribution feeder (*Source*: EPRI (2006)).

coordinating time interval (CTI) T_1 is sufficient. The fault current increases beyond design value $T_2 < CTI$. The breaker opens at the same time as Fuse 1, causing a wider outage.

11.6   PROTECTIVE RELAY RATINGS AND SETTINGS

Protective relays (EPRI, 2004b) have a reputation for providing reliable service for many years. Nonetheless, electromechanical protective relays are delicate instruments that are susceptible to the degradation of components that may affect performance. Owing to their design, numerical relays, Figure 11.9, have eliminated the degradation that can be expected from the mechanical components of electromechanical relays, Figure 11.10. Numerical relays also use minimal electronic components when compared to electronic relays. The failure of a protective relay to contain and isolate an electrical problem can have severe plantwide repercussions. When an expected protective action does not occur, the end result of an electrical abnormality may be catastrophic equipment damage and prolonged downtime instead of localized minor damage. Because of the severe consequences of a failure, protective relays should be maintained in a high state of readiness. Critical applications should be carefully evaluated for redundant protection.

Protective relaying is an integral part of any electrical power system. The fundamental objective of system protection is to quickly isolate a problem so that the unaffected portions of a system can continue to function. Protective relays are the decision-making device in the protection scheme. They monitor circuit conditions and initiate protective action when an undesired condition is detected. Protective relays work in concert with sensing and control devices to accomplish this function. There are several reasons to use protective relaying:

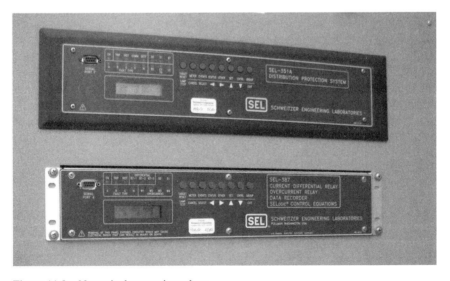

Figure 11.9   Numerical protective relays.

Figure 11.10   Electromechanical protective relay.

1. To provide alarms when measured process limits are exceeded, thereby allowing operators an opportunity to intervene with corrective actions.
2. To isolate faulted circuits or equipment from the remainder of the system so that the system can continue to function.
3. To limit damage to faulted equipment.
4. To minimize the possibility of fire or catastrophic damage to adjacent equipment.
5. To minimize hazards to personnel.
6. To provide post-fault information to help analyze the root cause.

Under normal power system operation, protective relays remain idle and serve no active function. However, when required to operate because of a faulted or undesirable condition, it is imperative that the relays function correctly. Another point of concern is the undesired operation of a protective relay during normal plant conditions or tolerable transients. Inadvertent relay operation can result in unnecessary system or plant downtime. A maintenance and surveillance program will help to ensure that the protective relays respond properly to normal and abnormal conditions. This frequency of testing can be extended to longer periods than in electromechanical devices. The number of tests can also be reduced owing to the design and construction of the relays. An effective maintenance program for protective relays accomplishes two primary goals. First, it provides a high degree of confidence that the electrical power protection system will respond to abnormal conditions as designed. Periodic assurance that protective relays are in an operable status is particularly important. Relay problems are generally detected by internal test routines and during operational checks via a human machine interface (HMI). Secondly, an effective maintenance program preserves the relay's readiness and helps to counteract normal and abnormal in-service deterioration that can affect a relay's electronic components over time. Even under normal conditions, electrical, thermal, and environmental stresses are continually at work, slowly degrading the relays. This deterioration is much slower in numerical relays than in electromechanical relays because numerical relays are not affected by mechanical deterioration. Routine maintenance checks help to identify any deterioration in the device. The life of numerical relays cannot be prolonged by recalibration, cleaning, and general maintenance because numerical relays are either functional or not. Failed devices must be removed, repaired, and/or replaced. Failed devices or device components such as printed circuit boards should be sent to the supplier for repair.

11.7 EFFECTS OF FAULT CURRENTS ON PROTECTIVE RELAYS

Numerical relays contain sensitive electrical components that are designed to operate at specific values of voltage and current. Voltages range from 110–480 VAC and 24–250 VDC. Currents are provided in either 1 A or 5 A (AC) rating. The maximum design-rated voltage is the highest rms alternating voltage or direct voltage. The maximum current is the highest rms alternating current or direct current. These maximum values are the limits at which the relay can operate continuously.

The operating coils of older electromechanical and solid-state relays typically determined the relay's rating. Today, numerical relays are supplied with multiple range or universal power supplies and binary inputs/outputs that operate at specific voltages that are set by using internal links or jumpers in the relay. The current and voltage sources can also have multiple operating ratings. Relays can be provided with secondary current selections of 1 or 5 A. Secondary voltages to the relay can be configured for a range of voltages and can typically be connected wye or delta by a setting parameter rather than a hardware configuration. Voltage inputs can be configured for phase-phase or phase-ground.

Relay burden is greatly reduced with the use of numerical relays. The power consumption of numerical relays can be expected in the 0.04–0.10 VA range. This means that in many applications, the relay burden is negligible. In older applications where the relay burden was much higher, intermediate CTs may have been required. Numerical relays normally no longer require these intermediate transformers.

Thermal overload capacities are also higher. For example, the effective thermal overload can reach a rating of 500 A (1 s) and a dynamic rating of 1250 A (half-cycle).

11.7.1   Examples

1. Electromechanical Relay (GE Multilin, 1997) GE IAC53 1.5–12 A time overcurrent taps; instantaneous range: 10–80 A; continuous current rating of the time overcurrent unit: 10–30.5 A; short-time current rating of the time overcurrent unit: $I^2t = 67,600$.

For example, with a 500 : 5 CT, 26 kA of fault current for 1 s will reach the thermal limit of this relay.

2. Digital (Schweitzer, 2003) SEL351 0.5–16 A time overcurrent range; instantaneous overcurrent range 0.25–100 A on 5 A CT; limits: 15 A continuous; 500 A @ 1 s, linear to 100 A; 1250 A @ 1 cycle.

For example, with 500 : 5 CT, 50 kA of fault current for 1 s is the limit.

11.8   METHODS FOR UPGRADING PROTECTION SYSTEMS

11.8.1   Update Short-Circuit Study

In order to obtain complete coordination of the protective equipment applied, it may be necessary to obtain some or all of the following information on short-circuit currents for each node or bus:

1. Maximum and minimum momentary (first cycle) short-circuit current.

2. Maximum and minimum interrupting duty short-circuit current.

3. Maximum and minimum ground-fault current.

The momentary currents are used to determine the maximum and minimum currents to which instantaneous and direct-acting trip devices respond. The maximum interrupting current is the value of the current at which the circuit protection coordination interval is established. The minimum interrupting current is needed to determine whether the protection sensitivity of the circuit is adequate.

The short-circuit study and coordination study should be updated when the available short circuit of the source to a plant is increased.

11.8.2   Update Protective Device Coordination Study

The objective of a coordination study is to determine the characteristics, ratings, and settings of overcurrent protective devices to ensure that the minimum unfaulted load is interrupted when the protective devices isolate a fault or an overload anywhere in the system. At the same time, the devices and the settings selected should provide satisfactory protection against overloads on the equipment and should interrupt short-circuit currents as rapidly as possible.

The coordination study provides data useful for the selection of instrument transformer ratios, protective relay characteristics, and settings and fuse ratings. It also provides other information pertinent to the provision of optimum protection and selectivity in the coordination of these devices. When plotting coordination curves, a certain time interval must be maintained between curves of the various protective devices in order to ensure correct sequential operation of the devices. This interval is called the *coordination time interval*.

When coordinating inverse curves, the time interval is usually 0.3–0.4 s. This interval is measured between relays in series at the instantaneous setting of the load side of the feeder relay or maximum short-circuit current which can flow through both devices simultaneously.

A basic understanding of time-current characteristics is essential in any study. Initial planning and power system data are also essential for any coordination study.

The initial planning process should include the following activities:

1. Develop a one-line diagram.
2. Determine the load flow.
3. Collect data.
4. Conduct a short-circuit study.
5. Determine time current coordination curves for all the devices in the system.

CHAPTER *12*

EFFECTS OF HIGH FAULT CURRENTS ON CIRCUIT BREAKERS

Circuit breakers (CBs) are an essential part of most electrical power systems, supplemented with other fault-interrupting and clearing equipment such as fuses, reclosers, and circuit switchers. The purpose of a circuit breaker is fundamentally to *break* a circuit. An electrical circuit is a closed path whereby current flows from a source and returns to the source. The circuit is broken when the path for the flow of current is opened and the current is reduced to a negligible or zero value. The circuit-breaking process begins with an initiating or trip signal, which may come from a protective relay or other device or from a control switch or automated control system. Some circuit breakers contain integral trip devices, whereas others require an external device to develop the trip signal. The second step is contact parting, whereby the closed contacts begin to open, spring pressure or other stored energy is released, and air, vacuum, or another insulating medium is introduced between the contacts. Once the contacts begin to open, arcing occurs, where current will flow through the intervening gap, both from vaporized metal particles from the contacts and from possible ionization of the insulating material. As the contacts move further apart, the arc will become attenuated and ultimately the voltage and current levels will be unable to support the arc, resulting in the arc being extinguished. The contacts will then come to rest in their fully open position and the fault will have been cleared. When a fault is cleared, there is no current flowing through the damaged circuit and no voltage should be present.

Circuit breakers are designed, built, and tested according to strict standards and procedures set forth by various standards-making bodies, and their application is regulated by electrical and building codes. These standards and procedures do vary in many ways throughout the world, but the essential purpose of protecting the public from electrical hazards, whether in their homes, workplaces, or any other location where electricity is used, is always paramount. Circuit breakers are not only designed for safety, but are used for the main purpose of making electrical systems safe for use. When a circuit breaker fails, it presents a danger to those working on or near the circuit breaker, as they may be injured by fire or explosion, or by shock from conductors which should be de-energized. A failed circuit breaker also may present a fire hazard

Principles of Electrical Safety, First Edition. Peter E. Sutherland.
© 2015 John Wiley & Sons, Inc. Published 2015 by John Wiley & Sons, Inc.

from the uninterrupted high fault currents. Power systems must be designed to with-stand safely the failure of the first level of protection, that is, the circuit breaker or fuse closest to the short circuit by means of backup protection and fault withstand capa-bility of cables, transformers, and other equipment. However, the conditions which cause circuit breaker failure may arise even in a system which was originally safely designed, and these failures must be prevented by regular maintenance, circuit anal-ysis, and testing. This chapter will examine some of the causes of circuit breaker failure, with examples, and also some of the factory testing procedures which help to ensure safe design and manufacture of circuit breakers. This chapter is adapted from EPRI (2006).

12.1 INSUFFICIENT INTERRUPTING CAPABILITY

For a variety of reasons, users might discover that available fault currents on their power systems exceed the interrupting rating of their existing circuit breakers. The increasing sophistication of tools for determining short-circuit (fault) currents, the addition of transformer capacity necessary to support plant life extensions, the addi-tion of motor loads as back-electromotive force (EMF) fault contributors, and the reconfiguring of systems, all contribute toward this situation. Safety considerations make the need to address the situation obvious.

12.2 HIGH VOLTAGE AIR CIRCUIT BREAKERS

Circuit breakers that are 25–40 years old can be considered to be approaching the end of their expected design life. It might not be possible to accurately predict the life of these circuit breakers. Problems affecting the operation of medium-voltage circuit breakers might not be detected during basic testing and maintenance.

Every year, the Doble Conference elicits responses from clients to a techni-cal questionnaire on various subjects. The following results on medium-voltage CB failures from 1991 to 1994 confirm the above assertion:

- 1991-133 replies, 54% return, 71 clients reported 165 CB failures
- 1992-158 replies, 61% return, 71 clients reported 180 CB failures
- 1993-125 replies, 48% return, 44 clients reported 117 CB failures
- 1994-109 replies, 42% return, 54 clients reported 109 CB failures.

Of particular interest in the Doble findings is the diversity of the nature of the failures. These include but are not restricted to the following:

1. Arc chute melting
2. Failure to clear faults
3. Insulation breakdown
4. Component deterioration
5. Bushing insulation destruction due to moisture and electrical stress

6. Arc chute carbon buildup and condensation

7. Interrupter disintegration

8. Fire destruction caused by failure to clear

9. Jamming of the contact blade by arc chute delamination preventing closure of contacts

10. Operating rod failure during an opening operation

11. Support arm failure

12. Motor drive and ratchet wheel failures

13. Operating coil and anti-pump relay failures.

Arc chute moisture contamination and reduced circuit breaker-operating velocities should be a major concern. Research in 1992 on more than 500 air circuit breakers at major Midwestern utilities showed that approximately 25% of all circuit breakers exhibited operating velocities below the manufacturers' specification, despite the fact that all circuit breakers had been timed at acceptance and at approximately 5-year intervals thereafter. The in-service dates of the circuit breakers ranged from the early 1950s to the mid-1980s (Genutis, 1992).

12.3   VACUUM CIRCUIT BREAKERS

Although the earliest attempts to make circuit-interrupting devices using contacts in a vacuum date back to the 1920s, it was not until the 1960s that the technology reached the point at which reliable vacuum interrupters capable of acceptable short-circuit-breaking current ratings could be manufactured economically.

Figures 12.1 and 12.2 provide an outline and the internal construction of a modern-day vacuum interrupter. Under working conditions, when contact separation occurs, arcing plasma that consists of metallic ions develops. The formation of metallic vapors sustains the arc. In proximity to the natural power frequency current zero, these vapors condense rapidly onto metal screens. Thus, a condition is created for ultrafast recovery of the dielectric strength.

At contact separation, where the arc is supported by ionized gas in other interrupters, the arc in vacuum circuit breakers is supported by ionized metal vapors that are generated by the contacts themselves. Thus, at contact zero, vapor condensation and the collapse of ionization is almost instantaneous.

Outwardly, the vacuum interrupter appears to be the ultimate in switching devices that are independent of the operating system, as its breaking capability is dependent only on the material and geometry of the contact structure and the quality of the vacuum. This is not strictly true, however, because care has to be taken to ensure the correct travel and force characteristics of the operating mechanism with respect to the interrupter.

The high dielectric strength of the vacuum permits a small contact gap in the open position, typically 1/4–3/4 in. (0.64-1.9 cm). As a result of this fact and the low resistance of the metal vapor arc, the arc energy is very low. Following arc quenching,

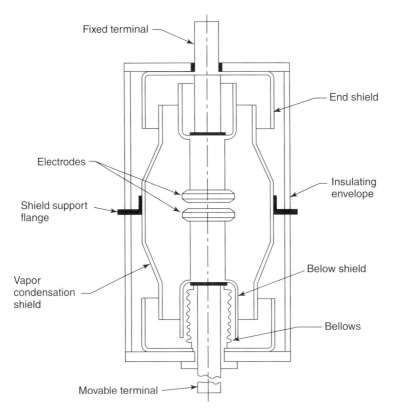

Figure 12.1   Schematic of vacuum interrupter (*Source*: EPRI (2006)).

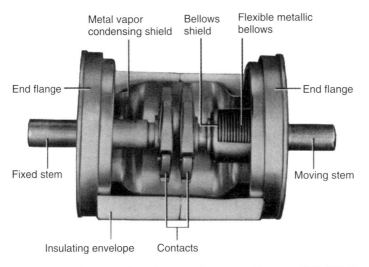

Figure 12.2   Cutaway view of vacuum interrupter (*Source*: EPRI (2006)).

most of the metal vapor condenses back onto the contacts and, by doing so, results in contact restoration; thus, contact erosion is extremely low.

At the moment of contact separation through which current is flowing, an arc discharge occurs. With currents approximating 10 kA, the arc burns diffusely on the whole contact surface between the contacts until the next current zero. Contact burn-off is low, as is the specific thermal load of the contact surface. For currents above 10 kA, the arc is severely pinched through the pressure of its own magnetic field. The concentrated burning arc occasioned by the high current density causes large amounts of contact material to vaporize very rapidly. The contact material, the shape of the contacts, and the manufacturing methods for the contacts play a significant role in the ultimate performance of a vacuum interrupter.

12.4   SF$_6$ CIRCUIT BREAKERS

Historically, the SF$_6$ circuit breaker found its application in transmission systems, where it was a simpler alternative to the air blast circuit breakers that were used in the 1960s. The original circuit breakers were of the two-pressure type, where the contacts and arc control devices were immersed in a metal tank full of SF$_6$, serving the dual purpose of an insulating and arc-extinguishing medium. Essentially, the gas used for arc extinction was drawn out of the tank, then compressed and pumped into a separate small reservoir. With the opening of the circuit breaker, the compressed gas was discharged through nozzles around the contacts by a valve, resulting in the extinguishing of the arc.

In distribution systems, the first technique used in SF$_6$ circuit breakers was that of a self-generated principle, referred to as the *puffer* system. This was a step toward simplifying the design with fewer moving parts. Eliminating compressors, high-pressure seals, and heating elements obtained a much higher degree of reliability. The design did not, however, overcome the requirement for relatively long strokes and high operating forces, resulting in the need for powerful operating mechanisms with large energy output requirements. A quantity of SF$_6$ gas is compressed with the commencement of a circuit-breaker-opening action prior to the separation of the arcing contacts. When the contacts part, an arc is drawn, and upon reaching current zero, the flow of gas rapidly cools the remaining plasma and sweeps away the arc products, leading to an extremely rapid increase in the dielectric strength of the gap, thus realizing a successful clearance.

An evolving design process developed self-blast interrupters, which use the arc energy to heat the gas and increase its pressure, allowing the gas to expand. This means that the pressure is raised by thermal means rather than by mechanical energy. It is a method that makes possible a lower-energy-operating mechanism than the puffer interrupter. The ensuing arc extinguishing process occurs in a similar manner to that of the puffer interrupters.

Figure 12.3 illustrates this thermal expansion method. From the figure, the travel of the gas can be observed. When heated by large arc currents, the gas rapidly expands and pressurizes the upper vessel chamber. Pressure differential between the

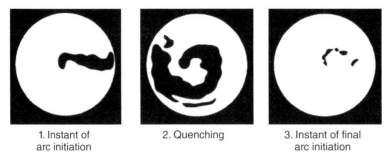

| 1. Instant of arc initiation | 2. Quenching | 3. Instant of final arc initiation |

Figure 12.3   Thermal expansion method (*Source*: EPRI (2006)).

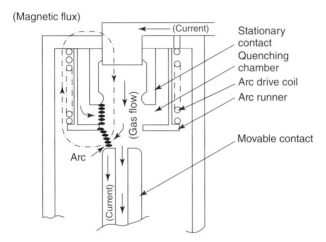

Figure 12.4   Rotary arc quenching method (*Source*: EPRI (2006)).

upper and lower vessel chambers creates a flow of gas, facilitating the cooling of the arc and the evacuation of the heat discharged by the arc. Thus, the gas flow effectively extinguishes the arc.

The rotary arc quenching method shown in Figure 12.4 is based on the interrupting current developing a rotating magnetic flux, which extinguishes the arc. Because the arc is rotated at approximately the speed of sound and is always exposed to the SF_6 gas, extinction is ensured. The purpose of rotating the arc is to bring the intense heat of the arc column into contact with as much of the gas as possible, thus raising its temperature and, consequently, its pressure.

By adopting the rotary arc and the thermal expansion method, a proportionate arc-quenching force is generated that is equivalent to the magnitude of the interrupting current. Figure 12.5 represents a plot of arc-quenching performance versus interrupting current. From the curves, the comparative performance for the different interrupting techniques as a function of the interrupting current can be readily observed.

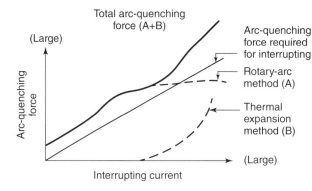

Figure 12.5   Arc quenching performance versus interrupting current (*Source*: EPRI (2006)).

With the advent of the latest technologies now available in SF_6, the closing energy has been reduced to levels comparable to vacuum. This translates to fewer parts in the circuit breaker mechanism. In fact, the difference is so minimal that one European manufacturer of both SF_6 and vacuum circuit breakers utilizes the same mechanism in each case.

12.5   LOSS OF INTERRUPTION MEDIUM

Both sealed vacuum and SF_6 interrupters are susceptible to the possible loss of their interrupting medium. Most users have shown a willingness to accept this susceptibility.

It should be noted that SF_6 circuit breakers can be monitored in service, while no practical monitoring for commercial purposes is currently available for vacuum circuit breakers.

Sophisticated methods akin to those used in a laboratory are available, but economics make their use prohibitive.

For the SF_6 circuit breaker, reliable pressure switches can be furnished within the circuit breakers poles, and these switches are wired into a two-stage alarm system to register any loss of the gas, facilitating action prior to a dangerous condition (Swindler, 1989). It should also be noted that, even with a total loss of SF_6 gas, a 30-kV withstand level is maintained, and load currents can still be safely interrupted on a one-time basis (Swindler, 1993).

The mechanical connection into the gas bottle and the pressure switch itself are two more potential leak paths. The manufacturer's history of gas bottle leakage should be a factor in choosing the claim of a higher reliability in gas integrity.

Because SF_6 gas is five times heavier than air, the possibility of some air replacing the SF_6, even after the pressure has equalized, cannot be ruled out. However, a large proportion of air is needed to significantly reduce the breakdown strength, and designers account for this remote possibility by designing the open-gap contact system to withstand the service voltage at atmospheric pressure. This should be contrasted with a vacuum interrupter suffering a leak, where the voltage strength of the gap falls to a minimum level of a few hundred volts in the glow discharge region of $0.1-10$ torr ($13.3-133.2$ Pa), recovering to around 30 kV/cm at atmospheric pressure. When the vacuum is leaking and the voltage drops, failure is possible (Blower, 1986).

The consequences of attempted interruption of fault currents by a vacuum interrupter (with loss of vacuum) and by an SF_6 circuit breaker (with loss of SF_6) have not been adequately addressed in the available technical literature. The following statement appears in a 1992 Doble Conference paper by a vacuum CB manufacturer:

> For load currents in an ungrounded system, the vacuum circuit breaker carried the current and interrupted it satisfactorily. In a 600 A grounded system, the phase in which the vacuum interrupter was at atmospheric pressure continued to arc for 15 seconds. All arcing was confined within the vacuum interrupter, although the envelope cracked. The SF_6 interrupter at atmospheric pressure switched the load current in both grounded and ungrounded circuits. For 10 kA and at 25 kA the phase that had the vacuum interrupter at atmospheric pressure again continued to arc for 30 cycles until the backup circuit breaker interrupted the circuit. All arcing was confined inside the interrupter's envelope. In these experiments, the envelope showed no evidence of rupture. In fact, the only evidence of thermal stress to the ceramic envelope was some cracking. The SF_6 puffer interrupter interrupted the 10 kA current, but at 25 kA the SF_6 did not interrupt the circuit and the continued burning of the arc caused the puffer to explode (Storms, 1992).

It should be noted that contemporary designs of both rotary arc and puffer SF_6 circuit breakers employ an overpressure diaphragm at the bottom of each interrupter. In the event that the interrupter fails to interrupt owing to loss of gas, the diaphragm ruptures, directing the pressurized gas downward.

In early designs of SF_6, problems were experienced with leaking seals. This is no longer the case with better contemporary varieties that use EP-type rubber O-rings. These O-rings have a longer life than the circuit breaker itself (Figure 12.6). For all production runs, the leakage rate of gas is tested by highly sensitive detecting equipment that can detect leakage rate values up to 10^{-9} cc/s.

12.6   INTERRUPTING RATINGS OF SWITCHING DEVICES

The most eminent reason for utilities to maintain or even reduce their fault current levels is to ensure proper functioning of circuit-interrupting devices such as circuit breakers and fuses. Interrupting capabilities of these devices will therefore be reviewed and provided in the guidebook.

Exceeding the interrupting rating will result in continued arcing, fire, explosions, and failure to interrupt the fault. Danger to any nearby personnel is likely.

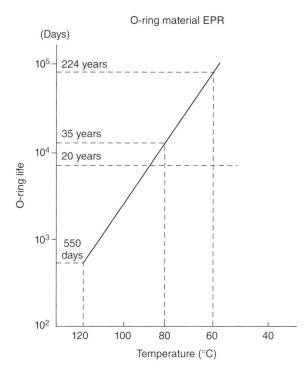

Figure 12.6   Relation between O-ring life and temperature (*Source*: EPRI (2006)).

Damage to substation structure, switchgear, buses, and other equipment is also likely. Backup equipment must interrupt the fault.

12.7   CIRCUIT BREAKERS

Circuit breakers in the United States are rated by the American National Standards Institute (ANSI) and the Institute of Electrical and Electronics Engineers (IEEE). Circuit breakers are classified by location, indoor or outdoor, and by voltage rating. Ratings for Class 1 circuit breakers, formerly classified as indoor circuit breakers, are shown in Table 12.1. Ratings for Class S2 circuit breakers, formerly classified as outdoor circuit breakers, are summarized in Table 12.2. Outdoor, high-voltage circuit breakers are summarized in Table 12.3. The maximum interrupting ratings for these various types are as follows:

- 63 kA for indoor distribution, 15 kV class
- 40 kA for outdoor distribution, 38 kV class
- 80 kA for outdoor transmission, 145 kV class.

TABLE 12.1 Typical Class S1 (formerly classified as indoor) Circuit Breakers for Cable Systems Rated Below 100 kV

Rated Maximum rms Voltage (kV)	Rated Continuous rms Current (A)	Rated rms Short-Circuit Current (kA)	Rated Peak Closing and Latching Current (kA)
4.76	1200, 2000	31.5	82
	1200, 2000	40	104
	1200, 2000, 3000	50	130
8.25	1200, 2000, 3000	40	104
15	1200, 2000	20	52
	1200, 2000	25	65
	1200, 2000	31.5	82
	1200, 2000, 3000	40	104
	1200, 2000, 3000	50	130
	1200, 2000, 3000	63	164
27	1200	16	42
	1200, 2000	25	65
38	1200	16	42
	1200, 2000	25	65
	1200, 2000, 3000	31.5	82
	1200, 2000, 3000	40	104

Source: IEEE (2009a).

12.8 FUSES

There are two major types of fuses: Current-limiting (Figure 12.7) and expulsion type (Figure 12.8). Fuse ratings are summarized in Table 12.4. While current-limiting fuses are enclosed in a sealed cylinder, and are usually contained in metal-enclosed switchgear, limiting the danger even if their short-circuit rating is exceeded, expulsion fuse links are mounted in open tubes on utility poles. These present several hazards when subjected to excessive short-circuit currents:

1. Exceeding the interrupting rating will result in continued arcing, fire, explosions, and failure to interrupt the fault.
2. Danger to public from falling arcing products if the fuse is on a pole along a street.
3. Damage to disconnecting switch, poles, enclosures, and other equipment is likely.
4. Backup equipment must interrupt fault, resulting in extended arcing time.

Current-limiting fuses generally have higher short-circuit ratings than expulsion fuses of the same size. Current-limiting fuses are available which fit in cutouts,

TABLE 12.2   Typical Class S2 (formerly classified as outdoor) Circuit Breakers for Line Systems Rated Below 100 kV

Rated Maximum rms Voltage (kV)	Rated Continuous rms Current (A)	Rated Short-Circuit rms Current (kA)	Rated Peak Closing and Latching Current (kA)
15.5	600, 1200	12.5	33
	1200, 2000	20	52
	1200, 2000	25	65
	1200, 2000, 3000	40	104
25.8	1200, 2000	12.5	33
	1200, 2000	25	65
38.0	1200, 2000	16	42
	1200, 2000	20	52
	1200, 2000	25	65
	1200, 2000	31.5	82
	1200, 2000, 3000	40	104
48.3	1200, 2000	20	52
	1200, 2000	31.5	82
	1200, 2000, 3000	40	104
72.5	1200, 2000	20	52
	1200, 2000	31.5	82
	1200, 2000, 3000	40	104

Source: IEEE (2009a).

replacing expulsion fuse links. Upgrading of expulsion fuses to current-limiting fuses and in general replacing fuses with units having higher interrupting ratings is the preferred solution to increasing available fault current levels.

12.9   CASE STUDIES

These case studies describe the replacement of air-magnetic circuit breakers whose interrupting ratings have been exceeded with new SF_6 or vacuum interrupters. Both of these cases are taken from nuclear power plants.

12.9.1   Example: Diablo Canyon

At Diablo Canyon (EPRI, 2003), the prime motivation for upgrading air-magnetic circuit breakers to SF_6 was an increase in the fault level and the elimination of a potentially hazardous design issue rather than the desire to save maintenance and other costs. Beginning in March 1994, a dedicated user team set out to evaluate the various options available. The team had extensive experience in preparing the type of quality assurance (QA)/dedication requirements that were to be imposed on the selected vendor for this project. The project consisted of 110 units.

TABLE 12.3 Typical Circuit Breakers Rated 100 kV and Above Including Circuit Breakers Applied in Gas-Insulated Substations

Rated Maximum rms Voltage (kV)	Rated Continuous rms Current (A)	Rated rms Short-Circuit Current (kA)	Rated Peak Closing and Latching Current (kA)
123	1200, 2000	31.5	82
	1600, 2000, 3000	40	104
	2000, 3000	63	164
145	1200, 2000	31.5	82
	1600, 2000, 3000	40	104
	2000, 3000	63	164
	2000, 3000	80	208
170	1600, 2000	31.5	82
	2000, 3000	40	104
	2000, 3000	50	130
	2000, 3000	63	164
245	1600, 2000, 3000	31.5	82
	2000, 3000	40	104
	2000, 3000	50	130
	2000, 3000	63	164
362	2000, 3000	40	104
	2000, 3000	50	130
	2000, 3000	63	164
550	2000, 3000	40	104
	3000, 4000	50	130
	3000, 4000	63	164
800	2000, 3000	40	104
	3000, 4000	50	130
	3000, 4000	63	164

Source: IEEE (2009a).

Maintaining the existing switchgear to provide a 350-MVA circuit breaker within the same footprint as the 250-MVA design necessitated a compact circuit breaker design. At the time, only one circuit breaker met this criterion. Others did not represent completed, tested designs. The fact that this element was SF_6 provided additional comfort because the user preferred monitoring the state of the interrupting medium and desired a surge-free performance. These criteria in conjunction with the QA requirements stipulated by Pacific Gas and Electric were the defining characteristics in selecting the vendor for this particular project. The Diablo project is a useful model for the nuclear industry for converting circuit breakers as the combination of high seismic levels, the qualification/dedication complexities, and the technical challenges are clearly defined.

The significant lessons learned from this project fall into two distinct areas that are unique to the nuclear environment. The first area concerns how to administer an

TABLE 12.4   Preferred Ratings for Fuses

Rated Maximum rms Voltage (kV)	Rated rms Symmetrical Interrupting Current (kA)
2.54–2.8	31.5–63
5.08–5.5	31.5–50
8.3	4.0–50
15-17.2	4.0–80
23–27	4.0–50
38	5.0–40
48.3	3.15–31.5
72.5	2.5–25
121	1.25–16
145	1.25–12.5
169	2.5–12.5

Source: IEEE (2009a).

Figure 12.7   Current-limiting fuses (Mersen).

Figure 12.8   Expulsion fuses are contained in a pole mounted fuse cutout (Cooper Power Systems).

adequate nuclear QA program in the dedication and building of circuit breaker conversions without losing sight of the primary mission of building reliable conversions in a timely and efficient manner. Many of the issues associated with the manufacturing process arose because of the intrusive nature of implementing the QA program. Careful consideration of QA needs was balanced with the requirements of a production facility that is geared to the non-nuclear aspects of building conversions.

The second area concerns the interface of the new design with the existing cubicles and the flexibility to undertake essential changes to cubicle interlock interfaces while being cognizant of nuclear constraints on the design of the conversion.

A further example of flexibility relates to a deviation in testing procedures. There is no economical method to field test the pressure switches contained within the poles of an SF_6 circuit breaker. After numerous meetings and detailed design reviews of the product and its associated manufacturing methods, acceptance was sanctioned as the lesser of two evils - in other words, accepting the pressure switch factory reliability test results and the track record of the product in the field as opposed to the incorporation of a vacuum element, which cannot be monitored.

To the users, this project is a success. A replacement circuit breaker was effectively integrated into the existing plant power system, significantly increasing fault protection and ensuring a considerable maintenance cost savings over the circuit breaker's installed life.

12.9.2   Example: Dresden and Quad Cities

There were several factors involved in the decision-making process at the Dresden and Quad Cities plants (EPRI, 2003). These factors extended from the problems associated with obsolescence and lack of spare parts; the required addition of cubicles;

the need to upgrade interruption levels; the objective of minimizing outage duration during modifications; and recognition of the state of the 20-year-old motors, transformers, and cables.

Initially, wholesale removal and replacement of the switchgear, with the inclusion of additional cubicles, was considered. The original equipment manufacturer (OEM) was favored in order to maintain continuity of products and maintenance services. The preferred technology was air magnetic because of its established acceptable overvoltage response, but this medium was no longer available from the OEM.

The costs for the subject proposal of full switchgear replacement were judged excessive given the extensive rigging effort, necessary extension in outage time, and the risk of damage to the old power and control cables (Shah *et al.,* 1988; Dinkel, Watts, and Langlois, 1986; Fish, 1994; Heath, Freeman, and Cochran, 1987). This can be readily appreciated in recognition of the fact that the total number of existing cubicles was 100 and 12 additional units had to be accounted for.

An alternative approach based on the concept of a 250-350 MVA conversion circuit breaker that would fit into the existing 26-in.-wide cubicle seemed appealing, as it became apparent that it was economically attractive and would eliminate the considerable risks pertaining to full replacement. At the time, IEEE Standard C37.59 was in draft form and could serve well as a guide for the planned scope of work.

The final selection of vacuum or SF_6 circuit breaker technology was made on the basis of cost, after factoring in the associated costs for surge arresters, spare parts, training, and vendor services. It was originally assumed that the existing 250 MVA bus rating was insufficient to withstand the 350 MVA momentary ratings. Testing in a high-power laboratory proved otherwise, and the approximately 20-year-old cubicle with its 20-year-old bus and bracing successfully withstood the momentary test stresses. This certification allowed the project to abandon the time-consuming and expensive change-out of the existing bus bracing.

12.10   LOW-VOLTAGE CIRCUIT BREAKERS

Low-voltage circuit breakers are operated at or below a nominal voltage of 1000 V AC or DC (IEEE, 1993). These can be classified into several types:

1. Power Circuit Breakers (LVPCBs) (Figure 12.9). A power circuit breaker is capable of making, breaking and carrying high fault currents for a specified time. Because they can carry a high current for a specified period of time, LVPCBs may be set up to operate with a short time delay tripping function, without an instantaneous trip. This property is helpful in coordinated systems, such as in main-tie-main Switchgear. However, the short time delay will increase the arc flash hazard due to the longer arcing time. LVPCBs generally are constructed on a metal frame. They are repairable, and all can be disassembled and refurbished when required.

2. Insulated-Case Circuit Breakers (ICCBs) (Figure 12.10). ICCBs are capable of making and breaking high fault currents, but are not rated to carry them for a specified time. Thus, ICCBs are required to have an instantaneous trip function.

Figure 12.9   Low-voltage power circuit breaker (General Electric Company).

Figure 12.10   Low-voltage insulated-case circuit breakers (General Electric Company).

ICCBs are usually constructed with enclosures of insulating materials and only selected components can be replaced.

3. Molded Case Circuit Breakers (MCCBs) (Figure 12.11). MCCBs are assembled as an integrated unit enclosed in a molded insulating (plastic) housing. MCCBs are capable of making and breaking high fault currents, but are not rated to carry them for a specified time. Thus, MCCBs are required to have an

Figure 12.11   Low-voltage molded-case circuit breakers (General Electric Company).

instantaneous trip function. MCCBs are not repairable, and must be replaced if they fail.

12.11   TESTING OF LOW-VOLTAGE CIRCUIT BREAKERS

In order for circuit breakers to be applied in a safe manner, samples of each type must be able to pass exhaustive testing for multiple switching operations (endurance tests), moderate degrees of overcurrent (overload tests), and high degrees of overcurrent (interrupting tests). The test procedures are specified in various standards published by standards bodies regulating the particular types and ratings of circuit breakers built according to those standards. Examples include the worldwide International Electrotechnical Commission (IEC) and the Institute of Electrical and Electronic Engineers (IEEE). Bodies specific to the United States include the ANSI, National Electrical Manufacturers Association (NEMA), and Underwriters Laboratories (UL). In many cases, the manufacturer's testing program must be observed by a representative of the standards-making body in order that the parts may be certified as being tested in conformance with the standards. The following standards are widely used in the testing of low-voltage (<1000 V) AC and DC circuit breakers:

1. UL489 Eleventh Edition (UL, 2009) September 1, 2009, UL Standard for Safety for Molded-Case Circuit Breakers, Molded-Case Switches and Circuit-Breaker Enclosures.

2. UL 1077 Sixth Edition (UL, 2005) July 14, 2005, UL Standard for Safety for Supplementary Protectors for Use in Electrical Equipment.

3. UL 1008 Sixth Edition (UL, 2011) April 15, 2011, UL Standard for Safety for Transfer Switch Equipment.

4. ANSI® C37.50-1989 (ANSI, 1989) American National Standard for Switchgear - Low-Voltage AC Power Circuit Breakers Used in Enclosures - Test Procedures.

5. IEC 60947

   a. IEC 60947-1 (IEC, 2011b) Edition 5.1 2011-03, Low-voltage switchgear and controlgear—Part 1: General rules.

   b. IEC 60947-2 (IEC, 2013a) Edition 4.2 2013-01, Low-voltage switchgear and controlgear—Part 2: Circuit-breakers.

6. IEC 60898

   a. IEC 60898-1 (IEC, 2003a) Edition 1.2, 2003-07, Electrical accessories— Circuit-breakers for overcurrent protection for household and similar installations—Part 1: Circuit-breakers for a.c. operation.

   b. IEC 60898-2 (IEC, 2003b) Edition 1.1, 2003-07, Edition 1:2000 consolidated with amendment 1:2003, Electrical accessories—Circuit-breakers for overcurrent protection for household and similar installations—Part 2:Circuit-breakers for a.c. and d.c. operation.

7. IEC 60934 (IEC, 2013b), Edition 3.2 2013-01, circuit-breakers for equipment (CBE).

Owing to the large number of types of circuit breakers and testing standards, the discussion which follows will only cover the highlights of some of the specified procedures. Low-voltage circuit breakers are tested for various types of duty and duty cycles appropriate for each type of breaker.

Due to the complexities of these test procedures, they are only summarized here. For anyone who intends to test circuit breakers or to evaluate tests of circuit breakers, the latest version of the applicable standard should be referred to.

12.11.1 Testing of Low-Voltage Molded-Case Circuit Breakers According to UL Standard 489

For MCCBs covered in UL 489-2009 (UL, 2009), the tests include the following:

1. Overload tests verify the circuit breaker can withstand repeated overloads at 600% of the rated currents.

2. Endurance tests verify that the circuit breaker can withstand multiple switching operations at 100% of the rated current.

3. Interrupting tests verify that the circuit breaker can safely interrupt its rated short-circuit current in the case of a short circuit.

Each type of test is conducted in specific test circuits according to the voltage and current ratings, number of poles (individual breaker units within the overall case), and circuit type (single-phase and three-phase). UL 489 provides diagrams of each

TABLE 12.5   Standard Ampere Ratings for Circuit Breakers

15	110	600
20	125	700
25	150	800
30	175	1000
35	200	1200
40	225	1600
45	250	2000
50	300	2500
60	350	3000
70	400	4000
80	450	5000
90	500	6000
100		

Source: NFPA (2014).

type of test circuit to be used. Each test also has a specific sequence of operations for repeated closings of the circuit breaker being tested.

UL 489 tests of AC MCCBs are designed for breakers rated 50-60 Hz, with the test frequency being in the range of 48-62 Hz. Circuit breakers have standard ampere ratings (continuous-current-carrying capability) as specified in the Article 240.6 (A) of the NEC (NFPA, 2014), listed in Table 12.5. The standard voltage ratings for MCCBs are listed in Table 12.6 for AC (both single and three-phase) and DC.

TABLE 12.6   Voltage Ratings for MCCBs from UL 489 Clause 8.2

AC (1-Phase) (V)	AC (3-Phase) (V)	DC (V)
		24
		48
		60
		65
		80
120		125
127		125/250
120/240		160
208	208Y/120	250
240		500
277		600
347		
480	480Y/277	
600	600Y/347	

Source: UL (2009).

TABLE 12.7 Applicable Current-Interrupting Ratings (AC or DC) From UL 489, Table 8-1

7500	25,000	65,000
10,000	30,000	85,000
14,000	35,000	100,000
18,000	42,000	125,000
20,000	50,000	150,000
22,000		200,000

Source: UL (2009).

TABLE 12.8 Overload Tests at 600% of Rated Amperes (UL 489, Clause. 7.1.3.7-8 and 7.1.3.14)

Rated Current (A)	Overload Test Current (A)	Rated Current (A)	Overload Test Current (A)	Rated Current (A)	Overload Test Current (A)
15	90	110	660	600	3600
20	120	125	750	700	4200
25	150	150	900	800	4800
30	180	175	1050	1000	6000
35	210	200	1200	1200	7200
40	240	225	1350	1600	9600
45	270	250	1500	2000	12,000
50	300	300	1800	2500	15,000
60	360	350	2100	3000	18,000
70	420	400	2400	4000	24,000
80	480	450	2700	5000	30,000
90	540	500	3000	6000	36,000
100	600				

Source: UL (2009).

The interrupting ratings of the circuit breakers, defined as "the highest RMS symmetrical current at rated voltage that a device is intended to interrupt under standard test conditions" are many times larger than their ampere ratings. For MCCBs rated under UL 489, the interrupting ratings range from 7.5 to 200 kA (Table 12.7).

The overload tests are performed at 600% of the rated current, with the values for various breaker ratings listed in Table 12.8. Test circuits for overload tests are shown in Figure 12.12. The no-load voltage of the test circuit should be in the range 100-105% of the rated voltage, while the overload voltage with 600% current should be in the range 85-115% of the rated voltage. If certain requirements for the recovery voltage (the voltage across the circuit breaker when it is opening) are met, then the ±15% requirement may be waived. Tests are performed at a lagging power factor from 45% to 50% for AC circuit breakers, as shown in Figure 12.13, except that some tests of two-pole and single-pole circuit breakers are performed at lagging a power

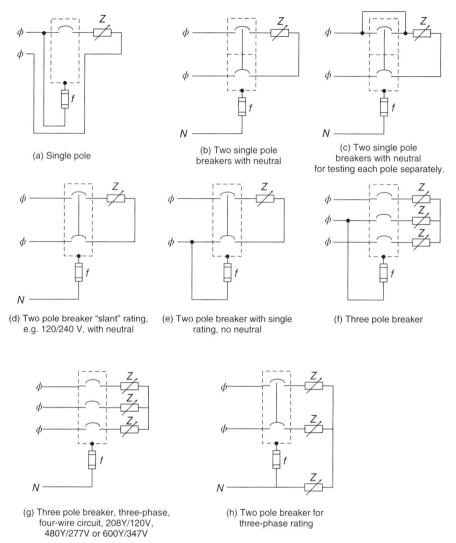

Figure 12.12   Overload test circuits based on UL 489 (UL, 2009). (a) Single pole; (b) two single-pole breakers with neutral for testing each pole separately; (c) two single-pole breakers with neutral; (d) two-pole breaker "slant" rating, for example, 120/240 V, with neutral; (e) two-pole breaker with single rating, no neutral; (f) three-pole breaker; (g) three-pole breaker, three-phase, four-wire circuit, 208Y/120 V, 480Y/277 V, or 600Y/347 V; (h) two-pole breaker for three-phase rating.

factor of 75-80%. For DC circuit breakers, the L/R time constant of the test circuit should not exceed 0.003 s, as shown in Figure 12.14. The test fuse should be rated 30 A with a voltage rating equal to or greater than that of the device under test, and a 5.3 mm^2 (#10 AWG) wire not longer than 1.8 m (6 ft). The timing sequences for these tests are listed in Table 12.9, and shown diagrammatically in Figures 12.15–12.19.

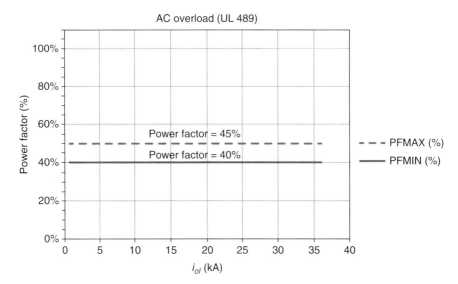

Figure 12.13   AC overload power factors.

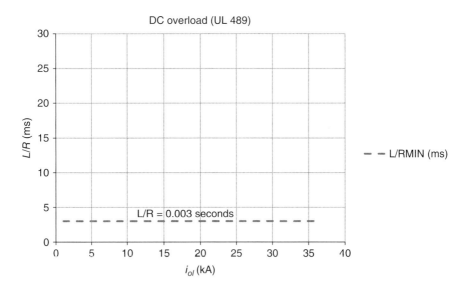

Figure 12.14   DC overload time constant.

The breakers are only switched on for a very short time in each test, as no particular duration is specified.

   Endurance tests are performed at 100% of the rated current, with the values for various breaker ratings listed in Table 12.5. The voltage of the test circuit with load should be greater than 97.5% of the rated voltage. Test circuits for endurance tests are shown in Figure 12.20. Tests are performed at a lagging power factor from

TABLE 12.9   Operations for Overload Test Operations (600% of Rated Current Unless Otherwise Noted)

Frame Size (A)	Test Current (A)	Voltage Rating (V)	Number of Operations	Operations Per Minute	Figure
≤100	600	All	50	6	Figure 12.15
101–150	900	All	50	5	Figure 12.16
151–225	1350	All	50	5	Figure 12.16
226–1600	9600	All	50	1[a]	Figure 12.17
1601–2500	15000	All	25	1[a]	Figure 12.18
2501–6000	36000	All	28	1[b]	Figure 12.19

[a]Operations may be in groups of 5, separated by 15 min.

[b]Three operations at 600% of rating at one operation per minute, followed by 25 operations at 200% of rating at one operation per minute, may be in groups of 5, separated by 15 min.

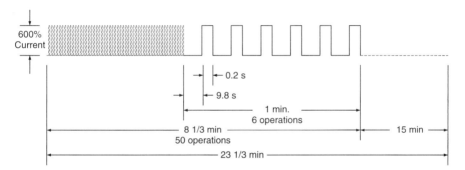

Figure 12.15   Overload test, ≤100 A frame (not to scale).

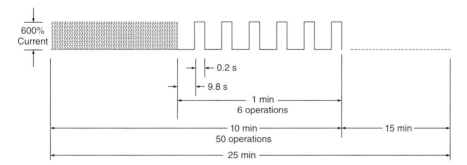

Figure 12.16   Overload test, 101-225 A frame (not to scale).

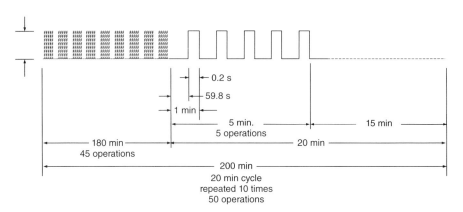

Figure 12.17   Overload test, 226-1600 A frame (not to scale).

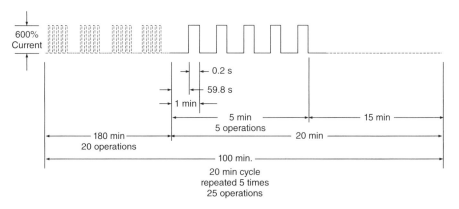

Figure 12.18   Overload test, 1601-2500 A frame (not to scale).

75% to 80% for AC circuit breakers, as shown in Figure 12.21. For DC circuit breakers, the L/R time constant of the test circuit should not exceed 0.003 s, as shown in Figure 12.22. The test fuse should be rated 30 A with a voltage rating equal to or greater than that of the device under test, and a 5.3 mm^2 (#10 AWG) wire not longer than 1.8 m (6 ft). The timing sequences for these tests listed in Table 12.10, and shown diagrammatically in Figures 12.23–12.28. The breakers are only switched on for a very short time in each test, as no particular duration is specified.

Interrupting tests are performed at high values of current, with the values for various breaker ratings listed in Table 12.11. Test circuits for interrupting tests are shown in Figure 12.29. The no-load voltage of the test circuit, both AC and DC tests, should be in the range 100-105% of the rated voltage. Tests are performed at a lagging power factor from 45% to 50% for AC circuit breakers tested at 10 kA or less, 25-30% for 10.001-20 kA, and 15-20% for test currents greater than 20 kA, as shown in Figure 12.30. For DC circuit breakers, the L/R time constant of the test circuit should be greater than or equal to 0.003 s if the test current is less than or equal to

TABLE 12.10   Operations for Endurance Test Operations (100% of Rated Current)

Maximum Frame Size (A)	Test Current (A)	Voltage Rating (V)	Number of Operations with Current	Operations Per Minute	Figure
100	100	All	6000	6	Figure 12.23
150	150	All	4000	5	Figure 12.24
225	225	All	4000	5	Figure 12.24
600	600	All	1000	4	Figure 12.25
800	800	All	500	1	Figure 12.26
1200	1200	All	500	1[a]	Figure 12.27
2500	2500	All	500	1[a]	Figure 12.27
6000	6000	All	400	1[a]	Figure 12.28

[a]May be in groups of 100 operations, separated by groups of no-load operation.

[b]Ten operations at one operation per minute, followed by 5 operations at one operation per minute, in groups of 5, separated by an interval (shown as 15 min in the figures).

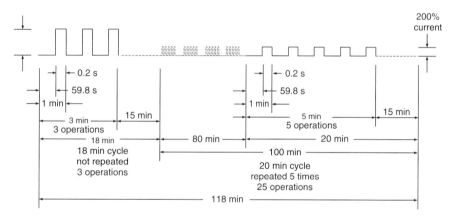

Figure 12.19   Overload test, 2501-6000 A frame (not to scale).

10 kA, as shown in Figure 12.31. If the test current exceeds 10 kA, the DC time constant should be greater than or equal to 0.008 s. The timing sequences for these tests are listed in Table 12.12, and shown diagrammatically in Figure 12.32. The test fuse should be rated 30 A with a voltage rating equal to or greater than that of the device under test, and a 5.3 mm² (#10 AWG) wire not longer than 1.8 m (6 ft).

12.11.2   Testing of Low-Voltage Molded-Case Circuit Breakers for Use With Uninterruptible Power Supplies According to UL Standard 489

Low-voltage MCCBs used with uninterruptable power supplies (UPS) are subject to special test conditions. All other aspects of these tests are as

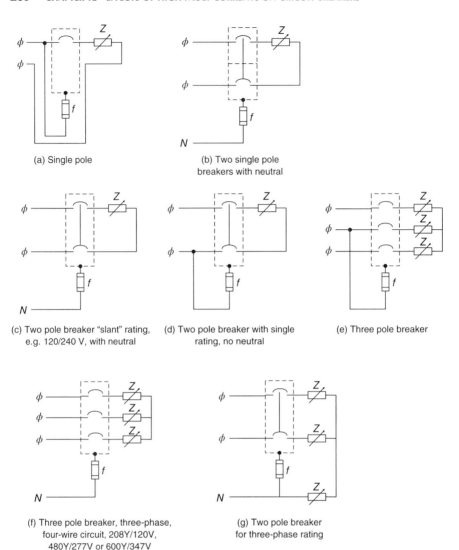

Figure 12.20 Endurance test circuits based on UL 489 (UL, 2009). (a) Single pole; (b) two single-pole breakers with neutral; (c) two-pole breaker "slant" rating, for example, 120/240 V, with neutral; (d) two-pole breaker with single rating, no neutral; (e) three-pole breaker; (f) three-pole breaker, three-phase, four-wire circuit, 208Y/120 V, 480Y/277 V, or 600Y/347 V; (g) two-pole breaker for three-phase rating.

previously described. The tests are to be applied at the maximum voltages shown in Table 12.13. In the overload test, the applied current is only 200% instead of 600% of the rated current. The specifications for the overload test are shown in Table 12.14. The specifications for the endurance test are shown in Table 12.15.

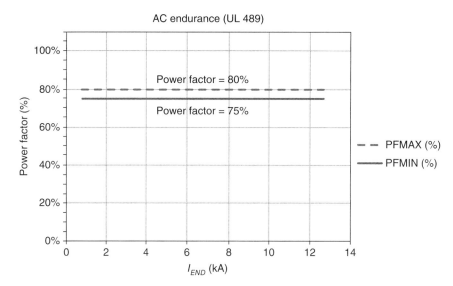

Figure 12.21   AC endurance power factors.

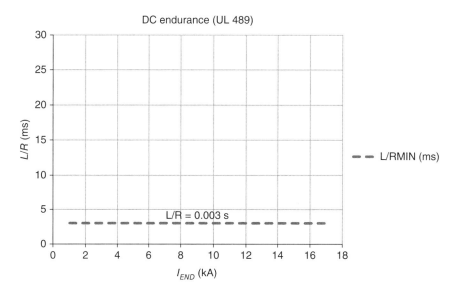

Figure 12.22   DC endurance time constant.

12.11.3   Testing of Supplementary Protectors for Use in Electrical Equipment According to UL Standard 1077

A supplementary protector, sometimes called a *miniature circuit breaker*, is similar to a molded-case circuit breaker, but is designed to be used inside an appliance or

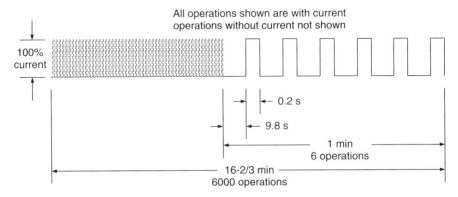

Figure 12.23   Endurance test, 100 A frame (not to scale).

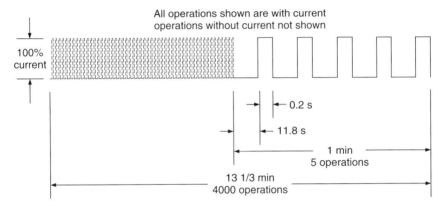

Figure 12.24   Endurance test, 150 A and 225 A frames (not to scale).

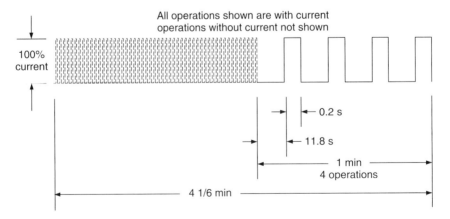

Figure 12.25   Endurance test, 600 A frame (not to scale).

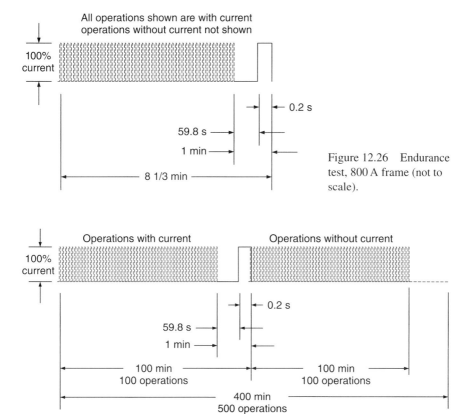

Figure 12.26   Endurance test, 800 A frame (not to scale).

Figure 12.27   Endurance test, 1200 A and 2500 A frames (not to scale).

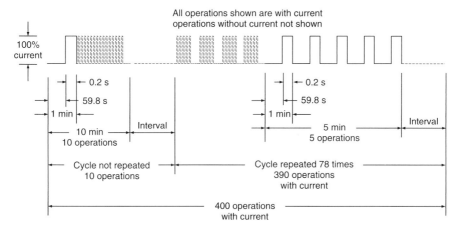

Figure 12.28   Endurance test, 6000 A frame (not to scale).

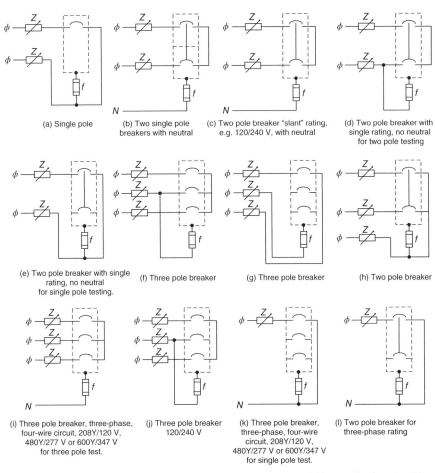

Figure 12.29   Interrupting test circuits based on UL 489 (UL, 2009). (a) Single pole; (b) two single-pole breakers with neutral; (c) two-pole breaker "slant" rating, for example, 120/240 V, with neutral; (d) two-pole breaker with single rating, no neutral for two pole testing; (e) two-pole breaker with single rating, no neutral for single pole testing; (f) three-pole breaker; (g) three-pole breaker; (h) two-pole breaker; (i) three-pole breaker, three-phase four-wire circuit, 208Y/120 V, 480Y/277 V, or 600Y/347 V for three pole test; (j) three-pole breaker 120/240 V; (k) three-pole breaker, three-phase, four-wire circuit, 208Y/120 V, 480Y/277 V, or 600Y/347 V for single pole test; (l) two pole breaker for three-phase rating.

other piece of equipment. Single-pole, two-pole, and three-pole supplementary protectors are shown in Figure 12.33 (General Electric Company, 2007). As such, they are important in protecting against appliance fires and other hazards to people and property. The definition of a supplementary protector is given in the *UL Standard for Safety for Supplementary Protectors for Use in Electrical Equipment*, UL Standard 1077 (UL, 2005):

"A manually resettable device designed to open the circuit automatically on a predetermined value of time versus current or voltage within an appliance or other

TABLE 12.11   Test Currents, Power Factors, and DC Time Constants for Interrupting Tests

Breaker Rating (A)	Available Amperes (1 Pole (≤1200 A), 2 Pole, Indiv.)	PF (%)	Available Amperes (3 Pole, Indiv.)	PF (%)	Available Amperes (2–3 Pole, Common)	PF (%)	DC Min Time Const (s)
≤100 (≤250 V)	5000	45–50	4330	45–50	5000	45–50	0.003
≤100 (251–600 V)	10,000	45–50	8660	45–50	10,000	45–50	0.003
101–800	10,000	45–50	8660	45–50	10,000	45–50	0.003
801–1200	14,000	25–30	12,120	25–30	14,000	25–30	0.008
1201–1600	14,000	25–30	14,000	25–30	20,000	25–30	0.008
1601–2000	14,000	25-30	14,000	25–30	25,000	15–20	0.008
2001–2500	20,000	25–30	20,000	25–30	30,000	15–20	0.008
2501–3000	25,000	15–20	25,000	15–20	35,000	15–20	0.008
3001–4000	30,000	15–20	30,000	15–20	45,000	15–20	0.008
4001–5000	40,000	15–20	40,000	15–20	60,000	15–20	0.008

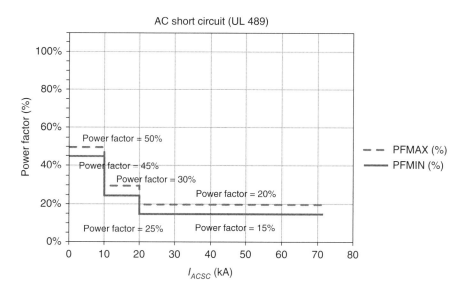

Figure 12.30   AC interrupting power factors.

electrical equipment. It is permitted to be provided with manual means for opening or closing the circuit."

Supplementary protectors are subjected to overload, endurance, and short circuit tests, similar to molded-case circuit breakers.

For supplementary protectors covered in UL 1077-2005 (UL, 2005), the tests include the following:

TABLE 12.12 Operations for Interrupting Test Operations

Poles	Frame Rating (A)	Circuit Breaker AC Voltage Rating (V)	Circuit Breaker DC Voltage Rating (V)	Number of Operations	Figure
1	All	120, 127, 208, 240, 277, 347, 480 or 600	250	3	Figure 12.32(a)
1	All	120/240 (Pairs)	125/250	3	Figure 12.32(a)
2	All	240, 480, 600	250	5	Figure 12.32(b)
2	All	120/240	125/250	3	Figure 12.32(a)
2	0–1200	208Y/120, 480Y/277 or 600Y/347	NA	5	Figure 12.32(b)
2	All	1φ–3φ	250	5	Figure 12.32(b)
3	0-1200	240, 480, 600	250	7	Figure 12.32(c)
3	1200-Up	240, 480, 600	250	7	Figure 12.32(c)
3	All	120/240	125/250	3	Figure 12.32(a)
3	0–1200	208Y/120, 480Y/277 or 600Y/347	NA	7	Figure 12.32(c)
3	1201-Up	208Y/120, 480Y/277 or 600Y/347	NA	7	Figure 12.32(c)

TABLE 12.13 Maximum Voltage for Tests to be Applied for Breakers Used in UPS

Level	Nominal	Maximum
1	384	500
2	400	600

TABLE 12.14 Overload Test for Breakers Used in UPS

Breaker Rating (A)	Number of Cycles	Test Current % of Rated	PF (All)	PF (1 and 2 pole only)	DC Min Time Const. (s)
≤249	50	200	45–50	75–80	0.003
250–6000	25	200	45–50	75–80	0.003

TABLE 12.15 Endurance Test for Breakers Used in UPS

Breaker Rating	Number of Cycles With Load	Number of Cycles Without Load
≤249	1000	1000
250–6000	400	400

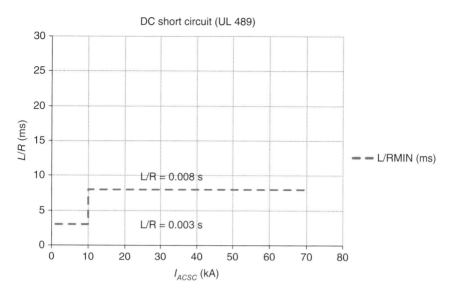

Figure 12.31   DC interrupting time constants.

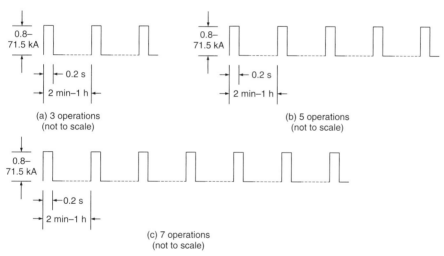

Figure 12.32   Interrupting tests (not to scale). (a) Three operations (not to scale); (b) five operations (not to scale); and (c) seven operations (not to scale).

1. Overload tests verify the supplementary protectors and can withstand repeated overloads at specified multiples of the rated current.
2. Endurance tests verify that the supplementary protectors can withstand multiple switching operations at 100% of the rated current.
3. Short circuit tests verify that the supplementary protector can safely open and close on its rated short circuit current in the case of a short circuit.

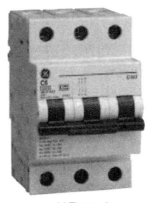

(a) Single pole
supplementary protector

(b) Two pole
supplementary protector

(c) Three pole
supplementary protector

Figure 12.33   Supplementary protectors for use in electrical equipment according to UL Standard 1077 (General Electric Company). (a) Single-pole supplementary protector; (b) two-pole supplementary protector; and (c) three-pole supplementary protector.

TABLE 12.16   Overload Test Currents and Power Factor for Supplementary Protectors

Supplementary Protector Type	Rating Basis	Test Current	Power Factor (%)
AC Motor starting (across the line)	HP	$6 \times$ motor FLA[a]	40–50
DC Motor starting (across the line)	HP	$10 \times$ motor FLA[a]	Resistive load
AC General use or incandescent lamp	Amperes (AC)	$1.5 \times$ rated	78–80% or Resistive load is so marked
DC General use or incandescent lamp	Amperes (DC)	$1.5 \times$ rated	Resistive load

[a]Full load amperes, from Tables 21.2 and 21.3 in UL (2005).

Each type of test is conducted in specific test circuits according to the voltage and current ratings, number of poles (individual supplementary protectors within the overall case), and circuit type (single-phase and three-phase). UL 1077 provides diagrams of each type of test circuit to be used. Each test also has a specific sequence of operations for repeated closings of the supplementary protector being tested.

UL 1077 tests of AC supplementary protectors are designed for devices rated 60 Hz, or the specified operating frequency of the device. For a 60 Hz device, the test frequency should be in the range of 60 ± 12 Hz. The short circuit current ratings of the supplementary protectors, if present, are marked on the device, along with code letters indicating whether the short circuit test was performed with or without series overcurrent protection.

The overload tests are performed at multiples of the rated current, with the values for various applications listed in Table 12.16. Test circuits for overload tests

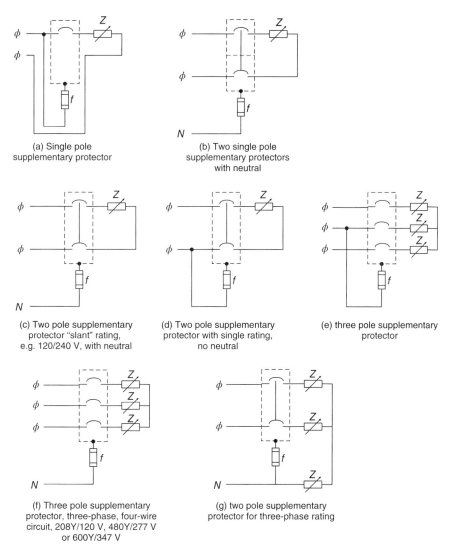

Figure 12.34   Overload and endurance test circuits for supplementary protectors based on UL 1077 (UL, 2005). (a) Single-pole supplementary protector; (b) two single-pole supplementary protectors with neutral; (c) two-pole supplementary protector "slant" rating, 120/240 V, with neutral; (d) two-pole supplementary protector with single rating, no neutral; (e) three-pole supplementary protector; (g) three-pole supplementary protector, three-phase, four-wire circuit, 208Y/120 V, 480Y/277 V or 600Y/347 V; (h) two-pole supplementary protector for three-phase rating.

are shown in Figure 12.34. The open-circuit voltage of the test circuit should be in the range of 100–110% of the rated voltage. The closed-circuit voltage of the test circuit should be in the range of 90–110% of the rated voltage, except in special circumstances where it may be as low as 85%. Tests are performed at a lagging power factor from 45% to 50% for supplementary protectors for AC motors, as shown in Table 12.16. The test fuse should be rated 30 A with a voltage rating equal to or greater than that of the device under test, and a 5.3 mm^2 (#10 AWG) wire not longer than 1.8 m (6 ft). The overload test occurs for 50 repetitions in a sequence of on for 1 s and off for the following 9 s, followed by 35 cycles of automatic closing and opening operation.

The endurance tests are performed at the rated current. Test circuits for endurance tests are shown in Figure 12.34. The open-circuit voltage of the test circuit should be in the range 100-105% of the rated voltage. The closed-circuit voltage of the test circuit should be in the range 90–110% of the rated voltage, except in special circumstances where it may be as low as 85%. Tests are performed at a lagging power factor from 75% to 80% for AC supplementary protectors. For DC supplementary protectors, a resistive load is used for testing. The test fuse should be rated 30 A with a voltage rating equal to or greater than that of the device under test and a 5.3 mm^2 (#10 AWG) wire not longer than 1.8 m (6 ft). The endurance test occurs for 6000 repetitions of a sequence of on for 1 s and off for the following 9 s, which is a rate of six cycles per minute.

The short-circuit tests are performed at high values of current, with the values for supplementary protector ratings listed in Table 12.17. The test circuit for short-circuit tests is shown in Figure 12.35. The circuit breaker "P" in the diagrams is the series overcurrent device (circuit breaker or fuse) which may be required. The supplementary protector should be protected by an appropriate fuse or circuit breaker during the testing. The no-load voltage of the test circuit, both AC and DC tests, should be in the range 99-105% of the rated voltage. AC tests are performed at a lagging power factor from 75% to 80%. For DC supplementary protectors, a resistive load should be used. The test fuse should be rated 30 A with a voltage rating equal to or greater than that of the device under test, and a 5.3 mm^2 (#10 AWG) wire not longer than 1.8 m (6 ft). The test is performed for three operations; the first operation is opening on a short circuit, while the subsequent two are closing on a short circuit.

TABLE 12.17 Short-Circuit Test Currents and Power Factor for Supplementary Protectors

Supplementary Protector Rating (HP)	Rated Voltage (V)	Test Current (A)	Power Factor (AC Protectors) (%)
HP ≤ ½	≤250	200	75–80
½ < HP ≤ 1	≤250	1000	75–80
HP ≤ 1	>250	1000	75–80
1 < HP ≤ 3	>250	2000	75–80
3 < HP ≤ 7-½	≤250	3500	75–80
HP > 7-½	≤250	5000	75–80
HP > 1	>250	5000	75–80

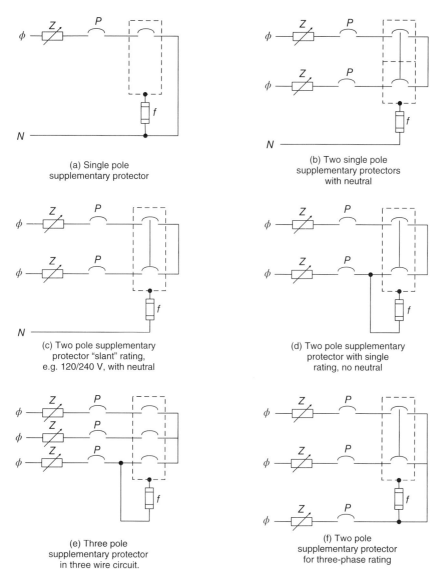

(a) Single pole
supplementary protector

(b) Two single pole
supplementary protectors
with neutral

(c) Two pole supplementary
protector "slant" rating,
e.g. 120/240 V, with neutral

(d) Two pole supplementary
protector with single
rating, no neutral

(e) Three pole
supplementary protector
in three wire circuit.

(f) Two pole
supplementary protector
for three-phase rating

Figure 12.35   Short-circuit test circuits for supplementary protectors based on UL 1077 (UL, 2005). (a) Single-pole supplementary protector; (b) two single-pole supplementary protectors with neutral; (c) two-pole supplementary protector "slant" rating, for example, 120/240 V, with neutral; (d) two-pole supplementary protector with single rating, no neutral; (e) three-pole supplementary protector in three wire circuit; (f) two-pole supplementary protector for three-phase rating; (g) three-pole supplementary protector in four wire circuit; (h) three-pole supplementary protector, 120/240 V; (i) three-pole supplementary protector, three-phase, four-wire circuit, 480Y/277 V or 600Y/347 V; (j) two-pole supplementary protector for three-phase, four-wire circuit, 480Y/277 V or 600Y/347 V.

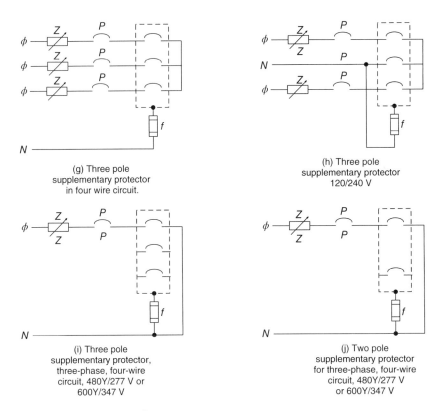

(g) Three pole
supplementary protector
in four wire circuit.

(h) Three pole
supplementary protector
120/240 V

(i) Three pole
supplementary protector,
three-phase, four-wire
circuit, 480Y/277 V or
600Y/347 V

(j) Two pole
supplementary protector
for three-phase, four-wire
circuit, 480Y/277 V
or 600Y/347 V

Figure 12.35   (*Continued*)

12.11.4   Testing of Transfer Switch Equipment According to UL Standard 1008

Transfer switch equipment (UL, 2011) is used to manually or automatically transfer the power to a load from one source to another. An automatic transfer switch (ATS) is shown in Figure 12.36 (General Electric Company, 2010). These switches are important in protecting against the disruptions caused by loss of power to emergency and critical loads. A transfer switch will typically switch power between a normal (utility) supply and a source of backup power, such as a generator. The load is often protected during the time necessary to start the generator and transfer power by an uninterruptible power supply (UPS). The definition of a transfer switch is given in the *National Electrical Code*, NFPA 70-2014 (NFPA, 2014):

"An automatic or nonautomatic device for transferring one or more load conductor connections from one power source to another."

Transfer switches are rated at a variety of low voltages for AC and DC operation, as listed in Table 12.18. The safety testing procedures for transfer switches is given in *Transfer Switch Equipment*, UL Standard 1008 (UL, 2011). Transfer switches are subjected to overload and short-circuit tests, similar to supplemental protectors.

Figure 12.36   Automatic transfer switch (*Source*: General Electric Company (2010)).

TABLE 12.18   Applicable Voltage Ratings for Transfer Switches

AC	DC
120	125
240	250
480	600
600	—

Source: UL (2011).

For transfer switches covered in UL 1008-2011 (UL, 2011), these tests include the following:

1. Overload tests verify the transfer switch can withstand repeated overloads at specified multiples of the rated currents.
2. Endurance tests verify that the transfer switch can withstand multiple switching operations at 100% of the rated current.

3. Short-circuit withstand tests verify that the transfer switch can withstand a short circuit on the load terminals until the designated overcurrent protective device clears the fault. For transfer switches used in emergency or standby systems, the switch must be able to transfer during a short circuit on the load terminals to a neutral position or to the opposite source. Tests are performed on both the normal and alternate sources circuits.

Each type of test is conducted in specific test circuits according to the voltage and current ratings, number of poles (individual switches within the overall case), and circuit type (single-phase and three-phase). UL 1008 provides diagrams of the test circuits to be used. Each test also has a specific sequence of operations for repeated closings of the transfer switch being tested.

UL 1008 tests of AC transfer switches are designed for devices rated 50/60 Hz, or the specified operating frequency of the device. For a 50 or 60 Hz device, the test frequency should be 60 Hz. The short-circuit current ratings of transfer switches are marked on the device. Standard short-circuit ratings are listed in Table 12.19.

Overload tests are performed at multiples of the rated current, with the values for various applications listed in Table 12.20. The open-circuit voltage of the test circuit should be in the range 100-110% of the rated voltage. The closed-circuit voltage of the test circuit should be in the range 100-110% of the rated voltage, except in special circumstances where it may be as low as 85%. The tests are performed at a lagging power factor from 40% to 50% for transfer switches for AC motors, as shown in Table 12.20. The overload test duty cycles are shown in Table 12.21.

TABLE 12.19   Short-Circuit Ratings for Transfer Switches

5000	25,000	65,000
7500	30,000	85,000
10,000	35,000	100,000
14,000	42,000	125,000
18,000	45,000	150,000
20,000	50,000	200,000
22,000	60,000	—

Source: UL (2011).

TABLE 12.20   Overload Test Currents and Power Factor for Transfer Switches

Protected Device	Protector Rating	Test Current	PF or DC Load
Motor or system loads	AC	6 × motor FLA	40–50%
	DC	10 × motor FLA	100% Resistive
Incandescent Lamp	AC	1.5 × rated current	75–80%
or resistive load	DC	1.5 × rated current	100% Resistive
Electric discharge lamp	AC	3 × rated current	40–50%

Source: UL (2011).

TABLE 12.21   Duty Cycles for Overload Tests of Transfer Switches

Transfer Switch Rating (A)	Number of Operations	Time Between Operations (min)
≤300	50	1
301–400	50	2
401–600	50	3
601–800	50	4
801–1600	50	5
1601–2500	25	5
≥2501	3	5

Source: UL (2011).

TABLE 12.22   Duty Cycles for Endurance Tests of Transfer Switches

Transfer Switch Rating (A)	Number of Operations with Current	Number of Operations Without Current	Time between Operations (min)
≤300	6000	0	1
301–400	4000	0	1
401–800	2000	1000	1
801–1600	1500	1500	2
≥1601	1000	2000	4

Source: UL (2011).

TABLE 12.23   Short-Circuit Test Currents and Power Factor for Transfer Switches

Transfer Switch Rating (A)	Short-Circuit Test Current (A)	PF (%)
≤100	5000	40–50
101–400	10,000	40–50
401–500	10,000	40–50
501–1000	10,001–20,000 (20× rating)	25–30
≥1001	>20,000 (20× rating)	≤20

Source: UL (2011).

Endurance tests are performed at 100% of the rated current. The open-circuit voltage of the test circuit should be in the range 100-105% of the rated voltage. The closed-circuit voltage of the test circuit should be in the range 90-110% of the rated voltage, except in special circumstances where it may be as low as 85%. Tests are performed at a lagging power factor from 75% to 80% for AC transfer switches. For DC transfer switches, a resistive load is used for testing. The endurance test duty cycles are shown in Table 12.22.

Short-circuit withstand tests are performed at high values of current, with the values for various transfer switch ratings listed in Table 12.23. The test circuit for

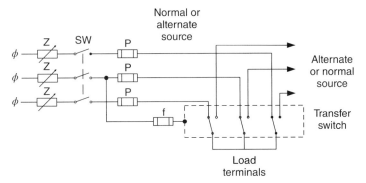

Figure 12.37 Short-circuit test circuit for transfer switches based on UL 1008 (UL, 2011). Three-pole supplementary protector.

short-circuit withstand tests is shown in Figure 12.37. The device should be protected by an appropriate fuse or circuit breaker during the testing. The fuse "P" in the diagram is the series overcurrent protective device which may be required for the tests. The no-load voltage of the test circuit, for both AC and DC tests, should be greater than or equal to the maximum rated voltage of the switch. The test fuse should be rated 30 A with a voltage rating equal to or greater than that of the device under test, and a 5.3 mm^2 (#10 AWG) wire not longer than 1.8 m (6 ft).

12.11.5 Testing of Low-Voltage AC Power Circuit Breakers According to ANSI Standard C37.50-1989

For AC LVPCBs covered in ANSI C37.50-1989 (ANSI, 1989), the tests include the following:

1. Overload switching tests verify the circuit breaker can withstand repeated overloads at 600% of the rated currents.

2. Endurance tests verify that the circuit breaker can withstand multiple switching operations at 100% of the rated current.

3. Short-circuit current tests verify that the circuit breaker can safely interrupt its rated short-circuit current in the case of a short circuit.

Each type of test is conducted in specific test circuits according to the voltage and current ratings, number of poles (individual breaker units within the overall case), and circuit type (single-phase and three-phase). Each test also has a specific sequence of operations for repeated closings of the circuit breaker being tested.

ANSI C37.50 tests of AC LVPCBs are designed for breakers rated 60 Hz, with the test frequency being in the range of 60 Hz ± 20%. AC circuit breakers have standard frame sizes (maximum ampere ratings or continuous-current-carrying capability) listed in Table 12.24 as specified in IEEE Standard C37.16-2009 (IEEE, 2009b). The standard fuse sizes are listed in Table 12.25. The nominal and maximum AC voltage ratings for LVPCBs are listed in Table 12.26. The interrupting ratings of the circuit breakers, defined as "the highest RMS symmetrical current at the rated voltage that a

TABLE 12.24   Standard Frame Sizes for AC and DC LVPCBs

Frame Size (A)
600
800
1600
2000
3000
3200
4000
5000
6000

Source: IEEE (2009a).

TABLE 12.25   Standard Fuse Sizes for AC LVPCBs

Fuse Size (A)
300
600
800
1000
1200
1600
2000
2500
3000
4000
5000
6000

Source: IEEE (2009a).

TABLE 12.26   Applicable Voltage Ratings for AC LVPCBs

Maximum AC Voltage (V)	Nominal AC Voltage (V)
254	240
508	480
635	600

Source: ANSI (1989) and IEEE (2009a).

TABLE 12.27   Overload Tests for AC LVPCBs at 600% of Rated Amperes

Breaker Rating (A)	Test Current (A)	PF (%)	Switching Operations for Opening and Closing
600	3600	≤50	50
800	4800	≤50	50
1600	9600	≤50	38
2000	12,000	≤50	38
3000	18,000	≤50	38
3200	19,200	≤50	38
4000	24,000	≤50	38
5000	30,000	≤50	38
6000	36,000	≤50	38

Source: ANSI (1989) and IEEE (2009a).

device is intended to interrupt under standard test conditions" are many times larger than their ampere ratings. For unfused AC LVPCBs rated under ANSI C37.16-2009, the interrupting ratings range from 22 to 200 kA. Fused AC LVPCBs have a 200 kA interrupting rating.

Overload tests are performed at 600% of the rated current, with the values for various breaker ratings listed in Table 12.27. The no-load voltage of the test circuit should not be less than 100% and greater than 105% of the rated voltage, while the overload voltage with 600% current should not be less than 65% of the rated voltage. The test frequency should be in the range of 60 Hz ±20%. Tests are performed at a lagging power factor of no more than 50% for AC circuit breakers. The test fuse should be rated 30 A with a voltage rating equal to or greater than that of the device under test, and a 5.3 mm^2 (#10 AWG) wire not longer than 1.8 m (6 ft). The breakers are tested at a rate of one operation per minute for 5 min, followed by an interval of 15 min before the next test sequence. The breakers are only switched on for a very short time in each test, not less than one cycle, and are opened by a shunt trip device.

Endurance tests are performed at 100% of the rated current, with the values for various breaker ratings listed in Table 12.28. The no-load voltage of the test circuit should not be less than 100% and greater than 105% of the rated voltage. The voltage of the test circuit with load should be greater than 80% of the rated voltage. The test frequency should be in the range of 60 Hz ± 20%. Tests are performed at a lagging power factor of not more than 85% for AC circuit breakers. The test fuse should be rated 30 A with a voltage rating equal to or greater than that of the device under test, and a 5.3 mm^2 (#10 AWG) wire not longer than 1.8 m (6 ft). The breakers are switched on and off at the rate of one operation every 2 min. Tests may be grouped in sequences of no fewer than 120 operations.

The short-circuit current tests are performed at high values of current, with the values for various breaker ratings listed in Table 12.29. The no-load voltage of the test circuit should not be larger than the rated maximum voltage of the breaker under

TABLE 12.28   Endurance Tests for AC LVPCBs at 600% of Rated Amperes

Breaker Rating (A)	Test Current (A)	PF (%)	Switching Operations for Opening and Closing (Energized)	Switching Operations for Opening and Closing (De-energized)
600	600	≤85	2800	9700
800	800	≤85	2800	9700
1600	1600	≤85	800	3200
2000	2000	≤85	800	3200
3000	3000	≤85	400	1100
3200	3200	≤85	400	1100
4000	4000	≤85	400	1100
5000	5000	≤85	400	1100
6000	6000	≤85	400	1100

Source: ANSI (1989) and IEEE (2009a).

TABLE 12.29   Maximum and Minimum Short-Circuit Ratings (kA) for AC LVPCBs

	Breakers with Instantaneous Trip Unit with Maximum Voltage (V)				Breakers Without Instantaneous Trip Unit with Maximum Voltage (V)	
	Minimum Short-Circuit Current (kA)	Maximum Short-Circuit Current (kA)			Minimum Short-Circuit Current (kA)	Maximum Short-Circuit Current (kA)
Frame Size (A)	254 V, 508 V, or 635 V	254 V	508 V	635 V	254 V, 508 V, or 635 V	254 V, 508 V, or 635 V
600	22	42	30	22	22	22
800	22	200	200	130	22	85
1600	42	200	200	130	42	85
2000	42	200	200	130	42	85
3000	65	85	65	65	65	65
3200	65	200	200	130	65	100
4000	85	200	200	130	85	100
5000	85	200	200	130	85	100
6000	85	200	200	130	85	100

Source: IEEE (2009a).

test. The test frequency should be in the range of 60 Hz ± 20%. The short-circuit current tests are performed at a lagging power factor not greater than 15% for unfused breakers, and 20% for fused breakers. The power frequency recovery voltage is the voltage across the breaker contacts while it is opening. For LVPCBs with instantaneous trip units, the power frequency recovery voltage should be greater than or equal to 95% of the rated maximum voltage of the breaker under test. For LVPCBs without instantaneous trip units, the power frequency recovery voltage should be greater than or equal to 80% of the rated maximum voltage of the breaker under test. The test fuse should be rated 30 A with a voltage rating equal to or greater than that of the device under test, and a 5.3 mm^2 (#10 AWG) wire not longer than 1.8 m (6 ft). The circuit breaker is tested with open-close-open sequences as specified in the standard.

12.11.6 Testing of Low-Voltage DC Power Circuit Breakers According to IEEE Standard C37.14-2002

For DC LVPCBs covered in IEEE C37.14-2002 (IEEE, 2002c), these tests include the following:

1. Endurance tests verify that the circuit breaker can withstand multiple switching operations at 100% of the rated current.

2. Short-circuit current tests verify that the circuit breaker can safely interrupt its rated short-circuit current in the case of a short circuit.

Each type of test is conducted in specific test circuits according to the voltage and current ratings, number of poles (individual breaker units within the overall case), and circuit type (single-pole and two-pole). Each test also has a specific sequence of operations for repeated closings of the circuit breaker being tested.

General-purpose DC circuit breakers have standard frame sizes (maximum ampere ratings or continuous-current-carrying capability) listed in Table 12.24 as specified in IEEE Standard C37.16-2009 (IEEE, 2009b). The nominal and maximum DC voltage ratings for LVPCBs are listed in Table 12.30. DC short circuits are

TABLE 12.30   Standard Nominal and Maximum Rated Voltage for DC LVPCBs

Rated Nominal Voltage (V)	Rated Maximum Voltage (V)
250	300
275	325
750	800
850	1000
1000	1200
1500	1600
3000	3200

Source: IEEE (2002c).

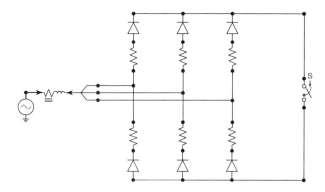

Figure 12.38   Circuit diagram for EMTP simulation of three-phase rectifier circuit.

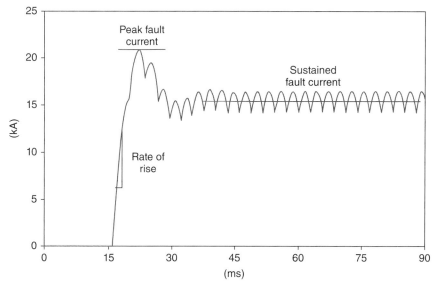

Figure 12.39   Current–time curve of rectifier short-circuit current for fault in DC system with appreciable inductance and resistance in the DC system, illustrating the components of the fault current waveform.

generally produced by faults across the load side of three-phase rectifiers, as shown in Figure 12.38. The DC short circuit waveform has a rise time, a peak, and a steady-state value (General Electric Company, 1972; Sutherland, 1999) as shown in Figure 12.39. The rated short-circuit current of the circuit breakers, defined as the "designated limit of available (prospective) rms current in amperes at which they shall be required to perform their short-circuit current duty cycle … at rated maximum voltage under the prescribed test conditions" (IEEE, 2002c) are many times larger than their ampere ratings. The rated peak current is defined as "the designated limit of nonrepetitive available (prospective) peak current in amperes that it shall be required to close into and still be able to open and then close" (IEEE,

2002c). For general-purpose DC LVPCBs rated under ANSI C37.16-2009, the interrupting ratings range from 25 to 100 kA.

The DC tests utilize a three-phase bridge rectifier. The characteristics of this device are that the open circuit DC output voltage V_{d0} is 1.35 times the AC line-line voltage. Thus for a 750 V AC source, approximately 1000 V of DC output voltage is produced. The current output of the rectifier has three modes, Modes 1–3. Mode 1 is the normal mode where load is supplied. Mode 3 is the short-circuit mode, and Mode 2 is a region of either high overloads or low short-circuit currents.

The load regulation curve for a 70 kA fault in Figure 12.40 shows that Mode 1 lasts until 75% voltage, Mode 2 as the voltage decreases from 75% to 43%, and Mode 3 is below 43% voltage. In short-circuit tests, the voltage is near zero in Mode 3. In overload and endurance tests, the voltage is in Mode 1, as indicated by the 1200 A load current shown. The load regulation curve in Figure 12.41 shows the response for a low fault current of 1.1 kA.

Endurance tests are performed at 100% of the rated current, with the values for various breaker ratings listed in Table 12.24. The no-load voltage of the test circuit

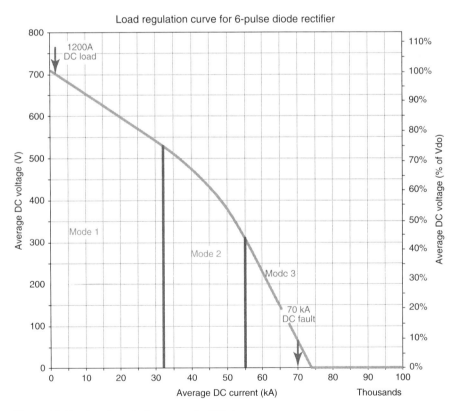

Figure 12.40   Load regulation curve for 70 kA output.

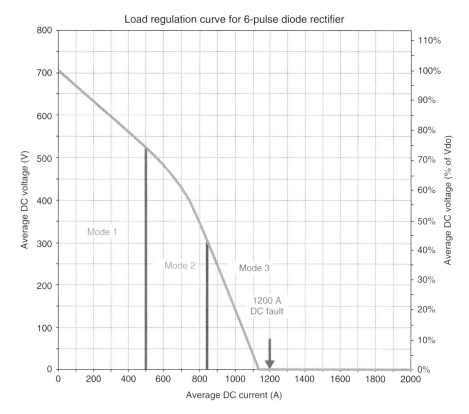

Figure 12.41   Load regulation curve for 1.1 kA output.

should not be less than 100% and greater than 105% of the rated voltage. The L/R time constant of the test circuit should not be less than 0.02 s or exceed 0.06 s. The test fuse should be rated 30 A with a voltage rating equal to or greater than that of the device under test, and a 5.3 mm^2 (#10 AWG) wire not longer than 1.8 m (6 ft). The breakers are switched on and off at the rate of one operation every 2 min. The circuit breaker must remain closed for at least 300 ms during each test. Tests may be grouped in sequences of no fewer than 120 operations. A mechanical endurance test must also be performed, with 20% of the operations listed for the energized endurance test.

   The short-circuit tests are performed at high values of current, with the values for various breaker ratings listed in Table 12.11. The short-circuit test for breakers used on solid-state rectifier applications also includes a test of peak current as part of the same operation. The no-load voltage of the test circuit should not be larger than the rated maximum voltage of the breaker under test. The L/R time constant of the test circuit is not specified directly, but specific values of R and L for different tests are given in IEEE (2009a). The test fuse should be rated 30 A with a voltage rating equal to or greater than that of the device under test, and a 5.3 mm^2 (#10 AWG) wire not longer than 1.8 m (6 ft).

12.11.7 Testing of Low-Voltage Switchgear and Controlgear According to IEC Standard 60947-1

IEC Standard 60947-1 (IEC, 2011b) covers all types of switchgear and controlgear up to 1000 V AC and 1500 V DC. The general rules of the switchgear and controlgear standard apply to circuit breakers with some modifications as detailed in the specific circuit breaker standards. IEC Standard 60947-2 (IEC, 2013a) covers all types of circuit breakers up to 1000 V AC and 1500 V DC. This standard covers all types of low-voltage circuit breakers except those covered in the following standards:

- IEC 60947-4-1 relating to circuit breakers used for across-the-line starters for motors.
- IEC 60898 relating to circuit breakers used in households and similar installations.
- IEC 60934 relating to circuit breakers used in equipment such as electrical appliances.

The standard IEC voltage ratings (IEC, 2009b) are listed in Tables 12.31 and 12.32. The standard IEC current ratings (IEC, 1999) are listed in Table 12.33. All

TABLE 12.31   Standard Nominal Rated AC Voltages for IEC Rated Equipment

Rated Nominal 50 Hz AC Voltage for Three-Phase, Three- or Four-Wire Systems (V)	Rated Nominal 60 Hz AC Voltage for Three-Phase, Three- or Four-Wire Systems (V)	Rated Nominal 50/60 Hz AC Voltage for Single-Phase, Three-Wire Systems (V)
	120/208	120/240
230	240	
230/400	230/400	
	277/480	
	480	
	347/600	
	600	
400/690		
1000		

Source: IEC (2009b).

TABLE 12.32   Standard DC Voltages for IEC Rated Traction Systems

Rated Minimum DC Voltage (V)	Rated Nominal DC Voltage (V)	Rated Maximum DC Voltage (V)
500	750	900
1000	1500	1800
2000	3000	3600

Source: IEC (2009b).

TABLE 12.33   Standard IEC Current Ratings for Equipment

1	1.25	1.6	2	2.5	3.15	4	5	6.3	8
10	12.5	16	20	25	31.5	40	50	63	80
100	125	160	200	250	315	400	500	630	800
1000	1250	1600	2000	2500	3150	4000	5000	6300	8000
10,000	12,500	16,000	20,000	25,000	31,500	40,000	50,000	63,000	80,000
100,000	125,000	160,000	200,000						

Source: IEC (1999)

rated currents both continuous-current and short-circuit-current ratings, are chosen from the standard currents in Table 12.33. IEC Standard 60947-1 tests of low-voltage AC switchgear and controlgear are to be performed with the test frequency being in the range ±25% of the rated frequency of the equipment.

For low-voltage circuit breakers covered under IEC Standard 60497-1 (IEC, 2011b), the tests include the following:

1. Overload performance tests verify the circuit breaker can withstand repeated overloads.

2. Operational performance capability (endurance) tests verify that the circuit breaker can withstand multiple switching operations at 100% of the rated current.

3. Short-circuit making and breaking capacity tests verify that the circuit breaker can safely close on (make) and interrupt (break) its rated short-circuit current in the case of a short circuit.

Each type of test is conducted in specific test circuits according to the voltage and current ratings, number of poles (individual breaker units within the overall case), and circuit type (single-phase and three-phase). Each test also has a specific sequence of operations for repeated closings of the circuit breaker being tested.

Test circuits are given in Figure 12.42 for overload and operational performance tests, and in Figure 12.43 for short-circuit tests. The fusible element, f, in Figure 12.42 and Figure 12.41 is rated for 1500 A ± 10%, and consists of a round copper conductor with a diameter of 0.8 mm, and a length of 50 mm, or equivalent. The resistor R_L limits the prospective fault current in fuse f to 1500 A, ±10%.

12.11.8   Testing of Low-Voltage AC and DC Circuit Breakers According to IEC Standard 60947-2

IEC Standard 60947-2 tests of low-voltage AC circuit breakers are to be performed with the test frequency being in the range of ±5% of the rated frequency of the circuit breaker. The overload tests are set up as shown in Figure 12.42 and the short-circuit tests as in Figure 12.43. The short-circuit and overload test currents, AC power factors, and DC time constants for circuit breakers tested in accordance with IEC 60947-2 are listed in Table 12.34, taken from Table 11 of (IEC, 2013a).

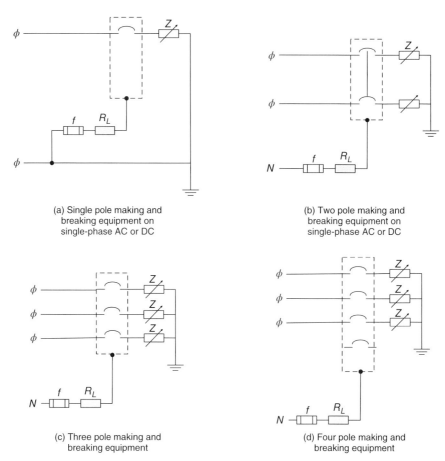

(a) Single pole making and breaking equipment on single-phase AC or DC

(b) Two pole making and breaking equipment on single-phase AC or DC

(c) Three pole making and breaking equipment

(d) Four pole making and breaking equipment

Figure 12.42 Overload and operational performance making and breaking test circuits based on IEC 60947-1 (IEC, 2011b). (a) Single-pole making and breaking equipment on single-phase AC or DC; (b) two-pole making and breaking equipment on single-phase AC or DC; (c) three-pole making and breaking equipment; (d) four-pole making and breaking equipment.

Overload performance tests for AC circuit breakers are performed 600% of the rated current for AC circuit breakers and 250% or the rated current for DC circuit breakers. The no-load voltage of the test circuit should not be less than 100% and greater than 105% of the rated voltage, while the voltage with the overload value of current should not be less than 65% of the rated voltage. Tests are performed at a lagging power factor of no more than 50% for AC circuit breakers, or a time constant of 2.5 ms for DC circuit breakers, as listed in Table 12.35. The test operating rates for the overload performance test are for a total of 12 operations, at the rates given in Table 12.36. The 12 operations consist of nine manual and three automatic openings, unless the circuit-breaker short-circuit trip setting is less than the test current, requiring that all openings be automatic. The breaker should be kept closed for long enough in each manual cycle for the full current to be reached, but not longer than 2 s.

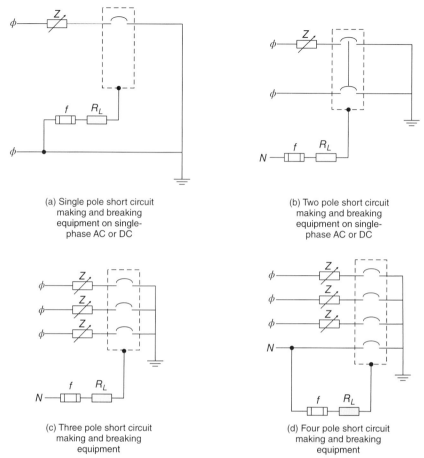

(a) Single pole short circuit
making and breaking
equipment on single-
phase AC or DC

(b) Two pole short circuit
making and breaking
equipment on single-
phase AC or DC

(c) Three pole short circuit
making and breaking
equipment

(d) Four pole short circuit
making and breaking
equipment

Figure 12.43   Short-circuit making and breaking test circuits. based on IEC 60947-1 (IEC, 2011b). (a) Single-pole, short-circuit making and breaking equipment on single-phase AC or DC; (b) two-pole, short-circuit making and breaking equipment on single-phase AC or DC; (c) three-pole, short-circuit making and breaking equipment; (d) four-pole, short-circuit making and breaking equipment.

The operational performance capability tests are performed at 100% of the rated current. The no-load voltage of the test circuit should not be less than 100% and greater than 105% of the rated voltage. The voltage of the test circuit with load should be greater than 80% of the rated voltage. The test frequency should be in the range 45-62 Hz. Tests are performed at a lagging power factor of not more than 80% for AC circuit breakers or a time constant of 2.0 ms for DC circuit breakers, as listed in Table 12.35. The test operating rates and number of operations for the operational performance test are given in Table 12.36.

The short-circuit current tests are performed at the rated values of short-circuit making and breaking current. The no-load voltage of the test circuit should not larger than the rated maximum voltage of the breaker under test. A tolerance of ±5% of

TABLE 12.34   IEC Short-Circuit Tests for Switchgear and Controlgear at Rated Short-Circuit Amperes

Test Current, Within ±5% (A)	PF for Short-Circuit Test (%)	DC Minimum Time Constant for Short-Circuit Test (ms)
$I \le 1500$	95	5
$1500 < I \le 3000$	90	5
$3000 < I \le 4500$	80	5
$4500 < I \le 6000$	70	5
$6000 < I \le 10{,}000$	50	5
$10{,}000 < I \le 20{,}000$	30	10
$20{,}000 < I \le 50{,}000$	25	15
$50{,}000 < I$	20	15

Source: IEC (2011b).

TABLE 12.35   IEC Short-Circuit Tests and Overload Test Currents for Circuit Breakers

Test Current, Within ±5% (A)	PF for Short Circuit Test (%)	PF for Operational Performance Test (%)	PF for Overload Test (%)	DC Minimum Time Constant for Short-Circuit Test (ms)	DC Minimum Time Constant for Operational Performance Test (ms)	DC Minimum Time Constant for Overload Test (ms)
$I \le 3000$	90%	80%	50%	5	2.0	2.5
$3000 < I \le 4500$	80%			5		
$4500 < I \le 6000$	70%			5		
$6000 < I \le\ < I \le 10{,}000$	50%			5		
$10{,}000 < I \le 20{,}000$	$\le 30\%$			10		
$20{,}000 < I \le 50{,}000$	$\le 25\%$			15		
$50{,}000 < I$	$\le 20\%$			15		

Source: IEC (2013a).

the rated frequency applies to the test frequency. The short-circuit current tests are performed at a lagging power factor as shown in Table 12.35. The power frequency recovery voltage is the voltage across the breaker contacts while it is opening. The power frequency recovery voltage should be 105% of the rated maximum voltage of the breaker under test.

12.11.9   Testing of Circuit Breakers Used for Across-the-Line Starters for Motors According to IEC Standard 60947-4-1

IEC Standard 60947-4-1 (IEC, 2009c) tests of low-voltage AC and DC contactors and AC motor starters are to be performed with the test frequency being 50 Hz for devices rated at 50 Hz or 60 Hz. While the standard covers many types of contactors

TABLE 12.36 Operating Rates for Overload Performance and Operational Performance Tests According to IEC 60947-2

Frame Size (A)	Minimum Operating Rate (cycles/h)	Number of Operations (Without Current)	Number of Operations (With Current)
$I \leq 100$	120	8500	1500
$100 < I \leq 315$	120	7000	1000
$315 < I \leq 630$	60	4000	1000
$630 < I \leq 2500$	20	2500	500
$2500 < I$	10	1500	500

Source: IEC (2013a).

and starters, only two categories of devices are considered here, AC-3 and AC-4 for operation of squirrel cage motors. Category AC-3 covers starting motors and disconnecting motors while running. Category AC-4 covers starting, plugging and inching.

The overload tests are set up as shown in Figure 12.40. The overload test currents are eight times the rated operational current for contactors and starters rated 630 A or less, and six times for those rated above 630 A. The duration of the overload test is 10 s and shall be performed one time.

The operational performance capability tests are performed at 200% of the rated current for category AC-3 and 600% for category AC-4. The no-load voltage of the test circuit should not be less than 100% and greater than 105% of the rated voltage. Tests are performed at a lagging power factor of 45% for devices rated 100 A or less and 35% for those rated above 100 A. The test is performed for 6000 operations, with intervals between tests as shown in Table 12.37.

TABLE 12.37 Time Interval Between Operations for Making and Breaking Tests of Contactors and Starters IEC 60947-4-1

Frame Size (A)	Interval (s)
$I \leq 100$	10
$100 < I \leq 200$	20
$200 < I \leq 300$	30
$300 < I \leq 400$	40
$400 < I \leq 600$	60
$600 < I \leq 800$	80
$800 < I \leq 1000$	100
$1000 < I \leq 1300$	140
$1300 < I \leq 1600$	180
$1600 < I$	240

Source: IEC (2009c).

The short-circuit current tests are performed at the rated values of short-circuit making and breaking current. The no-load voltage of the test circuit should not be larger than the rated maximum voltage of the breaker under test. The short-circuit current tests are performed at a lagging power factor as shown in Table 12.34. The power frequency recovery voltage is the voltage across the breaker contacts while it is opening. The power frequency recovery voltage should be 105% of the rated maximum voltage of the breaker under test. The fusible element, f, in Figures 12.42 and 12.43 consists of a round copper conductor with a cross-sectional area of 6 mm^2 and a length of $1.2 - 1.8$ m, connected to the neutral. The resistor R_L limits the prospective fault current in fuse f to 1500 A, $\pm 10\%$.

12.11.10 Testing of Circuit Breakers Used in Households and Similar Installations According to IEC Standard 60898-1 and -2

IEC Standard 60898-1 (IEC, 2003a) and 60898-2 (IEC, 2003b) cover tests of low-voltage AC and DC circuit breakers for overcurrent protection for household and similar installations. Circuit breakers under this standard with more than one pole are not required to have overcurrent tripping mechanisms on each pole. A pole with an overcurrent tripping mechanism is referred to as a *protected pole*. The tests may be performed at "any convenient voltage." These standards cover circuit breakers of the following voltage ratings for AC:

- 120, 230, 240, and 400 V, single- or three-phase
- 120/240 V, single-phase
- 230/400 V, single- or three-phase.

These standards cover circuit breakers of the following voltage ratings for DC:

- 125 or 220 V, single pole
- 220/440 or 125/250 V, two pole.

Preferred values of the rated current are listed in Table 12.38.

Mechanical and electrical endurance tests are performed at 100% of the rated current. The test frequency should be in the range 45-62 Hz. Tests are performed at a lagging power factor between 85% and 90% for AC circuit breakers, or a time constant of 4 ms $\pm$ 10% for DC circuit breakers, except that DC breakers marked T15 are tested at a time constant of 15 ms. The DC test current should have a ripple (ω) of no more than 5%. The endurance test consists of 4000 making and breaking operations (operating cycle). Circuit breakers less than or equal to 32 A are tested at a rate of

TABLE 12.38   Preferred Values of Rated Current (A) Per IEC 60898-1

6	8	10	13	16	20
25	32	40	50	63	80
100	125				

Source: IEC (2003a).

240 per hour, spaced at least 13 s apart. Circuit breakers greater than 32 A are tested at a rate of 120 per hour, spaced at least 28 s apart.

The-short circuit current tests are performed at either 500 A, or 10 times the rated current I_n. and at 1500 A. Circuit breakers rated above 1500 A short-circuit capacity, I_{cn}, are subject to testing of the "service short-circuit capacity" and at the "rated short-circuit capacity." Standard and preferred values of rated short-circuit capacity are listed in Table 12.39. The service short-circuit capacity, I_{cs}, is the breaking capacity when the circuit breaker is loaded at 85% of its "non-tripping current," I_{nt}, which is the highest value of current where the circuit breaker will not trip. The ratio between I_{cs} and I_{cn} is called "k" and is listed in Table 12.40. Short-circuit test currents have a tolerance of ±5% of the specified current. The frequency tolerance for the short circuit test is ±5% of the rated frequency.

Circuits for testing of one-, two-, three-, and four pole circuit breakers are shown in Figure 12.44. The following circuit elements are identified:

- A is the making switch.
- Z is the adjustable impedance to set the short-circuit current at the rated value.
- Z_1 is an adjustable impedance to reduce the short-circuit current below its rated value.
- R_1 is a resistor selected for a current of 10 A at the applied voltage.
- H is the "hot" wire of the source.
- N is the neutral wire of the source.
- f is a test fuse, consisting of a copper wire 50 mm or longer, having a diameter of 0.1 mm for a circuit breaker tested in free air, or 0.3 mm for a circuit breaker tested in an enclosure.
- R_2 is a 0.5 Ω resistor.

TABLE 12.39   Standard and Preferred (*) Values of Rated Short Circuit Capacity (A) per IEC 60898-1

1500	3000	4500	6000	10,000	20,000*

Source: IEC (2003a).

TABLE 12.40   Ratio "k" Between Service Short-Circuit Capacity and Rated Short-Circuit Capacity, Per IEC 60898-1

Short-Circuit Capacity Greater Than (A)	Maximum Short-Circuit Capacity (A)	Minimum Service Short-Circuit Capacity (A)	k
NA	6000	NA	1.0
6000	10,000	6000	0.75
10,000		7500	0.5

Source: IEC (2003a).

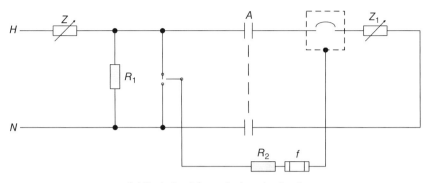

(a) Test circuit for a single pole circuit
breaker

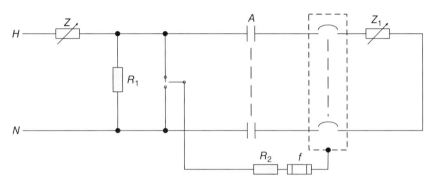

(b) Test circuit for a two pole circuit breaker
with one protected pole

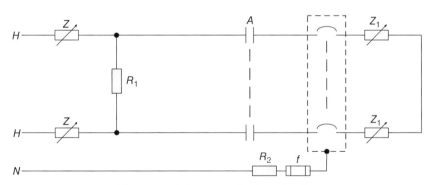

(c) Test circuit for a two pole circuit breaker
with two protected poles

Figure 12.44   Test circuits based on IEC 60898 (IEC, 2011b) ac circuit breakers for
households and similar installations. (a) Test circuit for a single-pole circuit breaker; (b) test
circuit for a two-pole circuit breaker with one protected pole; (c) test circuit for a two-pole
circuit breaker with two protected poles; (d) test circuit for a three-pole circuit breaker; (e)
test circuit for a four-pole circuit breaker.

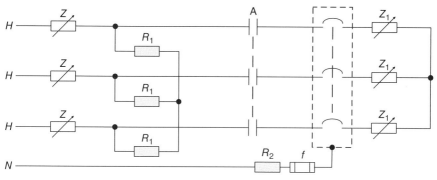

(d) Test circuit for a three pole circuit breaker

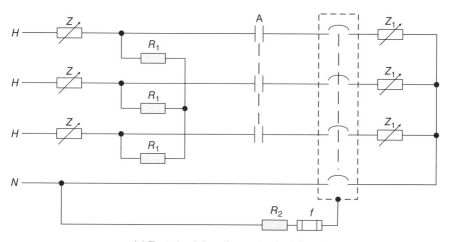

(e) Test circuit for a four pole circuit breaker

Figure 12.44   (*Continued*)

The power frequency recovery voltage is the voltage across the breaker contacts while it is opening. The power frequency recovery voltage should be 105%, with a tolerance of ±10%, of the rated maximum voltage of the breaker under test.

Short-circuit tests are performed at a lagging power factor between 20% and 98% for AC circuit breakers, as listed in Table 12.41, or a time constant of 4 ms ± 10% for DC circuit breakers, except that DC breakers marked T15 are tested at a time constant of 15 ms. The DC test current should have a ripple (ω) of no more than 5%. Tests are performed for a variety of make and break sequences and repetition rates as specified in the standard.

12.11.11   Testing of Circuit Breakers Used in Equipment such as Electrical Appliances According to IEC Standard 60934

IEC Standard 60934 (IEC, 2013b) covers tests of low-voltage AC and DC circuit breakers for overcurrent protection for use within equipment for household and

TABLE 12.41   Test Power Factor for AC Short-Circuit Tests, Per IEC 60898-1

Test Current Greater Than (A)	Maximum Test Current (A)	Minimum Power Factor (%)	Maximum Power Factor (%)
NA	1500	93	98
1500	3000	85	90
3000	4500	75	80
4500	6000	65	70
6000	10,000	45	50
10,000	25,000	20	25

Source: IEC (2003a).

TABLE 12.42   Rated Voltages for Circuit Breakers Under IEC 60934

AC	AC	DC
60	400/230	12
120	415/240	24
240/120	380	48
220	400	60
230	415	120
240	440	240
380/220		250

Source: IEC (2013b).

similar installations. These devices are similar to the supplementary protectors discussed under UL Standard 1077. Under this standard, circuit breakers with more than one pole are not required to have overcurrent tripping mechanisms on each pole. A pole with an overcurrent tripping mechanism is referred to as a *protected pole*. The preferred rated voltages are listed in Table 12.42. The rated currents are up to and including 125 A. The rated frequencies for the devices covered here are 50, 60, and 400 Hz.

Tests are performed with tolerances of +5% on frequency and voltage, and −0%/+5% on test current. Endurance (rated current), overload, and short-circuit tests are performed for a wide variety of operations, AC power factors, and DC time constants, as specified in the standard.

Short-circuit current tests are performed at the "rated switching capacity" and the "rated short circuit capacity." The rated switching capacity is the short-circuit making and breaking current. The rated short-circuit capacity is six times the rated current for AC and Four times for DC, except that it shall not be larger than 3000 A. Short-circuit current tests are also performed at the "rated conditional short-circuit current," which is the short-circuit withstand current when the device is protected by a short-circuit protective device, such as an upstream fuse or circuit breaker.

Circuits for testing of one-, two-, three-, and four-pole circuit breakers are shown in Figure 12.45. The circuit elements are identified as in Figure 12.44, except for the fuse, f:

- f is a test fuse, consisting of a copper wire 50 mm or longer, having a diameter of 0.1 mm.

The power frequency recovery voltage is the voltage across the breaker contacts while it is opening. The power frequency recovery voltage should be 105%, with a tolerance of $\pm 10\%$, of the rated maximum voltage of the breaker under test.

Tests at the rated conditional short-circuit current are performed at a lagging power factor between 85% and 98% for AC circuit breakers as listed in Table 12.43, or a time constant of 2.5 or 5 ms for DC circuit breakers as listed in Table 12.44. Tests at currents over 3000 A are performed according to IEC Standard 60898 (IEC, 2003a, 2003b). Tests are performed for a variety of make and break sequences and repetition rates as specified in the standard.

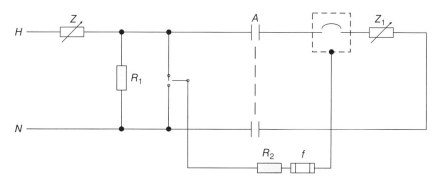

(a) Test circuit for a single pole circuit breaker

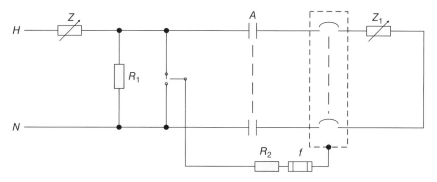

(b) Test circuit for a two pole circuit breaker with one protected pole

Figure 12.45   Test circuits based on IEC 60934 (IEC, 2013b). (a) Test circuit for a single-pole circuit breaker; (b) test circuit for a two-pole circuit breaker with one protected pole; (d) test circuit for a three-pole circuit breaker; (e) test circuit for a four-pole circuit breaker.

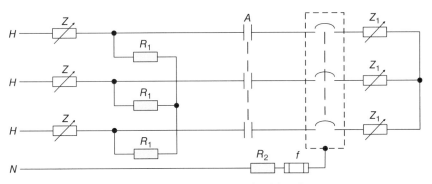

(d) Test circuit for a three pole circuit breaker

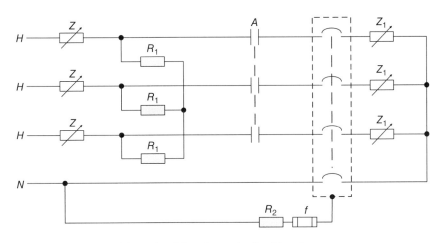

(e)test circuit for a four pole circuit breaker

Figure 12.45 *(Continued)*

12.12 TESTING OF HIGH-VOLTAGE CIRCUIT BREAKERS

High-voltage circuit breakers, greater than 1 kV, may be tested according IEEE Standard C37.09-1999 (IEEE, 1999a) or a similar IEC standard. The ratings of the high-voltage circuit breakers tested under C37.09 are defined in IEEE Standard C37.04 (IEEE, 1999b). Three types of tests are set forth:

Design tests are performed to demonstrate that a circuit breaker design can meet the requirements specified in C37.04.

Production tests are factory tests, which are part of the manufacturing process.

Conformance tests are generally performed on-site to demonstrate that a circuit breaker meets certain requirements.

TABLE 12.43   Test Power Factor for AC Conditional Short-Circuit Current Tests, Per IEC 60934

Test Current Greater Than (A)	Maximum Test Current (A)	Minimum Power Factor (%)	Maximum Power Factor (%)
300	1500	93	98
1500	3000	85	90

Source: IEC (2013b).

TABLE 12.44   Test Time Constants for DC Conditional Short-Circuit Current Tests, per IEC 60934

Test Current Greater Than (A)	Maximum Test Current (A)	Minimum Time Constant (ms)	Nominal Time Constant (ms)	Maximum Time Constant (ms)
NA	1000	2.0	2.5	3.0
1000	NA	4.0	5.0	6.0

Source: IEC (2013b).

Production and conformance tests will not be described here. The design tests include the following:

1. *Maximum-voltage tests* are considered to be passed if the circuit breaker is successfully tested for short-circuit and other tests at its rated maximum voltage.

2. *Power frequency tests* are considered to be passed if the circuit breaker is successfully tested at its rated frequency ±10%, or at the rated frequency if allowed.

3. *Continuous-current-carrying tests* require that the circuit breaker temperature should not increase by more than 1 °C over an ambient temperature between 10 °C and 40 °C for three separate 30 min tests. The resistance of the closed circuit breaker is also measured, using a test current of at least 1 A, up to the rated continuous current.

4. *Dielectric withstand tests* are for power frequency dielectric withstand, lightning impulse, and chopped wave tests, as specified in IEEE Standard C37.06 (IEEE, 2009a).

5. *Standard operating duty (standard duty cycle) tests* are performed at the rated current for selected duty cycles.

6. *Interrupting time tests* measure the total of the contact parting time plus the arcing time.

7. *Transient recovery voltage (TRV) tests* determine whether a circuit breaker can withstand a certain voltage while the contacts are opening as specified in IEEE Standard C37.06. This is done as part of the short circuit tests.

8. *Short-circuit current interrupting tests* are conducted over a wide variety of test conditions to ensure that the circuit breaker can interrupt short circuits as specified in IEEE Standard C37.04.

9. *Load current switching tests* demonstrate the ability of the circuit breaker to switch specified load currents over given duty cycles.

10. *Capacitor switching current tests* demonstrate the ability of the circuit breaker to switch specified capacitor banks or highly capacitive lines and cables over given duty cycles.

CHAPTER 13

MECHANICAL FORCES AND THERMAL EFFECTS IN SUBSTATION EQUIPMENT DUE TO HIGH FAULT CURRENTS

13.1 INTRODUCTION

Mechanical forces and thermal effects produced by high fault currents can damage or destroy substation equipment. This damage is a serious safety issue for a number of reasons. All substation equipment can suffer damage from mechanical forces caused by short circuits. Examples include bending of bus work, sudden expansion of transformer coils, and breaking of insulators and bushings. In addition, even faults with rather moderate magnitude may cause long-term effects such as accelerated aging of dielectric insulation as a result of repetitive mechanical stresses. Solving the problem of increased fault currents means repeating portions of the original design process. Because substation design has become an automated procedure (Anders *et al.*, 1992), the uprating should be in the nature of a design review. This chapter is adapted from EPRI (2006).

13.2 DEFINITIONS

Fixed bus bar end. The end of a rigid bus bar that is not free to rotate.

Pinned bus bar end. The end of a rigid bus bar that is free to rotate.

SCADA Systems. Supervisory controls and data acquisition systems.

Strain bus. A bus made of flexible conductors, which hangs from suspension insulators, such that

$$l = l_c + 2l_i \tag{13.1}$$

where

l = distance between supports, in m

l_c = length of bus conductor, in m

l_i = length of one insulator chain, in m.

Principles of Electrical Safety, First Edition. Peter E. Sutherland.

Slack bus. A bus made of flexible conductors, which hangs from post insulators, such that

$$l = l_c \tag{13.2}$$

13.3   SHORT-CIRCUIT MECHANICAL FORCES ON RIGID BUS BARS

13.3.1   Short-Circuit Mechanical Forces on Rigid Bus Bars — Circular Cross Section

The short-circuit forces on three, rigid, side-by-side busbars are shown in Figure 13.1 and are calculated according to IEEE Standard 605 (IEEE, 2008):

$$F_{SC} = K_f \frac{\mu_0}{2\pi} \frac{8\Gamma(D_f I_{SC})^2}{D} \tag{13.3}$$

where

F_{SC} = fault current force in N/m

I_{SC} = symmetrical rms fault current in A

D = conductor center-to-center spacing in m

Γ = a constant whose value is

1.00 for phase-phase faults on either conductor,

0.866 for three-phase faults on conductor B,

0.808 for three-phase faults on conductor A or C.

D_f is the decrement factor. Where fault durations are less than 1 s or the X/R ratio is greater than 5, the asymmetry of fault current waveforms produces additional heating, which must be taken into account:

$$D_f = \sqrt{1 + \frac{T_a}{t_f}\left(1 - e^{-2t_f/T_a}\right)} \tag{13.4}$$

where

t_f = fault duration in s

T_a = time constant $X/2\pi f R$ in s

K_f = mounting structure flexibility factor. This is 1.0 unless the heights of the mounting structures are greater than 3 m (IEEE, 1998, Figure 4).

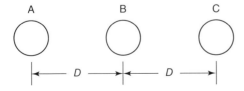

Figure 13.1   Three-phase side-by-side bus bar configuration.

TABLE 13.1   Allowable Stress for Common Conductor Materials

Material	Minimum Yield Stress (kPa^2)
Al alloy 6063-T6 or 6101-T6	172,375
Al alloy 6061-T6	241,325
Al alloy 6061-T6	103,452
Cu No. 110 hard drawn	275,800

IEEE (2008).

The short-circuit forces on a conductor are added to the other forces to produce a total force, which must be less than the minimum yield stress of the conductor material (Table 13.1) (IEEE, 2008, Section 11). The magnitude of the total force is

$$F_T = \sqrt{\left(F_w + K_H F_{SC}\right)^2 + \left(F_G + K_V F_{SC}\right)^2} \tag{13.5}$$

The angle of the force below the horizontal is

$$\theta = \tan^{-1}\left[\frac{F_G + K_V F_{SC}}{F_w + K_H F_{SC}}\right] \tag{13.6}$$

where

F_T = total vector force on the bus, in N/m

F_w = wind force in N/m

F_G = total bus unit weight, including ice loading and connectors, in N/m

K_V = 1 if the bus conductors are vertical, otherwise 0

K_H = 1 if the bus conductors are horizontal, otherwise 0

The maximum allowable length between spans is limited by the maximum of the vertical deflection or the span length for fiber stress. The vertical deflection is primarily an aesthetic concern, which is not affected by short-circuit forces. The maximum allowable length based on fiber stress is calculated from the maximum allowable stress in Table 13.1:

$$L_S = 3.16\sqrt{\frac{K_S F_A S}{F_T}} \tag{13.7}$$

where

L_S = maximum length of the bus in cm

K_S = constant based on the number of spans and end types (Table 13.2)

F_A = maximum yield stress in kPa^2

S = section modulus in cm^3.

When increased forces are anticipated owing to increased short-circuit levels, supporting them with additional insulators can protect substation bus bars.

TABLE 13.2   Conductor Span Constant K_S

Number of Spans	Fixed Ends	Pinned Ends	K_S
1	0	2	8
1	2	0	12
1	1	1	8
2	N/A	N/A	8
3	N/A	N/A	10
4+	N/A	N/A	28

13.3.2   Short-Circuit Mechanical Forces — Rectangular Cross Section

In the case of substations with rectangular cross-section bus bars, a proximity factor K is used (Copper Development Association, 2001):

$$F_{SC} = K \frac{\mu_0}{2\pi} \frac{8\Gamma(D_f I_{SC})^2}{D} \tag{13.8}$$

where K is taken from Figure 13.2. K is equal to 1.0 for a round conductor and is almost 1.0 for a square conductor. Conductor shape is most significant for thin, strip conductors.

13.4   DYNAMIC EFFECTS OF SHORT CIRCUITS

When excited by a displacement force, a rigid conductor will vibrate at its natural frequency, subject to damping forces (IEEE, 2008). The stimulus of a short circuit will provide a twice power frequency periodic force which may be amplified if the natural frequency of the bus bar is greater than or equal to the power frequency. If additional supports are being added to stiffen the bus owing to increased fault currents, the dynamic effects should be checked as well. The natural frequency of the bus bar is

$$f_b = \frac{\pi K^2}{20 L^2} \sqrt{\frac{EJ}{m}} \tag{13.9}$$

where

f_b = natural frequency of the bus, in Hz

K = pinning factor:

    1.00 if both ends are pinned,

    1.22 if one end is fixed and one end is pinned,

    1.51 if both ends are fixed.

E = modulus of elasticity (Table 13.3), in kPa

J = moment of inertia of the crosssection, in cm^4

m = mass per unit length, in kg/m.

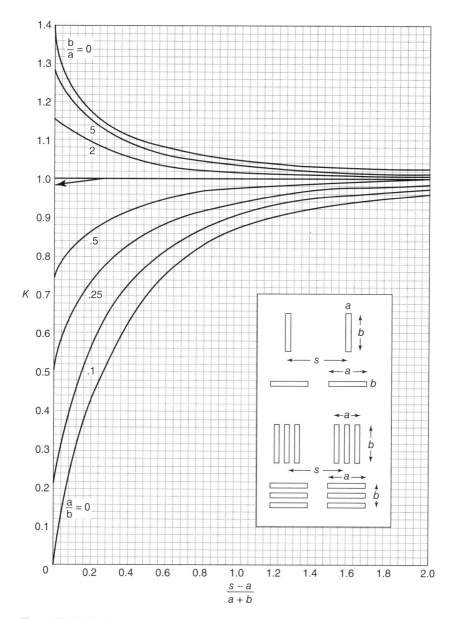

Figure 13.2 Dwight curves for proximity factor (Dwight, 1945).

If the resonant frequency calculation leads to the suspicion of a possible resonance problem, a dynamic (Bergeron *et al.*, 1999) or static (Bergeron and Trahan,
1999) finite element analysis should be performed.

In the case of hollow bus bars, internal weights or stiffeners may be added to
dampen vibration modes.

TABLE 13.3   Modulus of Elasticity for Common Conductor Materials

Material	Modulus of Elasticity (kPa)
Al alloy 6061-T6	
Al alloy 6063-T6	6.895×10^7
Al alloy 6101-T6	
Cu	11.03×10^7

IEEE (2008).

13.5   SHORT-CIRCUIT THERMAL EFFECTS

Heating of bus bars can cause annealing, thermal expansion or damage to attached equipment. The limit for thermal expansion adopted during the design of the substation is generally used. Annealing can occur at temperatures of 100°C or more.

The amount of current required to heat a conductor from the ambient to a given final temperature during the duration of a fault can be calculated (IEEE, 2008, Section 5.2):

$$I = C \times 10^6 A \sqrt{\frac{1}{t} \log_{10} \frac{T_f - 20 + (K/G)}{T_i - 20 + (K/G)}} \tag{13.10}$$

where

I = maximum allowable rms symmetrical fault current in A

C = constant: 92.9 for aluminum conductors, 142 for copper conductors

t = fault duration in s

T_f = final conductor temperature in °C

T_i = initial conductor temperature in °C

K = constant: 15150 for aluminum, 24500 for copper

G = International Annealed Copper Standard (IACS) conductivity as a percentage.

When the ends of a bus bar are not fixed, heating will result in thermal expansion, which may cause damage to attached equipment, such as switches, insulators, and other devices. The amount of expansion (IEEE, 2008, Section 13) is

$$\frac{\Delta L}{L_i} = \frac{\alpha(T_f - T_i)}{1 + \alpha T_i} \tag{13.11}$$

where

L_i = initial bus bar length in m

ΔL = change in bus bar length in m

α = coefficient of thermal expansion in 1/°C.

If both ends of the bus bar are fixed, then the resulting a force results may damage attached equipment:

$$F_{TE} = 0.1AE\alpha(T_i - T_f) \qquad (13.12)$$

where

A = cross-sectional area in cm^2

Where thermal expansion is anticipated owing to increased fault currents, expansion fittings can be added to long bus structures.

13.6   FLEXIBLE CONDUCTOR BUSES

Flexible conductor buses may be constructed as strain buses, suspended from insulator strings (Figure 13.3). This type of construction is usually used for substation main buses at high voltages (>100 kV). The slack bus construction (Figure 13.4) with post insulators is normally used for connections between equipment within a substation. When high current-carrying capability is needed, conductors are often bundled (Figure 13.5) separated from 8 to 60 cm with spacers at regular intervals of 2 to 30 m (CIGRE, 1996, p. 12).

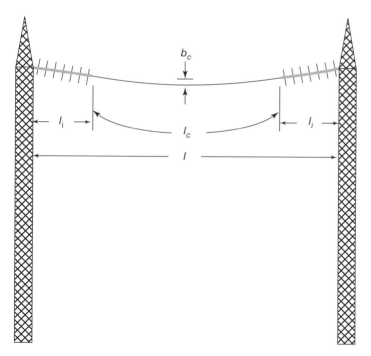

Figure 13.3   Strain bus from suspension insulators (EPRI, 2006).

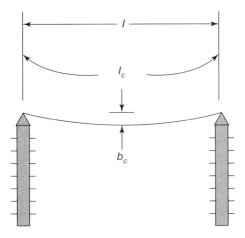

Figure 13.4 Slack bus from post insulators (EPRI, 2006).

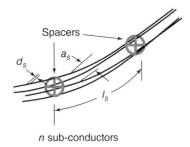

n sub-conductors

Figure 13.5 Details of flexible conductor bundle with spacers (EPRI, 2006).

The following are the effects of high fault currents on flexible conductor buses:

1. Increased tension on the conductors
2. Increased tension on insulators
3. Unwanted forces on support structures
4. Increased thermal stress on the conductors
5. Possibility of arcing due to decreased minimum clearance between conductors during swing
6. Possibility of damage due to increased drop force
7. Pinch-effect damage to conductors due to clashing, to spacers due to compression, and to suspension insulators and supports due to impulse tension.

Flexible conductor substation buses are discussed in detail in IEC Standard 60865 (IEC, 1994; IEC, 2011a) with further explanations in (CIGRE, 1996). The design standard is intended for horizontal buses up to 60 m long in a temperature range of −20 to +60°C and maximum sag of 8%. Automatic reclosing does not increase the effect of short circuits on flexible conductors. The simplified calculation procedure used in the standards has been verified by tests and detailed finite element method (FEM) simulations (Stein, Miri, and Meyer, 2000; Miri and Stein, 2002; Herrmann, Stein, and Keißling 1989).

13.6.1 Conductor Motion During a Fault

The short-circuit force on flexible conductors for a phase-to-phase or three-phase fault is approximately

$$F_{SC} = \frac{\mu_0}{2\pi} \frac{0.75 I_{SC}^2}{D} \frac{l_c}{l} \tag{13.13}$$

The forces will cause the conductors to separate (swing out), gravity will then bring them together (drop force), and they will oscillate with a characteristic period. The forces will also cause an elastic expansion of the conductor material, while the high currents will cause thermal expansion. If the bus is constructed from bundled conductors instead of single conductors, then the short circuit will force them together through the pinch effect, which produces tension on the conductor.

The IEC standard (IEC, 2011a) defines the ratio of electromagnetic force from the short circuit to the weight of the conductor:

$$r = \frac{F_{SC}}{nm_s'g} \tag{13.14}$$

where

n = number of sub-conductors

m_s' = mass of the sub-conductors in kg/m

g = gravitational constant in m/s^2

then the direction of the resultant force is

$$\delta = \arctan r \tag{13.15}$$

Without current flow, the static conductor sag is

$$b_c = \frac{nm_s'gl^2}{8F_{st}} \tag{13.16}$$

where

F_{st} = is the static force on the conductors in N.

If the conductor oscillates at small swing-out angles, again, with no current flow, the period of oscillation is

$$T = 2\pi\sqrt{0.8\frac{b_c}{g}} \tag{13.17}$$

With a short circuit, the period is

$$T_{SC} = \frac{T}{\sqrt[4]{1+r^2}\left[1-\frac{\delta^2}{16}\right]} \tag{13.18}$$

The maximum angle which the conductor can swing out for a given short-circuit current magnitude and fault duration can now be calculated. The swing-out angle at the end of the short circuit is

$$
\delta_k =
\begin{cases}
\delta \left[1 - \cos \left(2\pi \dfrac{T_{k1}}{T_{SC}} \right) \right] & \text{for} \quad 0 \le \dfrac{T_{k1}}{T_{SC}} \le 0.5 \\[2ex]
2\delta & \text{for} \quad \dfrac{T_{k1}}{T_{SC}} > 0.5
\end{cases}
\tag{13.19}
$$

where

T_{kl} = duration of the fault, in s.

Using r, we calculate a quantity χ,

$$
\chi =
\begin{cases}
1 - r \sin \delta_k & \text{for} \quad 0 \le \delta_k \le \dfrac{\pi}{2} \\[2ex]
1 - r & \text{for} \quad \delta_k > \dfrac{\pi}{2}
\end{cases}
\tag{13.20}
$$

and the maximum swing-out δ_m,

$$
\delta_m =
\begin{cases}
1.25 \arccos \chi & \text{for} \quad 0.766 \le \chi \le 1 \\
\pi/18 + \arccos \chi & \text{for} \quad -0.985 \le \chi \le 0.766 \\
\pi & \text{for} \quad \chi < -0.985
\end{cases}
\tag{13.21}
$$

Once the angle of displacement is known, the objective is to determine the minimum conductor clearance during the fault. This will require calculating the conductor tensile force and the thermal and elastic expansion.

Short-circuit tensile force on the conductors may be calculated as follows:

Calculate the stiffness norm N:

$$
N = \frac{1}{Sl} + \frac{1}{nE_S A_S}
\tag{13.22}
$$

where

S = spring constant of both supports in N/m

E_S = actual Young's modulus in N/m^2

A_S − cross section of one sub-conductor in m^2

Calculate the stress factor ζ:

$$
\zeta = \frac{(ngm_s'l)^2}{24F_{st}^3 N}
\tag{13.23}
$$

Calculate the load parameter φ:

$$
\varphi =
\begin{cases}
3 \left(\sqrt{1 + r^2} - 1 \right) & \text{for} \quad 0 \le \delta \le \dfrac{\pi}{2} \\[2ex]
3(r \sin \delta + \cos \delta - 1) & \text{for} \quad \delta > \dfrac{\pi}{2}
\end{cases}
\tag{13.24}
$$

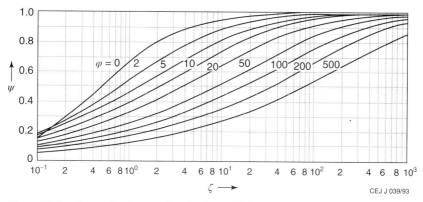

Figure 13.6   Curves for determining the factor Ψ from IEC Standard 60865, Figure 7 (EPRI, 2006).

Determine the factor ψ from Figure 13.6 (IEC, 2011a), or calculate from a real solution of

$$\begin{cases} \varphi^2\psi^3 + \varphi\,(2 + \zeta)\,\psi^2 + (1 + 2\zeta)\psi - \zeta(2 + \varphi) = 0 \\ 0 \le \psi \le 1 \end{cases} \tag{13.25}$$

Calculate the conductor bundling factor, K_B:

$$K_B = \begin{cases} 1.0 & \text{for} \quad n = 1 \\ 1.1 & \text{for} \quad n \ge 2 \end{cases} \tag{13.26}$$

Then

$$F_t = K_B F_{st}(1 + \varphi\psi) \tag{13.27}$$

The thermal expansion of the conductors is given by

$$\varepsilon_{th} = c_{th}\left[\frac{I''_{k3}}{nA_S}\right]^2 T \tag{13.28}$$

where

$$T = \max\left\{\frac{T_{sc}}{4}, T_{k1}\right\} \tag{13.29}$$

I''_{k3} = initial three-phase symmetrical rms short-circuit current in A.

$$c_{th} = \begin{cases} 0.27 \times 10^{-18}\,\text{m}^4/(\text{A}^2\text{s}) & \text{for} \quad \text{Al, Al alloy, and Al/steel conductors} \\ & \quad\;\; \text{with a cross-section ratio of Al/St} > 6 \\ 0.17 \times 10^{-18} & \text{for} \quad \text{Al/St} \le 6 \\ 0.088 \times 10^{-18} & \text{for} \quad \text{Cu} \end{cases}$$

$$\tag{13.30}$$

The elastic expansion of the conductors is given by

$$\varepsilon_{ela} = N(F_t - F_{st}) \tag{13.31}$$

The thermal and elastic expansions are combined into an expansion factor C_D:

$$C_D = \sqrt{1 + \frac{3}{8}\left[\frac{l}{b_c}\right]^2 (\varepsilon_{els} + \varepsilon_{th})} \tag{13.32}$$

The actual shape of the conductor sag can vary from a catenary at low current to a triangle at high short-circuit currents. A form factor C_F is used:

$$C_F = \begin{cases} 1.05 & \text{for} \quad r \le 0.8 \\ 0.97 + 0.1r & \text{for} \quad 0.8 \le r < 1.8 \\ 1.15 & \text{for} \quad r \ge 1.8 \end{cases} \tag{13.33}$$

The maximum horizontal displacement (Figure 13.7) for a slack bus ($l_c = l$) is

$$b_h = \begin{cases} C_F C_D b_c & \text{for} \quad \delta_m \ge \dfrac{\pi}{2} \\ C_F C_D b_c \sin \delta_m & \text{for} \quad \delta_m < \dfrac{\pi}{2} \end{cases} \tag{13.34}$$

For a strain bus ($l_c = l - 2l_i$),

$$b_h = \begin{cases} C_F C_D b_c \sin \delta & \text{for} \quad \delta_m \ge \delta \\ C_F C_D b_c \sin \delta_m & \text{for} \quad \delta_m < \delta \end{cases} \tag{13.35}$$

The closest approach of the conductors is

$$a_{min} = a - 2b_h \tag{13.36}$$

where

a = distance between the centers of the conductors while at rest in m.

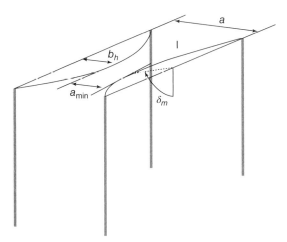

Figure 13.7   Horizontal displacement and distance between midpoints of a slack bus (EPRI, 2006).

Once the conductors have reached their maximum height, they will fall, experiencing the drop force:

$$F_f = 1.2F_{st}\sqrt{1 + 8\zeta\frac{\delta_m}{\pi}} \quad \text{for} \quad r > 0.6 \quad \text{and} \quad \delta_m \geq 70° \tag{13.37}$$

13.6.2   Pinch Forces on Bundled Conductors

In bundled conductor configurations (Figure 13.5), short-circuit forces cause the sub-conductors to come together rapidly. This discussion applies to n sub-conductors arranged in a circular configuration, separated by a distance a_S.

Sub-conductors are said to clash when

$$\begin{cases} \dfrac{a_s}{d_s} \leq 2.0 \quad \text{and} \quad l_s \geq 50a_s \\[1em] \qquad\qquad \text{or} \\[1em] \dfrac{a_s}{d_s} \leq 2.5 \quad \text{and} \quad l_s \geq 70a_s \end{cases} \tag{13.38}$$

If there are no more than four sub-conductors, which clash according to the above definition, then the force F_t, described above, may be used. Otherwise, the force F_{pi}, as described below, should be calculated.

In order to calculate F_{pi}, the following quantities are required:

1. The maximum of the single-phase to ground and the three-phase short-circuit current

$$I'' = \max(I''_{k1}, I''_{k3}) \tag{13.39}$$

2. The short-circuit current force is

$$F_v = (n-1)\frac{\mu_0}{2\pi}\left(\frac{I''}{n}\right)^2\frac{l_s}{a_s}\frac{v_2}{v_3} \tag{13.40}$$

3. The factor v_1 is given by

$$v_1 = \frac{f}{\sin\dfrac{\pi}{n}}\sqrt{\frac{(a_s - d_s)m'_s}{\dfrac{\mu_0}{2\pi}\left(\dfrac{I''}{n}\right)^2\dfrac{n-1}{a_s}}} \tag{13.41}$$

where

v_1 and $v_2 =$ given by Figure 13.8 (IEC, 2011a) and

$\quad v_3 =$ given by Figure 13.9 (IEC, 2011a).

4. The first strain factor for bundle contraction is

$$\varepsilon_{st} = 1.5\frac{F_{st}l_s^2N}{(a_s - d_s)^2}\sin^2\left(\frac{\pi}{n}\right) \tag{13.42}$$

5. The second strain factor for bundle contraction is

$$\varepsilon_{pi} = 0.375\frac{F_v l_s^3N}{(a_s - d_s)3}\sin^3\left(\frac{\pi}{n}\right) \tag{13.43}$$

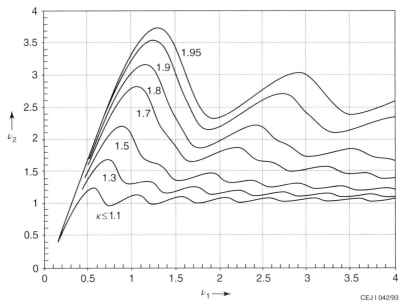

Figure 13.8 Curves for determining the factors v_1 and v_2 from IEC Standard 60865, Figure 9 (EPRI, 2006).

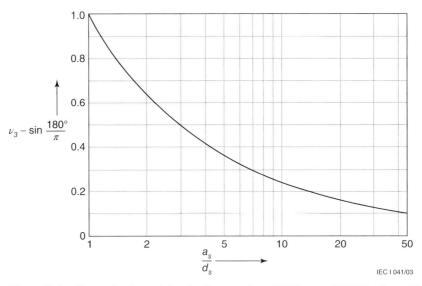

Figure 13.9 Curves for determining the factor v_3 from IEC Standard 60865, Figure 10 (EPRI, 2006).

6. The clashing factor is

$$j = \sqrt{\frac{\varepsilon_{pi}}{1 + \varepsilon_{st}}} \qquad (13.44)$$

7. If $j \geq 1$, the sub-conductors clash, go to step 8.

8. If $j < 1$ the sub-conductors do not clash, go to step 10.

9. The overswing factor for clashing of sub-conductors is

$$v_e = \frac{1}{2} + \left[\frac{9}{8} n (n-1) \frac{\mu_0}{2\pi} \left(\frac{I''}{n} \right)^2 N v_2 \left(\frac{l_s}{a_s - d_s} \right)^4 \right.$$

$$\left. \frac{\sin^4 \left(\frac{\pi}{n} \right)}{\xi^3} \left\{ 1 - \frac{\arctan \sqrt{v_4}}{\sqrt{v_4}} \right\} - \frac{1}{4} \right]^{1/2} \qquad (13.45)$$

where

ξ = is the pinch-force factor taken from Figure 13.10 (IEC, 2011a)

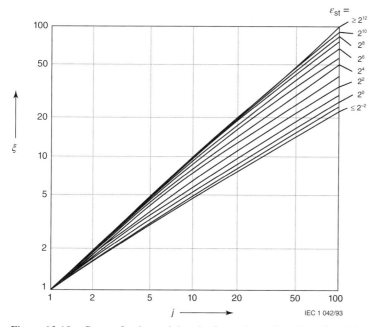

Figure 13.10   Curves for determining the factor ζ as a function of j and from IEC Standard 60865, Figure 11 (EPRI, 2006).

and the factor v_4 is given by:

$$v_4 = \frac{a_s - d_s}{d_s} \qquad (13.46)$$

10. The tensile force is then

$$F_{pi} = F_{st}\left(1 + \frac{v_e}{\varepsilon_{st}}\xi\right) \qquad (13.47)$$

11. The factor v_4 is given by

$$v_4 = \eta\frac{a_s - d_s}{a_s - \eta(a_s - d_s)} \qquad (13.48)$$

where

η = nonclashing sub-conductor factor taken from Figures 13.11–13.13 (IEC, 2011a).

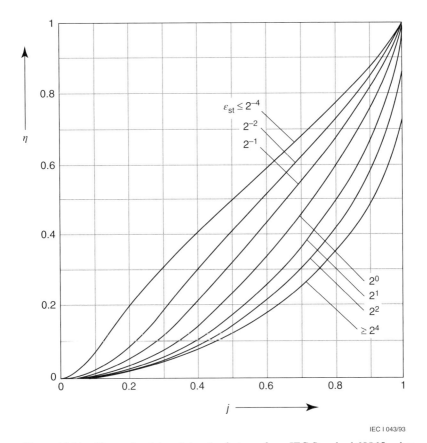

IEC I 043/93

Figure 13.11   Curves for determining the factor η from IEC Standard 60865, when $2.5 < a_s/d_s \leq 5.0$, Figure 12a (EPRI, 2006).

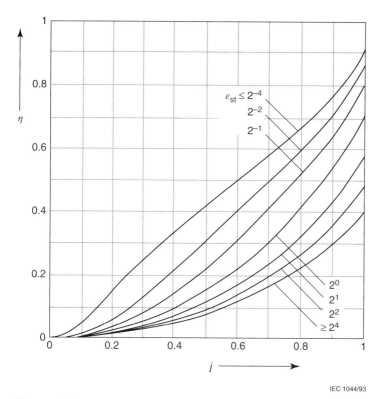

IEC 1044/93

Figure 13.12   Curves for determining the factor η from IEC Standard 60865, when $5.0 < a_s/d_s \leq 10.0$, Figure 12b (EPRI, 2006).

12. The overswing factor for nonclashing sub-conductors is given by

$$v_e = \frac{1}{2} + \left[\frac{9}{8} n (n-1) \frac{\mu_0}{2\pi} \left(\frac{I''}{n} \right)^2 N v_2 \left(\frac{l_s}{a_s - d_s} \right)^4 \right.$$

$$\left. \frac{\sin^4 \left(\frac{\pi}{n} \right)}{\eta^4} \left\{ 1 - \frac{\arctan \sqrt{v_4}}{\sqrt{v_4}} \right\} - \frac{1}{4} \right]^{1/2} \qquad (13.49)$$

13. The tensile force is then

$$F_{pi} = F_{st} \left(1 + \frac{v_e}{\varepsilon_{st}} \eta^2 \right) \qquad (13.50)$$

Spacer compression may be calculated with the Manuzio formula (Lilien *et al.* 2000):

$$P_{max} = 1.45 I'' \sqrt{F_{st} \log \left(\frac{a_S}{d_S} \right)} \qquad (13.51)$$

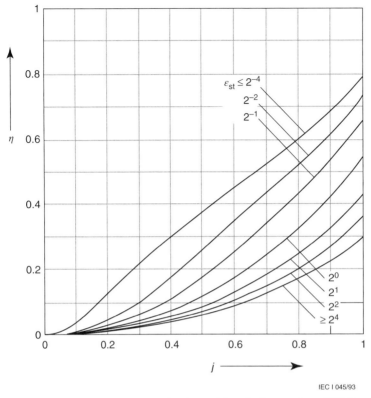

IEC I 045/93

Figure 13.13   Additional curves for determining the factor η from IEC Standard 60865, when $10.0 < a_s/d_s \leq 15.0$, Figure 11 (EPRI, 2006).

where

P_{max} = compression force on the spacer in N

F_{st} = initial static tension on the conductor bundle, in N

a_s = conductor spacing in mm

d_s = conductor diameter in mm.

Tests by Lillien *et al.*, showed that the Manuzio formula underestimated the stress by 50%. Better results (within $\pm 10\%$) were obtained using finite element analysis.

13.7   FORCE SAFETY DEVICES

When short-circuit forces increase, force safety devices (FSDs) can mitigate pinch-force effects. A FSD (Miroshnik, 2003) is a deformable mechanical link which can be placed in series with a flexible substation bus to limit damage due to

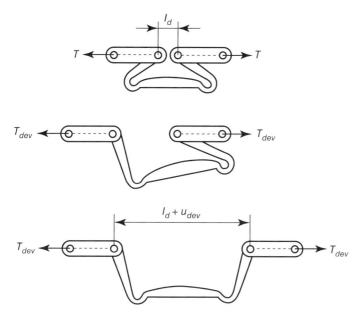

Figure 13.14   Operation of force safety device (EPRI, 2006).

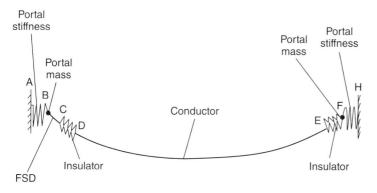

Figure 13.15   Connection of FSD to flexible substation bus structure (EPRI, 2006).

short-circuit forces. It is similar to a fuse, in that it is nonrecoverable, and must be replaced after a short circuit. The principle of operation (Figure 13.14) is that of a metallic cramp, having two weakened cross-sectional areas which are calibrated for an actuation force of P_{dev}. The FSD is connected between the support structure and the suspension insulator (Figure 13.15). When the total force F_t exceeds P_{dev}, the FSD will be deformed, limiting the force. A graph of force limitation is plotted in Figure 13.16.

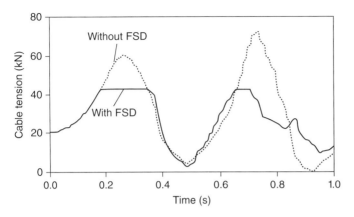

Figure 13.16   Limitation of bus tension by FSD (EPRI, 2006).

13.8   SUBSTATION CABLE AND CONDUCTOR SYSTEMS

There are many types of cables and conductors used in substations (IEEE, 2007c). These include the following:

1. High-voltage power cables, defined as >1000 V. These may connect to other substations, to substation equipment, or to customer loads.

2. Low-voltage power cables, defined as >1000 V. These supply auxiliary power to substation equipment.

3. Control cables. These include instrument transformer secondary cables.

4. Instrumentation cables. These are used primarily for SCADA systems.

5. Overhead secondary conductors. In distribution substations, these medium voltage open-wire lines are the termination points of distribution feeders.

13.8.1   Cable Thermal Limits

Cables are subject to thermal damage from prolonged exposure to short-circuit currents. Protective relay operating times and circuit-breaker clearing times must be fast enough to prevent prolonged overheating (IEEE, 1993, Section 5.6.2). Although the protection requirements of the National Electrical Code (NEC) (NFPA, 2014) do not apply in most substations, they should be considered when evaluating cable protection systems.

In addition to the NEC requirements, it is recommended that cable protection adhere to the $I^2 t$ limits of the cable damage curve for the insulation type as published by the cable manufacturers.

$$\left(\frac{I}{A}\right)^2 t = 0.0297 \log_{10} \frac{T_2 + T_k}{T_1 + T_k} \tag{13.52}$$

where

I = symmetrical short-circuit current in A

A = conductor cross-section in circular mils

T_1 = initial conductor temperature in °C, the maximum continuous conductor temperature for the insulation system is used, typically 75 or 90°C for low-voltage cables.

T_2 = final conductor temperature in °C, the short-circuit temperature limit of the insulation system is used, typically 250°C for low-voltage cables.

T_k = heating temperature constant for the conductor material, 234°C for Cu and 228°C for Al.

Similar limits are available for the sheaths of medium-voltage cables. They should be used for ground fault currents. In the case of increased fault current levels, protective relay settings should be changed, as necessary, to protect the cables. If this is not possible, resizing of the cable may be necessary.

13.8.2   Cable Mechanical Limits

When a short duration fault bends a cable, the mechanical effect is more significant than the thermal effect (Rüger, 1989). Permanent deformation may occur in plastic-insulated single-core cables. When cleats confine a cable such that short-circuit forces create outward bows with small bending radii, the cable may be damaged. Friction between the cleats and the cable may damage the outer sheath. Softening of the insulation by simultaneous heating further increases the damage caused by bending of the conductors. Proper support of cables can prevent this type of damage from occurring.

13.9   DISTRIBUTION LINE CONDUCTOR MOTION

When overhead distribution lines enter substations, the opportunity exists for the substation to be exposed to damage from distribution faults (Ward, 2003). A fault on the distribution line causes the overhead conductors to swing side-to-side closer to the substation. As a result, the conductors may move close enough to arc (0.1 m) or even touch. This causes a second fault, which may cause increased stress on the substation transformer and cause backup protective devices to operate. Ward prepared a computer program that calculates critical clearing-time curves for overhead distribution conductors, based on conductor motion. Critical clearing time increases as spans decrease. The following are possible solutions to the problem of damage caused by distribution conductor motion:

Using faster recloser time curves. This is the preferred solution, if it is possible.

Installing fiberglass spacers at mid span to shorten the effective span. This is fairly inexpensive.

Adding intermediate poles to shorten spans. This is expensive.

Increasing the phase spacing. This requires replacing cross arms, and is expensive.

Removing slack in the lines to reduce conductor motion. This is time consuming and expensive.

The first two options, faster reclosing times and fiberglass spacers at mid span, are the best alternatives if increased levels of fault current result in added stress to the substation due to overhead distribution line faults.

13.10 EFFECTS OF HIGH FAULT CURRENTS ON SUBSTATION INSULATORS

Brittle fracture of non-ceramic insulators (NCIs) have mostly occurred in polymer suspension insulators (Burnham *et al.*, 2002), however, the same problems could occur in post-type insulators. In terms of short-circuit stresses, exceeding the mechanical loading limits could result in cracks or splits in the rod or in damaged seals. Water intrusion leads to brittle fracture, through the leaching of acids in combination with tensile stress (de Tourreil *et al.*, 2000).

13.10.1 Station Post Insulators for Rigid Bus Bars

High fault current forces on rigid bus bars are transmitted to supporting insulators, which will be subject to forces that may exceed their design limits. The effects on the insulators could be cracks, fractures, or breakage. These, in turn, will weaken the support structure of the bus, resulting in greater damage should a second fault occur before the damage is repaired. The action of reclosers is of particular concern here.

The short-circuit force on a bus bar is transmitted to the insulator (IEEE, 2008, Section 12) through the bus-support fitting (Figures 13.17 and 13.18):

$$F_{SB} = L_E F_{SC} \qquad (13.53)$$

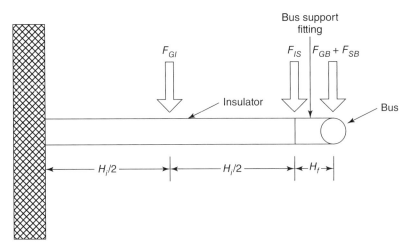

Figure 13.17   Insulator configuration for vertical bus (EPRI, 2006).

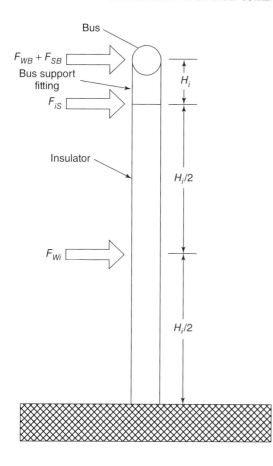

Figure 13.18   Insulator configuration for horizontal bus (EPRI, 2006).

where

F_{SB} = bus short-circuit force transmitted to the bus support fitting in N

L_E = effective length of the bus span in m.

Similarly, the gravitational forces are

$$F_{GB} = L_E F_G$$

where

F_{GB} = gravitational force transmitted to the bus support fitting in N

F_G = weight of the bus in N.

The cantilever force on the insulator is then

$$F_{IS} = K_V K_1 \left[\frac{F_{WI}}{2} + \frac{(H_i + H_f) F_{WB}}{H_i} \right]$$
$$+ K_H K_3 \left[\frac{F_{GII}}{2} + \frac{(H_i + H_f) F_{GB}}{H_i} \right]$$
$$+ K_2 \left[\frac{(H_i + H_f) F_{SB}}{H_i} \right] \tag{13.54}$$

TABLE 13.4   Insulator Force Multipliers

Force	Post Insulator	Chain Insulator
Tension	1.5	1.0
Drop	1.0	1.0
Pinch	1.0	1.0

where

K_1 = overload factor for wind forces, typically 2.5

K_2 = overload factor for fault current forces; this should also be 2.5, unless certain resonance criteria are met.

K_3 = overload factor for gravitational forces, typically 2.5

F_{WI} = wind force on the insulator in N

H_i = height of the insulator in cm

H_f = height of the bus centerline above the insulator in cm.

If the cantilever force is exceeded as prospective fault currents increase, two possible solutions are (i) to increase the number of insulators, thus decreasing L_E or (ii) to replace the insulators with units having greater cantilever strength. If insulator spacing is changed, the mechanical resonant frequencies will have to be recalculated, and a dynamic study may need to be performed. Experimental results (Barrett *et al.*, 2003) show that the IEEE method is conservative, and it is unlikely that increased fault currents will damage an IEEE-designed insulator structure.

13.10.2   Suspension Insulators for Flexible Conductor Buses

In accordance with (IEC, 2011a), the forces calculated shall be used with multipliers according to the type of insulator, as shown in Table 13.4.

13.11   EFFECTS OF HIGH FAULT CURRENTS ON GAS-INSULATED SUBSTATIONS (GIS)

Gas-insulated substations (GIS) are designed and tested in accordance to (IEEE, 2011), and have short-circuit ratings as listed in Table 13.5. These are for external faults, where the GIS is tested in the same manner as circuit breakers.

TABLE 13.5   GIS Short-Circuit Ratings

Short-Time Current-Carrying Capability (kA, rms) for a Specified Time of 1 or 3 s	
20	50
25	63
31.5	80
40	100

IEEE (2011).

TABLE 13.6   GIS Phase-to-Ground Burn-Through Times

Phase-to-Ground Burn-Through Times	
RMS Current (kA)	Time (s)
<40	0.2
≥40	0.1

IEEE (2011).

The internal arcing fault-withstand capability of GIS is based on the thickness of the metal walls and the gas pressure (IEEE, 2011), and is thus not easy to upgrade. The withstand times are listed in Table 13.6. IEC standards are similar in regard to short-circuit ratings as well as burn-through times. The time to puncture an aluminum plate is approximately (Boeck and Krüger, 1992)

$$t = C\frac{d^2}{I} \quad C = (60 \ldots 500) \ \mathrm{kA\,ms\,mm^{-2}} \tag{13.55}$$

where

t = time in ms

C = constant

d = thickness of the aluminum in mm

I = current in kA.

A rotating arc can puncture a GIS wall in two different ways (Boeck, 2003). If there is an oblique arc, which rotates, the burst will be similar to that shown in Figure 13.19. If an insulating barrier stops the moving arc, the vertical arc will

Figure 13.19   GIS enclosure punctured by a rotating arc (EPRI, 2006).

puncture a hole in a much shorter time. The internal design of GIS is intended to keep the arcs moving and to prevent them from sweeping over the same locations more than once.

A statistical analysis (Trinh, 1992) shows that "An increase in the mean fault current from 20 to 30 kA raises the risk of burn-through from 0.34 to 0.78, which illustrates the importance of designing the GIS in terms of the distribution of the local fault current." This is expressed as a probability formula:

$$R = \int_0^\infty p(ic) \int_0^\infty P(tc)p(tc)dtc\,dic \qquad (13.56)$$

where

R = risk of burn-through of a GIS having an envelope thickness d associated with a fault at a certain location on the transmission system

$p(ic)$ = probability density of the local fault current ic

$P(tc)$ = probability that the burn-through time will not exceed tc

$p(tc)$ = probability density of the fault-clearing times.

When a GIS unit is inspected and maintained, or replaced after a fault, very specific safety procedures should be followed (IEEE, 2011). Sulfur hexafluoride (SF_6) is nontoxic, but produces numerous toxic by-products during arcing and burning.

CHAPTER **14**

EFFECT OF HIGH FAULT CURRENTS ON TRANSMISSION LINES

14.1 INTRODUCTION

The purpose of this chapter is to provide information on the effects on transmission lines of the increased fault currents which often accompany increased power flows. These effects can include conductor motion and the breakage of conductors and insulators, all of which can be hazardous to personnel. The proper design of transmission lines and equipment will include ensuring that high fault currents do not cause hazards to operating personnel, line workers, and passersby. The standard reference works on transmission lines, such as the EPRI "Red Book" (EPRI, 1975, 1982, 2004c) and the *Westinghouse Transmission and Distribution Book* (Central Station Engineers, 1964), include very little data on short circuits and none on the effects of short circuit currents on transmission lines themselves. The exception to this is the *Compact Line Design Reference Book* (EPRI, 1978). To compile data for this chapter, technical papers and IEC standards relating to flexible conductor substation buses (covered elsewhere in this book) have been referred to. Some information has been found in an EPRI report on non-ceramic insulators (NCI) (EPRI, 1998). This chapter is adapted from EPRI (2006).

14.2 EFFECT OF HIGH FAULT CURRENT ON NON-CERAMIC INSULATORS (NCI)

Three main types of NCI are used on overhead transmission lines:

- Suspension insulators (Figure 14.1).
- Post insulators.
- Phase-to-phase insulators.

Principles of Electrical Safety, First Edition. Peter E. Sutherland.
© 2015 John Wiley & Sons, Inc. Published 2015 by John Wiley & Sons, Inc.

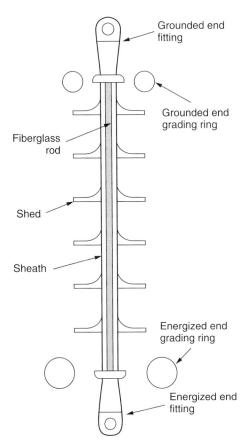

Figure 14.1   Cross section of a typical transmission non-ceramic insulator (*Source*: EPRI (2006)).

Each of the above may be applied in numerous ways:

- In general, suspension insulators are intended primarily to carry tension loads. Suspension insulators can be applied in I-string, Vee-string, and dead-end applications.

- Post insulators are intended to be loaded in tension, bending, or compression. The most common application is horizontal posts. These post insulators may be applied either alone or together with a suspension insulator in a braced post configuration. Post insulators are also applied in substations as bus support or in disconnect switch applications.

- Phase-to-phase insulators are intended to be loaded in tension, torsion, bending, or compression. Phase-to-phase insulators couple two phases together in order to control conductor spacing during galloping.

This chapter deals mainly with suspension insulators although many areas may be relevant to both post and phase-to-phase insulators.

Modern suspension and strain insulators generally consist of the following main elements:

- Energized metallic end fitting.
- Energized end grading ring (need depends on application).
- Fiberglass-reinforced plastic rod (FRP).
- Polymeric weather shed system, consisting of weather sheds and sheath.
- Grounded end grading ring (need depends on application).
- Grounded metallic end fitting.

Grading rings, also called *corona rings*, are not installed at all voltage levels or on all applications, and often only a single grading ring is installed at the energized end. The method of construction and materials used depends on the manufacturer and application.

Concern has been raised over end-fitting damage caused by fault currents flowing during a flashover. Although tests have indicated that the critical tensile strength of an NCI may be reduced to 80% of its specified mechanical load (SML) rating during a power arc test, the insulator's tensile strength recovers to a level above its SML after the test. Since NCI are usually applied at less than 50% of their SML, this is not a significant concern. However, during a power arc, the galvanization of the end fitting may be damaged, making the fitting susceptible to corrosion. The long-term effects of localized heating on the end fitting, FRP rod, and weather shed system due to power arcs still require further research.

If a flashover occurred across the insulator because of a transient event such as lightning, the rubber weather-shed system or FRP rod may or may not have sustained damage. If the weather-shed system or FRP rod has been damaged (such damage is usually obvious), then the insulator should be removed (see Figure 14.2). In some cases, the flashover may have caused no obvious damage to the insulator apart from

Figure 14.2   Flashover damage sustained by an NCI (*Source*: EPRI (2006)).

areas of degalvanization of the end-fitting hardware. This may be especially true if the power arc terminated on the grading rings. Insulators whose external appearance indicates they have only sustained this type of damage are still of concern. Owing to the nature of grading ring attachments and end-fitting design, power fault currents usually flow through the end fitting, which is in direct contact with the FRP and rubber weather-shed system. Whether the increased end-fitting temperatures associated with the power arc current result in long-term degradation of either the polymeric rubber material or the FRP rod is unknown. Hence, it is advisable to remove such insulators from service.

14.3 CONDUCTOR MOTION DUE TO FAULT CURRENTS

While normal transmission line construction, with widely separated phases, does not appear to be significantly impacted by conductor motion due to fault currents, it is an important consideration for compact transmission lines (EPRI, 1978).

Two parallel conductors, each carrying current, will be subject to a force of attraction or repulsion, depending on current direction. The magnitude of the force on each conductor is

$$F = \frac{\mu_0}{2\pi} \frac{I^2 \ell}{d} \tag{14.1}$$

where

$I =$ current in each conductor

$\ell =$ length of each conductor, and

$d =$ distance between conductors.

If the current flow in each conductor is in the same direction, the force will cause attraction; and, conversely, if the current is in opposite directions, the force will cause repulsion. For short-circuit currents, these forces may be sufficient to cause significant conductor movement, particularly where conductors are closely spaced (such as in Extra High Voltage (EHV) (EHV is generally considered to be 345 kV and above) conductor bundles or in adjacent phases of a compact line), because of the inverse relationship of conductor spacing and the resultant force. The actual movement of a conductor, considering inertia, is a function of both the magnitude of the current and the time it is applied, and is therefore dependent on circuit-breaker interrupting time.

A phase-to-phase fault will cause current in the two affected phases to flow in opposite directions. The two conductors will then be repelled and, on interruption of the fault current, will swing together.

If the current is due to a fault on the line section under consideration, the electrical consequences of conductor motion (even clashing) are generally unimportant as that section will be tripped out anyway. However, if the fault is on an adjacent line section, the motion may be serious as it might cause interruption of the unfaulted section. "Through-fault" currents, or those supplied through a nonfaulted line to a fault elsewhere on the system, can be an important design consideration for compact circuits. Even though such currents on the compact line may be less than the maximum fault currents attainable on the system, they may be sufficient to be determining in the selection of phase-to-phase spacing or in establishing the need for insulating spacers.

The motion of conductors subjected to electromagnetic forces is similar to that of weighted, stretched strings, with the complication that the string is usually a compound conductor, such as an aluminum conductor, steel reinforced (ACSR). Relatively simple analyses of conductor motion of both vertically and horizontally spaced conductors can be shown to give results well within line design accuracy requirements.

14.4 CALCULATION OF FAULT CURRENT MOTION FOR HORIZONTALLY SPACED CONDUCTORS

Figure 14.3 illustrates the conductor configuration used as a basis for calculations. It is assumed that the forces to which each catenary span of the conductor is subjected will cause the span to swing in a plane, as shown in Figure 14.3(a). The plan projection of each catenary is again a catenary (Figure 14.3(b) and (c)). This assumption, supported by experimental results, simplifies the calculation technique. The most severe fault is phase to phase on adjacent phases, which impresses a cyclic separating electromagnetic force.

Since all spans of a line contributing to a through-fault will behave similarly, the net pole-top force along the span, and therefore motion, will be zero. Consequently, each span can be assumed to be rigidly terminated.

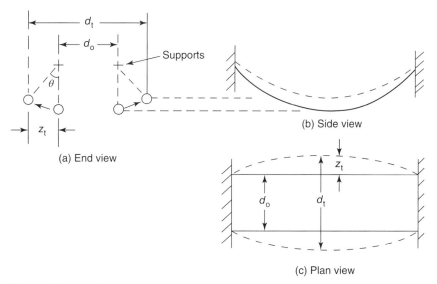

Figure 14.3   Horizontal conductor motion during through-fault. (a) end view; (b) side view; and (c) plan view (*Source*: EPRI (2006)).

14.5 EFFECT OF CONDUCTOR SHAPE

The true shape of each conductor is a catenary, as shown in Figure 14.4. For a catenary,

$$y = k\left[\cosh\left(x/k\right) - 1\right] \qquad (14.2)$$

At $x = S/2$, where S is the span length (i.e., at the conductor support), let $y = y_0$, where y_0 is the conductor sag, to define k.

$$y_0 = k\left[\cosh\left(\frac{S}{2k}\right) - 1\right]$$
$$= \frac{S^2}{8k} \qquad (14.3)$$

where

$$k = \frac{S^2}{8y_0} \qquad (14.4)$$

and

$$y_x = \frac{S^2}{8y_0}\left[\cosh\left(\frac{8y_0 x}{S^2}\right) - 1\right] \qquad (14.5)$$

$$y_{\text{avg}} = \frac{S}{4y_0}\left[\frac{S^2}{8y_0}\sinh\left(\frac{4y_0}{S}\right) - \frac{S}{2}\right] \qquad (14.6)$$

By substitution in the above equation, it can be shown that for transmission lines,

$$y_{\text{avg}} = \frac{2}{3}y_0 \qquad (14.7)$$

is an almost exact solution

Since the force between two parallel conductors is given by equation (14.1), and

$$d_t = d_0 + 2y_t \qquad (14.8)$$

where

d_t = conductor spacing at time t

d_0 = initial conductor spacing

y_t = conductor sag at time t

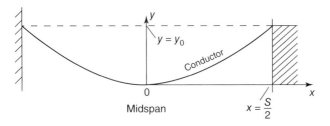

Figure 14.4   Conductor geometry (*Source*: EPRI (2006)).

then the average force on one conductor is

$$F_{avg} = \frac{\mu_0}{2\pi} \frac{I^2 \ell}{d_0 + \frac{2}{3} y_0}$$ (14.9)

The conductor mass is modeled as a pendulum, which swings at a distance $\frac{2}{3} y_0$ below the support points.

14.6   CONDUCTOR EQUATIONS OF MOTION

From the previous equations, the forces on a conductor are represented in Figure 14.5, where

F = electromagnetic force of repulsion

W = conductor weight

P_0 = actual conductor position at midspan when at rest

P_t = actual conductor position at midspan, time t

Q = effective position of mass and forces

F_c = normal component of conductor tension.

Resolving tangentially, the force accelerating conductor swing is

$$F_{\tan g} = F \cos \theta - W \sin \theta$$ (14.10)

The conductor acceleration point Q is then

$$Q_{\tan g} = F_{\tan g} \left(\frac{g}{W} \right)$$ (14.11)

where g is the gravitational constant.

Using a step-by-step analysis, the conductor velocity

$$v_{\tan g(t)} = v_{(t-\delta t)} + \frac{a_{(t)} - a_{(t-\delta t)}}{2} dt$$ (14.12)

then

$$\theta_{(t)} = \theta_{(t-\delta t)} + \frac{v_{\tan g(t)} + v_{\tan g(t-\delta t)}}{2r} dt$$ (14.13)

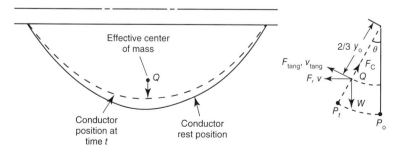

Figure 14.5   Forces on a conductor (*Source*: EPRI (2006)).

where

$$r = \frac{2}{3}y_0 \tag{14.14}$$

and the true horizontal midspan displacement of point P is then

$$z_t = y_0 \sin \theta \tag{14.15}$$

14.7 EFFECT OF CONDUCTOR STRETCH

As the conductor deflects under load, the effective weight per unit length changes. Resolving perpendicular to the conductor in the conductor plane, using the terminology of Figure 14.5,

$$F_c = W \cos \theta + F \sin \theta \tag{14.16}$$

If $\theta = 0$ (i.e., the conductor is in vertical rest position),

$$F_c = W \tag{14.17}$$

and

$$sag = y_0 \cong \frac{WS^2}{8H} \tag{14.18}$$

where H is the conductor tension.
If $\theta \neq 0$, then

$$
\begin{aligned}
y_0 &= \frac{F_c S^2}{8H} \\
&= \frac{W \cos \theta + F \sin \theta}{8H} S^2
\end{aligned}
\tag{14.19}
$$

14.8 CALCULATION OF FAULT CURRENT MOTION FOR VERTICALLY SPACED CONDUCTORS

As a conductor span moves upward owing to fault current forces, the tension is reduced and the acceleration is restrained by the increase in the effective conductor weight. Conversely, as a conductor moves downward owing to these forces, the acceleration is inhibited by an increase in conductor tension. Because of these effects, it is important that the modulus of elasticity be considered in calculations.

Figure 14.6 illustrates a typical vertical conductor arrangement. As a simplifying approximation, it is assumed that the forces to which each conductor is subjected will cause an increase or reduction in sag, but that the conductor will retain a catenary shape. This assumption is supported by experimental results for low currents applied for long durations. The assumption is even more accurate for high fault current levels and short durations, where most of the kinetic energy is imparted to the conductor before the conductor can move appreciably.

The terminology used in analyzing the vertical case is the same as for the horizontal case. The configuration used as a basis for calculations is illustrated in Figure 14.7.

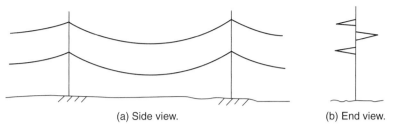

(a) Side view.   (b) End view.

Figure 14.6   Typical vertical conductor arrangement. (a) Side view and (b) end view (*Source*: EPRI (2006)).

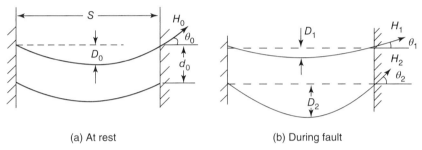

(a) At rest   (b) During fault

Figure 14.7   Vertical displacement during fault. (a) At rest and (b) during fault (*Source*: EPRI (2006)).

14.9   CALCULATION PROCEDURE

From the rest position, $D_0 = D_1 = D_2$, that is, all sags are equal. For any other position, assuming both conductors are a catenary, the average separation distance can be expressed:

$$d_{avg} = D_0 + \frac{2}{3}(D_2 - D_1) \tag{14.20}$$

so that the average electromagnetic force is

$$F = \frac{\mu_0}{2\pi} \frac{I^2 \ell}{d_{avg}} \tag{14.21}$$

Note that the electromagnetic force has the effect of changing the effective conductor weight. The net accelerating force on each span of the bottom conductor is:

$$F_{net} = 2H_2 \sin \theta_2 - S(W + F) \tag{14.22}$$

and for the top conductor,

$$F_{net} = 2H_1 \sin \theta_1 - S(W - F) \tag{14.23}$$

Using $F_{net} = ma$, conductor motion can be expressed as a function of time. The rates of acceleration and the period of oscillation of each conductor need not be the same.

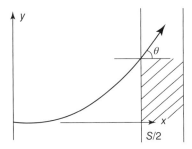

Figure 14.8 Conductor angle at support (*Source*: EPRI (2006)).

Figure 14.8 illustrates the basis on which the angle of the conductor at the support is calculated.

For the conductor catenary,

$$y = \frac{H}{W}\left[\cosh\left(\frac{Wx}{H}\right) + 1\right] \tag{14.24}$$

that is,

$$\frac{\partial y}{\partial x} = \sinh\frac{Wx}{H} \tag{14.25}$$

At $x = S/2$ (i.e., at the conductor support),

$$\frac{\partial y}{\partial x} = \sinh\frac{WS}{2H} = \sinh\frac{4D}{S} \tag{14.26}$$

14.10 CALCULATION OF TENSION CHANGE WITH MOTION

If an initial conductor length L_1 and an initial conductor tension T_1 are assumed, then for any subsequent motion resulting in L_2 and T_2,

$$L_2 - L_1 = \frac{H_1 - H_2}{aE}L \tag{14.27}$$

or

$$\delta H = \frac{\delta L}{L}aE \tag{14.28}$$

where the conductor length, L, can be approximated as

$$L = S + \frac{8D^2}{3S} \tag{14.29}$$

δL can be expressed as

$$\delta L = \frac{8D_2^2}{3S} - \frac{8D_1^2}{3S} = \frac{8}{3S}(\delta D - 2D_1) \tag{14.30}$$

where

$$\delta D = D_1 - D_2 \tag{14.31}$$

The change in tension can be expressed in terms of the change of vertical displacement as follows:

$$\delta H = \frac{8aE}{3SL_1} \delta D(\delta D - 2D_1) \tag{14.32}$$

14.11   CALCULATION OF MECHANICAL LOADING ON PHASE-TO-PHASE SPACERS

The electromagnetic forces during a phase-to-phase fault will act to move the conductors apart, placing phase-to-phase spacers in tension. After the fault is cleared, the conductors will swing together, compressing the spacers. These forces can be analyzed using the diagram of Figure 14.9. Using the previously defined terminology, for any subspan swing angle, θ,

$$F_c = F \sin \theta + W \cos \theta \tag{14.33}$$

$$\begin{aligned} F_{spacer} &= 2F_c \cos \theta \\ &= 2(F \sin \theta + W \cos \theta) \sin \theta \end{aligned} \tag{14.34}$$

and

$$\sin \theta = \frac{z_t}{D} \tag{14.35}$$

$$\cos \theta = \sqrt{1 - \left(\frac{z_t}{D}\right)^2} \tag{14.36}$$

In the simple case where F_{out} reduces to zero before maximum swing is reached (i.e., the fault clears),

$$F_{spacer} = 2SW \frac{z_t}{D} \sqrt{1 - \left(\frac{z_t}{D}\right)^2} \tag{14.37}$$

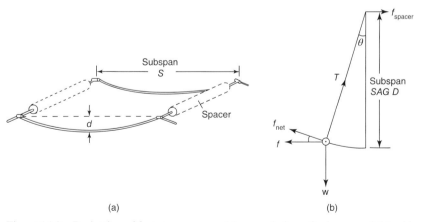

(a)   (b)

Figure 14.9   Derivation of forces on spacers. (a) general view of subspan and (b) midspan cross section of subspan (*Source*: EPRI (2006)).

Since

$$D = \frac{WS^2}{8H} \tag{14.38}$$

the maximum spacer force is

$$F_{\text{spacer}} = 16H\frac{z_t}{S}\sqrt{1 - \left(\frac{z_t}{D}\right)^2} \cong \frac{16Hz_t}{S} \tag{14.39}$$

This will be the maximum spacer force in both tension and compression for the usual case where the fault has cleared before maximum conductor deflection has occurred.

14.12   EFFECT OF BUNDLE PINCH ON CONDUCTORS AND SPACERS

Transmission line spacers (Figure 14.10) are designed to withstand the compressive force on bundled conductors caused by short-circuit forces. Spacer compression may be calculated with the Manuzio formula (Lilien *et al.*, 2000), originally for flexible bus substation design:

$$P_{\max} = 1.45I''\sqrt{F_{st} \log\left(\frac{a_S}{d_S}\right)} \tag{14.40}$$

where

$P_{\max}$ = compression force on the spacer in N

F_{st} = initial static tension on the conductor bundle in N

a_s = conductor spacing in mm

d_s = conductor diameter in mm.

Tests by Lillien *et al.* showed that the Manuzio formula underestimated the stress by 50%. Better results (within $\pm10\%$) were obtained using finite element analysis. These results were extended (Lilien and Papailiou, 2000) to transmission line design. The Manuzio approach neglects

1. the pinch effect, which results in an increase in tension of the subconductors during the short circuit;

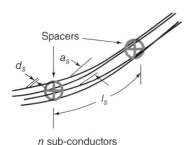

Figure 14.10   Details of transmission line conductor bundle with spacers (*Source*: EPRI (2006)).

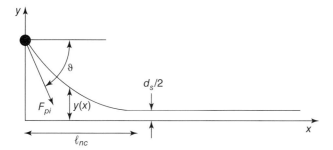

Figure 14.11   Parabolic model of subconductor pinch forces (*Source*: EPRI (2006)).

2. the asymmetry of the fault current;

3. the length of subspan between spacers, ℓ_S.

All of these causes result in higher compression forces on spacers during faults. Lilien and Papaliou stress that the Manuzio formula should no longer be used, especially as fault current levels are increasing in transmission systems. After an extensive series of tests and computer simulations, they recommend a new calculation method based on the IEC 60865-1 approach (IEC, 1994, 2011a) The subconductor (Figure 14.11) is assumed to take a parabolic shape between the spacer and the point where all the subconductors touch:

$$y(x) = \frac{a_s - d_s}{2} \cdot \left(\frac{x}{\ell_{nc}}\right)^2 - (a_s - d_s) \cdot \left(\frac{x}{\ell_{nc}}\right) + \frac{a_s}{2} \qquad (14.41)$$

where ℓ_{nc} is the noncontact length, which must be calculated.

Simultaneous numerical solution of the following two equations is required:

$$F_c = \frac{2F_{pi}}{\sqrt{1 + (\ell_{nc}/(a_s - d_s))^2}} \qquad (14.42)$$

and

$$F_c = \frac{\mu_0}{2\pi} I^2 \int_0^{\ell_{nc}} \frac{\cos(\vartheta(x))}{2y(x)} dx \qquad (14.43)$$

where

$$\cos(\vartheta) = \frac{1}{\sqrt{1 + (dy/dx)^2}} \qquad (14.44)$$

and

ϑ = deviation from the horizontal (Figure 14.11).

I = time-average short circuit current (CIGRE, 1996).

Solution of these equations results in values for F_C, the compression force on the spacer and the ℓ_{nc} length of subconductor before the pinch occurs. Finite element method simulations can also be used for this problem (Kruse and Pearce, 2000).

CHAPTER *15*

LIGHTNING AND SURGE PROTECTION

15.1   SURGE VOLTAGE SOURCES AND WAVESHAPES

Surge voltages are a significant hazard in power transmission and distribution systems (IEEE, 1993; Greenwood, 1991). Surges are the causes of much equipment failure, such as downed conductors, fires, and step and touch potentials, which may injure people nearby. The design manufacture and installation of electrical power equipment and systems must be fundamentally based on safety both for users of the equipment and for those who operate and maintain it. As with other electrical hazards, there are extensive sets of industry standards, governmental regulations, and electrical and building codes which must be followed. The first hazard of high voltages, as we have seen, is the threat of shock, which with high voltage transients may jump across the air to a person standing near an energized conductor. Even if a person is touching a conductor, the high voltage causes a higher current to flow through them than would otherwise happen. However, the short duration and high frequency of transients may mitigate these effects to some extent. High transient voltages can cause the failure of insulation, the failure of fuses and circuit breakers to interrupt faults, the destruction of load equipment, and resulting explosions, fires, and other hazards.

Surges may be caused by a variety of sources:

Lightning strokes may hit a variety of locations within the power system. Shielding failures may cause a direct stroke to hit a phase conductor. When a lightning stroke hits an overhead ground wire, it may cause the voltage on the tower to exceed the voltage on the conductors, causing a back flashover across the insulator string. When there is a lightning strike to the earth near the tower, an induced flashover can result.

Power equipment itself may cause transient overvoltages, even during normal operation. Forced interruption of current during switching (current chop) can cause significant overvoltages in some cases (Sutherland, 2012). Switching of thyristors, vacuum switches, and fast circuit breakers can cause energization transients. When the current changes rapidly, Ldi/dt causes voltage surges due to sudden change of current. An example of this is the blowing of current-limiting fuses (CLFs). Restriking is arcing across the contacts

Principles of Electrical Safety, First Edition. Peter E. Sutherland.
© 2015 John Wiley & Sons, Inc. Published 2015 by John Wiley & Sons, Inc.

during fault interruption by circuit-switching devices. This causes voltage increases similar to that due to arcing ground faults on an ungrounded LV system. Another cause of overvoltage transients is sudden contact between MV and LV system conductors.

While voltage transients have very fast rise times when generated, they may be considerably slowed by passing through power system conductors and equipment. Rise times may be limited by the surge passing through isolation transformers, passing through a system with many stub taps, surge capacitor protection, and snubber circuits.

The surge voltage waveshapes are defined in standard test procedures (IEEE, 2012b). The equation for the surge voltage is called the *impulse function* (Dommel, 1987):

$$f(t) = k(e^{-\alpha_1 t} - e^{-\alpha_2 t}) \tag{15.1}$$

The full-wave impulse test, shown in Figure 15.1 has a rise time of $1.2\,\mu s$ and decays to 50% of the peak value in $50\,\mu s$. This is called the *1.2/50 impulse wave*. Using (15.1), $\alpha_1 = 1.47 \times 10^4\ 1/s$, while $\alpha_2 = 2.47 \times 10^6\ 1/s$. The peak or crest voltage used in the test is the basic lightning impulse level (BIL) of the system. This approximates the shape of the surge from lightning.

The chopped-wave impulse test (Figure 15.2) has a 1.2 μs rise time, and the voltage is chopped after suitable time. The crest voltage is set to $110-115\%$ of the system BIL.

The switching surge test uses a slow front waveform with one of the following characteristics:

1. Time to crest $30-60\,\mu s$
2. Time to crest $150-300\,\mu s$
3. Time to crest $1000-2000\,\mu s$.

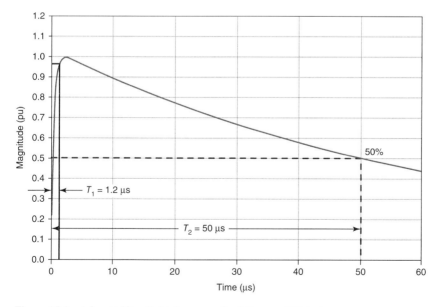

Figure 15.1 $1.2\,\mu s \times 50\,\mu s$ lightning surge for full-wave BIL test.

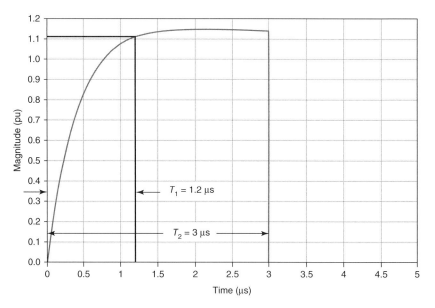

Figure 15.2   1.2 µs waveform for chopped-wave test.

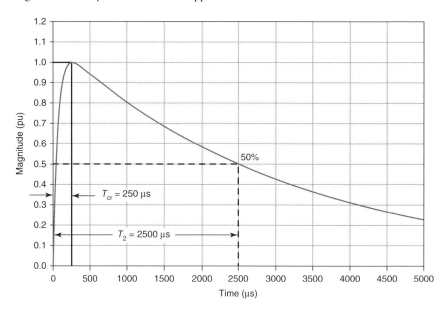

Figure 15.3   250 µs × 2500 µs switching surge for BSL test.

For a 250/2500 impulse, an example of which is shown in Figure 15.3, $\alpha_1 = 3.17 \times 10^2$ 1/s and $\alpha_2 = 1.60 \times 10^4$ 1/s in (15.1). The switching surge caused by sudden interruption of fault current by a CLF) develops when the available (prospective) fault current is reduced to less than its normal value by the melting of the fusible element. In a CLF, the links are usually silver, buried in fine quartz sand inside

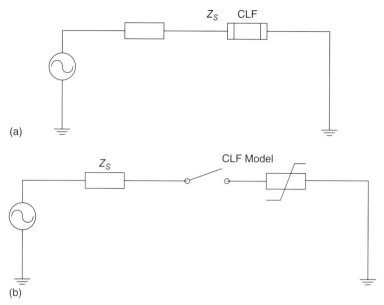

Figure 15.4   Modeling of current limiting fuse. (a) One-line diagram of current-limiting fuse (CLF); (b) simplified simulation model of CLF using switch and nonlinear resistance.

the fuse's cylindrical casing. The melting fuse link also melts the sand, forming a glass-like insulating barrier to form, rapidly interrupting the arcing fault current. The reduced current is called the *let-through* current. The *di/dt* caused a rapid increase in voltage producing the transient voltage spike. The CLF can be modeled as one or more nonlinear resistances (Petit, St-Jean, and Fecteau, 1989). A simplified version of this model, using a time-varying resistance, is shown in Figure 15.4. Figures 15.5 and 15.6 are plots of the simulation results, showing the voltage spike produced by the current-limiting action of the fuse.

Opening of circuit breakers on capacitors may produce overvoltages resulting in restrike. The restrike is caused by the voltage across the opening contacts to exceed the breakdown voltage of the gap. The restrike phenomenon can be modeled using the concept that an arc may be considered as a resistance. Each restrike is considered as a switch rapidly closing and then opening on a resistance. This model is shown in Figure 15.7, and the simulation results in Figure 15.8. Once the contacts open, the capacitor holds the voltage on the capacitor side constant, while the voltage on the line side continues on its sinusoidal path. If the circuit breaker contacts have not opened sufficiently as the peak voltage approaches, a restrike may occur. This is the "first restrike" in the figure. This results in a high frequency, high voltage transient oscillation and momentary interruption of the current. The voltage on the capacitor side is now held at the opposite peak voltage, and the scenario repeats on the next half-cycle. The resulting transient can reach three times the peak voltage.

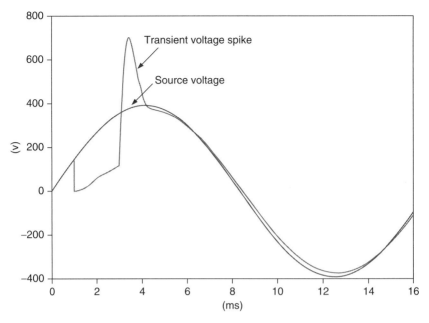

Figure 15.5   Overvoltage caused by CLF opening.

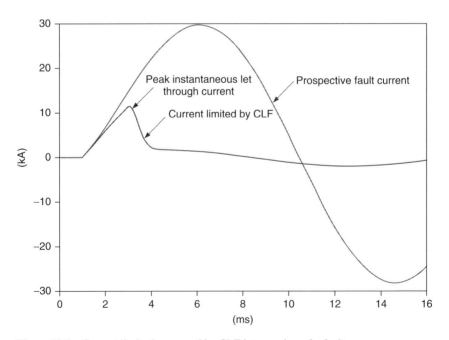

Figure 15.6   Current limitation caused by CLF interruption of a fault.

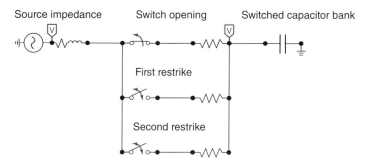

Figure 15.7   Equivalent circuit for capacitor switching restrike phenomena.

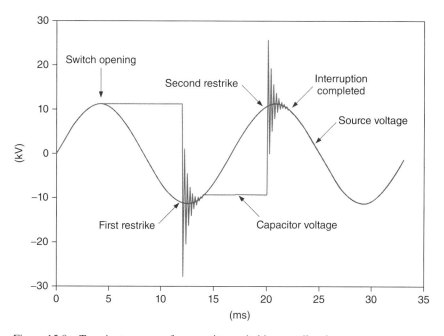

Figure 15.8   Transient response for capacitor switching restrike phenomena.

15.2   SURGE PROPAGATION, REFRACTION, AND REFLECTION

Surges are high frequency waves which travel on a system designed for low frequency waves. The surge waveform may have only one peak. A transmission line carries the wave by the constant interchange of energy between the distributed capacitance and inductance. This can be modeled as a lumped line of discrete inductors ΔL and capacitors ΔC (Figure 15.9). The velocity of the wave (Figure 15.10) is

$$v = \frac{1}{\sqrt{LC}} \tag{15.2}$$

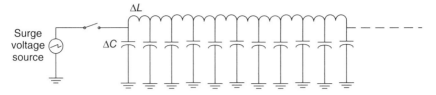

Figure 15.9   Cable conductor distributed constant equivalent circuit.

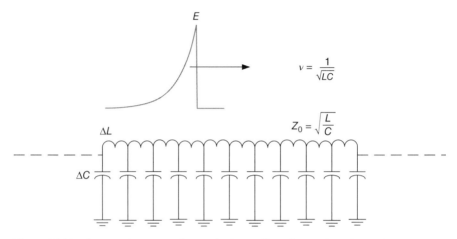

$$v = \frac{1}{\sqrt{LC}}$$

$$Z_0 = \sqrt{\frac{L}{C}}$$

Figure 15.10   Surge-voltage wave in transit along a line of surge impedance Z_0.

where

v = velocity in m/s

L = inductance in II/m

C = capacitance in F/m.

The ratio of voltage to current is called the *surge impedance*, Z_0, which is entirely resistive, and is also determined by L and C:

$$Z_0 = \sqrt{\frac{L}{C}} \qquad (15.3)$$

Typical surge impedance values are 400 Ω for an open-wire overhead line and 40 Ω for a solid dielectric metal-sheathed cable. Typical surge velocities are 300 m/μs for an open-wire overhead line and 150 m/μs for a solid dielectric metal-sheathed cable.

When a surge meets a junction between two circuits having different surge impedances (Z_{0-1} and Z_{0-2}), a reflection and refraction of the incident wave occurs. This is similar to the reflection and refraction of light passing from one medium to another. The magnitude of the reflected wave may be shown to be

$$E_R = E \times \frac{(Z_{0-2} - Z_{0-1})}{(Z_{0-2} + Z_{0-1})} \qquad (15.4)$$

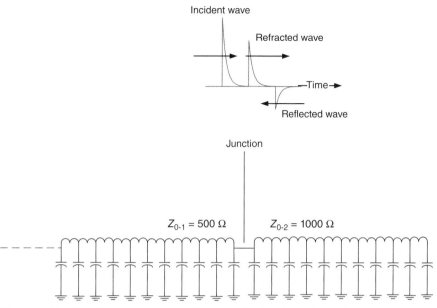

Figure 15.11   Traveling current wave reflection and refraction at low to high impedance junction.

The magnitude of the refracted wave may be shown to be

$$E_F = E \times \frac{(2 \times Z_{0\text{-}2})}{(Z_{0\text{-}2} + Z_{0\text{-}1})} \qquad (15.5)$$

Thus, given a positive incident wave, the reflected wave may have positive or negative polarities while the refracted wave is always positive. If the reflected wave is positive, because the wave is in transition from low to high impedance, it will add to the incident wave, thereby increasing the total transient voltage seen on the incident side of the junction (Figure 15.11). If the reflected wave is negative, because the wave is in transition from high to low impedance, it will subtract from the incident wave, thereby decreasing the total transient voltage seen on the incident side of the junction (Figure 15.12).

On encountering a junction terminated with a higher surge impedance, a surge voltage wave traveling along a transmission line will increase in voltage to as much as double if the junction is terminated in an open circuit. The terminal voltage will not exceed two times the traveling-wave value under all possible conditions. Even with a surge arrester placed ahead of the junction at a finite distance, the resultant junction voltage rise will depend on

1. the steepness of the surge voltage wave;
2. the propagation velocity along the line;
3. the distance of the line extension ΔD;
4. the magnitude of surge impedance connected to the junction.

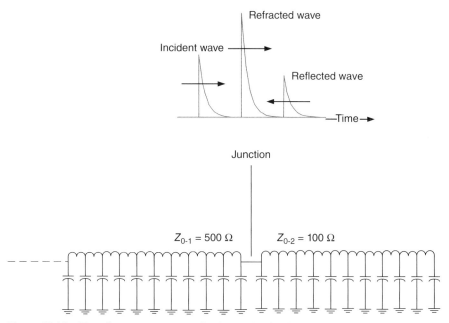

Figure 15.12   Traveling current wave reflection and refraction at high to low impedance junction.

The rise in terminal voltage will be aggravated by

1. the steeper front wave;

2. greater ΔD;

3. slower line propagation velocity;

4. greater magnitude of the terminating surge impedance.

15.3   INSULATION WITHSTAND CHARACTERISTICS AND PROTECTION

The high potential, low frequency test lasts for 1 min, and is an elevated voltage, 60 Hz insulation test. For transformer testing it is a 1 min, elevated voltage, constant V−Hz test. The full-wave impulse test crest voltage establishes equipment BIL using the $1.2 \times 50\,\mu s$ wave. This test is most stressful on line-to-ground insulation. The chopped-wave impulse test crest voltage is 10−15% higher than the BIL and follows a positive wave, steep negative gradient. The chopped-wave test is most stressful on turn-to-turn insulation. The front of wave impulse test crest voltage is higher than for the chopped-wave test. The impulse is chopped on the rising front of the wave before it reaches the crest voltage. The switching surge impulse test crest voltage is lower than the BIL. The test rise time is slower than for the impulse test; however, switching surges can be fast front waves. The BILs for common electric power equipment, according to ANSI standards, are shown in Tables 15.1–15.5. It can be seen that dry-type transformers typically have lower BIL than liquid-filled ones, which has

TABLE 15.1   Basic Impulse Insulation Levels (BILs) of Power Circuit Breakers, Switchgear Assemblies, and Metal-Enclosed Buses

Voltage Rating (kV)	Basic Impulse Insulation Level (BIL) (kV)
2.4	45
4.16	60
7.2	75 (95[a])
13.8	95
14.4	110
23	150
34.5	200
46	250
69	350
92	450
115	550
138	650
161	750
230	900
345	1300

[a]For power circuit breakers in metal clad switchgear.

Source: IEEE (1993)

TABLE 15.2   Impulse Test Levels for Liquid-Immersed Power Transformers

Rated Insulation Class, rms	Winding					Bushing		
	High Potential Test Value, rms	Chopped-Wave Test Minimum Time to Flashover, Crest		BIL Full Wave, $1.2 \times 50\,\mu s$ Crest	Switching Surge Level, Crest	60 Hz, 1 m Dry, rms	60 Hz, 1 m Wet, rms	BIL Full Wave, $1.2 \times 50\,\mu s$ Crest
(kV)	(kV)	(kV)	(μs)	(kV)	(kV)	(kV)	(kV)	(kV)
1.2	10	54	1.5	45	20	15	13	45
2.5	15	69	1.5	60	35	21	20	60
5.0	19	88	1.6	75	38	27	24	75
8.7	26	110	1.8	95	55	35	30	95
15.0	34	130	2.0	110	75	50	45	110
25.0	50	175	3.0	150	100	70	70	150
34.5	70	230	3.0	200	140	95	95	200
46.0	95	290	3.0	250	190	120	120	250
69.0	140	400	3.0	350	280	175	175	350
92.0	185	520	3.0	450	375	225	190	450
115.0	230	630	3.0	550	460	280	230	550
161.0	325	865	3.0	750	620	385	315	750

Source: IEEE (1993)

TABLE 15.3 Impulse Test Levels for Liquid-Immersed Distribution Transformers

Rated Insulation Class, rms	Winding					Bushing		
	High Potential Test Value, rms	Chopped-Wave Test Minimum Time To Flashover, Crest		BIL Full Wave, $1.2 \times 50\,\mu s$ Crest	Switching Surge Level, Crest	60 Hz, 1 m Dry, rms	60 Hz, 1 m Wet, rms	BIL Full Wave, $1.2 \times 50\,\mu s$ Crest
(kV)	(kV)	(kV)	(µs)	(kV)	(kV)	(kV)	(kV)	(kV)
1.2	10	36	1.0	30	20	10	6	30
2.5	15	54	1.25	45	35	15	13	45
5.0	19	69	1.5	60	38	21	20	60
8.7	26	88	1.6	75	55	27	24	75
15.0	34	110	1.8	95	75	35	30	95

Source: IEEE (1993)

TABLE 15.4 Impulse Test Levels for Dry-Type Transformers, Delta or Ungrounded Wye

Nominal Winding Voltage (kV)	High Potential Test, rms (kV)	Standard Basic Impulse Insulation Level (BIL), Crest (kV)
0.12–1.2	4	10
2.52	10	20
4.16–7.2	12	30
8.32	19	45
12.0–13.8	31	60
18.0	34	95
23.0	37	110
27.6	40	125
34.5	50	150

Source: IEEE (1993)

TABLE 15.5 Impulse Test Levels for Dry-Type Transformers, Grounded Wye

Nominal Winding Voltage (kV)	High Potential Test, rms (kV)	Standard Basic Impulse Insulation Level (BIL), Crest (kV)
1.2Y/0.693	4	10
24.36Y/2.52	10	20
8.72Y/5.04	10	30
13.8Y/7.97	10	60
22.86Y/13.2	10	95
29.95Y/14.4	10	110
34.5Y/14.4	10	125

Source: IEEE (1993)

TABLE 15.6   Rotating Machine 60 Hz, 1 min High Potential Test Voltages, Phase to Ground

Rated Voltage (kV)	Rated Voltage Crest Value (kV)	High Potential Test Crest Value (kV)	High Potential Test Crest Value (pu of Normal Crest)
0.46	0.38	2.71	7.21
2.3	1.88	7.92	4.22
4.0	3.27	12.73	3.9
4.6	3.76	14.43	3.84
6.6	5.39	20.1	3.73
13.2	10.78	38.8	3.6

Source: IEEE (1993)

TABLE 15.7   Rotating-Machine 60 Hz Winding Impulse Voltages, Phase to Ground

Rated Voltage (kV)	Rated Voltage Crest Value (kV)	Impulse Voltage Crest Value (kV)	Impulse Voltage Crest Value (pu of Normal Crest)
0.46	0.38	3.39	9.01
2.3	1.88	9.9	5.27
4.0	3.27	15.91	4.87
4.6	3.76	18.0	4.8
6.6	5.39	25.1	4.66
13.2	10.78	48.5	4.5

Source: IEEE (1993)

been the cause of numerous failures of dry-type transformers with inadequate surge protection. Note that rotating machines do not have BILs, but have an impulse-level time characteristic shown in Figure 15.10. When surge capacitors are used for motor protection (Sutherland, 1997), the steep wavefront of a surge is slowed as shown in curve V_C. The high potential test and impulse voltage requirements are shown in Tables 15.6 and 15.7.

A thorough knowledge of the insulation system withstand capabilities and endurance qualities is required. The likely sources of overvoltage exposure and character, magnitude, duration, and repetition rate that is likely to be impressed on the apparatus and circuits need to be identified. Appropriate application of surge protection devices is the most effective tool for achieving the desired insulation security. The problem of protection is complicated by the fact that insulation failure results from excessive magnitude of impressed voltage and from the aggregate sum total duration of such overvoltages.

15.4   SURGE ARRESTER CHARACTERISTICS

All arresters are based on using a nonlinear voltage/current characteristic to reduce overvoltages, as shown in Figure 15.11. Surge arrester design has evolved over the years from gapped, pellet, and expulsion type arresters. The silicon carbide (SiC) valve-type arrester was the most widely used type of arrester for many decades in the twentieth century. The valve element consists of nonlinear resistance. The parallel gap element provides isolation from normal line-to-ground voltage and reseals after-surge discharge to prevent flow of follow current. Application difficulty for valve-type arresters stemmed from variable arc characteristics with gapped designs.

All modern arresters are gapless metal oxide (ZnO) type. The ZnO valve element is sufficiently nonlinear to eliminate the need for gap elements. The nonlinear element provides a more constant voltage throughout the range of application. ZnO arresters initiate and complete their surge protective cycle more ideally than do gapped valve-type arresters.

On the basis of their durability, surge arresters come in three classes for a range of applications. These are

1. Distribution class
2. Intermediate class
3. Station class.

Distribution-class arresters are the least expensive, least durable, and have the lowest repetitive duty capability. They are capable of discharging a 75 A × 1000 μs rectangular current wave.

Intermediate-class arresters are of medium expense and durability, having a medium repetitive duty capability. They are capable of discharging up to 161 km (100 miles) of transmission line.

Station-class arresters have the greatest expense and durability, with the maximum repetitive duty capability of all classes. They are capable of discharging 240–320 km (150–200 miles) of transmission line. Some arresters have discharge capabilities in excess of minimum requirements. Manufacturer data should be consulted when selecting arresters for particular applications.

15.5 SURGE ARRESTER APPLICATION

Selection of arrester rating is by a number of published characteristics (IEEE, 2012b). The maximum continuous overvoltage (MCOV) should be greater than or equal to the maximum line-to-neutral voltage. The temporary overvoltage (TOV) rating should be greater than or equal to the maximum line-to-ground voltage during a single-line to ground fault (SLGF). The line-to-ground voltage can be calculated as the line-to-line voltage multiplied by the coefficient of grounding.

The TOV is reduced by previous energy.

While selecting the arrester class,

- switching impulse energy capability ≥ switching surge duty imposed by the system;
- discharge currents in effectively shielded (in-plant) systems are relatively low and usually have no bearing on arrester selection.

Generally,

- use station class to protect components of 7.5 MVA or more and essential rotating machines;
- use intermediate class. to protect 1-20 MVA substations and other rotating machines;
- use distribution class to protect distribution class apparatus, small rotating machines and dry-type transformers.

For selection of arrester location,

- place arrester as close as possible to the equipment being protected;
- reflected wave from protected equipment should participate in the clamping process.

The process of sizing arresters to protect an insulation system is called *insulation coordination*. The following factors need to be considered:

PM = (insulation withstand level/voltage @ equip — 1) 100%

CWW = chopped wave withstand level

BIL = basic impulse insulation level

SSL = switching surge level

FOW = front of wave protective (sparkover) level

LPL = lightning protective (sparkover) level

SPL = switching surge protective (sparkover) level

The recommended protective margins are

$PM(1)$ = (CCW/FOW· 1) 100% ~ 20%

$PM(2)$ = (BIL/LPL· 1) 100% ~ 20%

$PM(1)$ = (SSL/SPL. 1) 100% ~ 15%

Distribution system coordination is usually based on only PM(1) and PM(2) as switching surges are normally less severe than lightning in such systems.

REFERENCES

AIHA (2012), ANSI/AIHA® Z10-2012, *Occupational Health and Safety Management Systems*, Falls Church, VA: American Industrial Hygiene Association.

Ammerman, R.F.; Sen, P.K.; and Nelson, J.P. (2009), "Electrical Arcing Phenomena," *IEEE Industry Applications Magazine*, Vol. *15*, No. 3, May/June 2009, pp. 36–41.

Anders, G.J.; Ford, G.L.; Vainberg, M.; Arnot, S.; and Germani, M. (1992). "Optimization of Tubular Rigid Bus Design," *IEEE Transactions on Power Delivery*, Vol. *7*, No. 3, pp. 1188–1195.

ANSI (1989), ANSI® C37.50-1989, *American National Standard for Switchgear—Low-Voltage AC Power Circuit Breakers Used in Enclosures—Test Procedures*, New York, NY: American National Standards Institute.

Barrett, J.S.; Chisholm, W.A.; Kuffel, J.; Ng, B.P.; Sahazizian, A-M.; and de Tourreil, C. (2003), "Testing and Modelling Hollow-Core Composite Station Post Insulators under Short-Circuit Conditions," *IEEE PES General Meeting*, Vol. *1*, 13–17 July 2003, pp. 211–218.

Bergeron, D.A. and Trahan, R. E. Jr. (1999), "A Static Finite Element Analysis of Substation Busbar Structures," *IEEE Transactions on Power Delivery*, Vol. *14*, No. 3, pp. 890–896.

Bergeron, D.A.; Trahan, R. E. Jr.; Budinich, M.D.; and Opsetmoen, A. (1999), "Verification of a Dynamic Finite Element Analysis of Substation Busbar Structures," *IEEE Transactions on Power Delivery*, Vol. *14*, No. 3, pp. 884–889.

Blackburn, J.L. (1998), *Protective Relaying Principles and Applications*, New York: Marcel Dekker.

Blower, R.W. (1986), *Distribution Switchgear*, London: William Collins and Co., Ltd.

Boeck, W. (2003), "Solutions of Essential Problems of Gas Insulated Systems for Substations (GIS) and Lines (GIL)." *ICPADM 2003, Nagoya, Japan.*

Boeck, W. and Krüger, K. (1992), "Arc Motion and Burn Through in GIS," *IEEE Transactions on Power Delivery*, Vol. *7*, No. 1, pp. 254–261.

Brown, W.A. and Shapiro, R. (2009), "Incident Energy Reduction Techniques," *IEEE Industry Applications Magazine,* Vol. *15*, No. 3, May/June 2009, pp. 36–41.

Burke, J. and Untiedt, C. (2009), "Stray Voltage," *IEEE Industry Applications Magazine,* Vol. *15*, No. 3, May/June 2009, pp. 36–41.

Burnham, J.T., *et al.* (2002), "IEEE Task Force Report: Brittle Fracture in Nonceramic Insulators," *IEEE Transactions on Power Delivery*, Vol. *17*, No. 3, pp. 848–856.

Cadick, J.; Capelli-Schellpheiffer, M.; and Neitzel, D. (2000), *Electrical Safety Handbook*, Second Edition, New York: McGraw-Hill.

Copper Development Association (2001), *Copper for Busbars*, Publication 22, June 1996, Reprinted 2001. Hemel Hempstead: Copper Development Association

Copper Development Association (undated), *Application Data Sheet*, Publication A6022-XX/78, New York, NY: Copper Development Association.

Principles of Electrical Safety, First Edition. Peter E. Sutherland.
© 2015 John Wiley & Sons, Inc. Published 2015 by John Wiley & Sons, Inc.

Central Station Engineers of the Westinghouse Electric Corp. (1964), *Electrical Transmission and Distribution Reference Book,* East Pittsburgh, PA: Westinghouse Electric Corp.

Christensen, D.A. (2009), *Introduction to Biomedical Engineering Biomechanics and Bioelectricity,* San Rafael, CA: Morgan & Claypool.

CIGRE (1996), "The mechanical effects of short-circuit currents in open-air substations rigid and flexible bus-bars," in *CIGRE Brochure No. 105*, Vols *1 and 2*, April 1996, Paris: CIGRE.

Crudele, F.; Short, T.; and Sutherland, P. (2006), "Lightning Protection of Distribution Capacitor Controllers," *2006 IEEE-PES Conference & Exposition*, Dallas, TX, May 22–24, 2006.

Dalziel, C.F. (1972), "Electric Shock Hazard," *IEEE Spectrum*, Vol. *9*, No. 2, February 1972, 41–50.

Dalziel, C.F. and Lee, W.R. (1969), "Lethal Electric Currents," *IEEE Spectrum*, Vol. *6*, No. 2, February 1969, 44–50.

Das, J.C. (2005), "Design Aspects of Industrial Distribution Systems to Limit Arc Flash Hazard," *IEEE Transactions on Industry Applications*, Vol. *41*, No. 6, Nov/Dec. 2005, pp. 1467–1475.

Dinkel, D.G.; Watts, W.G.; and Langlois, J.R., (1986), "13.8 KV Switchgear Uprate," *Tappi Annual Engineering Conference*. 1986.

Doan, R. and Sweigart, R. (2003), "A Summary of Arc-Flash Energy Calculations," *IEEE Transactions on Industry Applications,* Vol. *39*, No. 4, July/Aug. 2003, pp. 1200–1204.

Dommel, H.W. *(1987), Electro-Magnetic Transients Program (EMTP) Theory Book*, Portland, OR: Can-Am EMTP Users Group.

Dwight, H.B. (1945), *Electrical Coils and Conductors—Their Electrical Characteristics and Theory*, New York: McGraw-Hill.

Elmore, W.A. (ed.) (1994), *Protective Relaying Theory and Applications*, New York: Marcel Dekker.

EPRI (1975), *Transmission Line Reference Book 345 kV and Above*, Palo Alto, CA: EPRI. EL-2500.

EPRI (1978), *Transmission Line Reference Book: 115–138 kV Compact Line Design*, Palo Alto, CA: EPRI. EL-100-V3.

EPRI (1982), *Transmission Line Reference Book: 345 kV and Above*, Second Edition, Revised, Palo Alto, CA: EPRI.

EPRI (1992), EL-2500-R1EPRI, *Seasonal Variations of Grounding Parameters by Field Tests.* Report No. TR-100863, Electric Power Research Institute, Palo Alto, CA.

EPRI (1998) *Application Guide for Transmission Line Non-Ceramic Insulators.* Palo Alto, CA: EPRI. TR-111566.

EPRI (1999*), Identifying, Diagnosing and Resolving Residential Shocking Incidents.* Report. No. TR0-113566, Electric Power Research Institute, Palo Alto, CA.

EPRI (2000), *Technical Brief "Pool shocking—Fun in the sun can be a shocking event,"* Palo Alto, CA: Electric Power Research Institute.

EPRI (2003) *Considerations for Conversion or Replacement of Medium-Voltage Air-Magnetic Circuit Breakers Using Vacuum or SF6 Technology: Revision to TR-106761*, Palo Alto, CA: EPRI. 2003. 1007912.

EPRI (2004a), EPRI *Guide for Transmission Line Grounding: A Roadmap for Design, Testing and Remediation*. Report No. 1002021. Electric Power Research Institute, Palo Alto, CA.

EPRI (2004b). *Protective Relays: Numerical Protective Relays*, Palo Alto, CA: EPRI. 2004. 1009704.

EPRI (2004c), *EPRI AC Transmission Line Reference Book - 200 kV and Above*, Third Edition, Palo Alto, CA: EPRI. 1008742.

EPRI (2005), *Grounding and Lightning Protection of Capacitor Controllers*, Palo Alto, CA: EPRI. 1008573.

EPRI (2006), *EPRI Fault Current Management Guidebook*, Palo Alto, CA: EPRI. 1010680.

Fish, M.W. (1994), "When You Have to Retrofit 15 kV Switchgear," *1994 IEEE Pulp and Paper Technical Conference,* pp. 188–193.

Freschi, F.; Guerrisi, A.; Tartaglia, M.; and Mitolo, M. (2013), "Numerical Simulation of Heart-Current Factors and Electrical Models of the Human Body" *IEEE Transactions on Industry Applications*, Vol. *49*, No. 5, September/October 2013, pp. 2290–2299.

Multilin, G.E. (1997), *Instructions Time Overcurrent Relays* IAC53A IAC53B IAC54A IAC54B Form 800, General Electric Co. Publication GEK-3054H.

General Electric Company (1972), "Short-circuit characteristics of electronic power converters," in *GE Industrial Power Systems Data Book*, Schenectady, NY: General Electric Co.

General Electric Company (2007), *E2000 Miniature Circuit Breakers, Catalog/Selection Guide*, GE Publication DEP130B, Plainville, CT: GE Consumer & Industrial.

General Electric Company (2008a), *GE Overcurrent Device Instantaneous Selectivity Tables*, GE Publication DET537B, Plainville, CT: GE Consumer & Industrial.

General Electric Company (2008b), *Guide to Low Voltage System Design and Selectivity*, GE Publication DET654, Plainville, CT: GE Consumer & Industrial.

General Electric Company (2010), *Zenith ZTSCT Closed Transition Transfer Switches*, GE Publication PB-5069, Chicago, IL: GE Energy—Digital Energy.

General Electric Company (undated), *Industrial Control Components ("Rainbow Catalog")*, GE Publication GEP-345D, Salem, VA: General Electric Company.

Genutis, D.A. (1992), "Problems with Medium Voltage Air Magnetic Circuit Breakers," *NETA World*, Spring 1992.

Gradwell, B. (2014) *"Arc Flash Mitigation Through the use of an Engineered Parallel High Speed Semi Conductor Fuse Assembly"*, 2014 IEEE/IAS 50th Industrial & Commercial Power Systems Technical Conference (I&CPS), Fort Worth, TX, May 20–23, 2014.

Greenwood, A. (1991), *Electrical Transients in Power Systems*, Second Edition, Hoboken, NJ: John Wiley & Sons, Inc.

Grover, F.W. (1946), *Inductance Calculations, Working Formulas and Tables*, New York: Dover Publications, Inc.

Gustavsen, B.; Martinez, J.A.; and Durbak, D. (2005), "Parameter Determination for Modeling System Transients-Part II: Insulated Cables" *IEEE Transactions on Power Delivery*, Vol. *20*, No. 3, July 2005, pp. 2045–2050.

Hall, J.R. (2012), *Home Electrical Fires*, Quincy, MA: National Fire Protection Association, Fire Analysis and Research Division, January 2012.

Heath, W.; Freeman, T.S.; and Cochran, F.R. (1987), "Replacement of Air Magnetic Breakers with Vacuum Breakers in Medium Voltage Switchgear Assemblies," presented at the Electrical Systems & Equipment Committee Meeting No. 58, Engineering and Operations Division, Electric Council of New England.

Herrmann, B.; Stein, N.; and Keißling, G. (1989), "Short-Circuit Effects in HV Substations with Strained Conductors Systematic Full Scale Tests and a Simple Calculation Method," *IEEE Transactions on Power Delivery*, Vol. *4*, No. 2, pp. 1021–1028.

Hodder, M.; Vilcheck, W.; Croyle, F.; and McCue, D. (2006), "Practical Arc-Flash Reduction," *IEEE Industry Applications Magazine*, Vol. *12*, No. 3, May/June 2006, pp. 22–29.

IEC (1994), International Standard 60865–2: 1994, *Short-Circuit Currents—Calculation of Effects. Part 2: Examples of Calculation,* Geneva, Switzerland: International Electrotechnical Commission.

IEC (1999), International Standard 60059, Edition 2.0, 1999–06, *IEC Standard Current Ratings,* Geneva, Switzerland: International Electrotechnical Commission.

IEC (2003a), International Standard 60898–1, Edition 1.2, 2003–07, *Electrical Accessories– Circuit-Breakers for Overcurrent Protection for Household and Similar Installations—*

Part 1: Circuit-Breakers for a.c. Operation, Geneva, Switzerland: International Electrotechnical Commission.

IEC (2003b), International Standard 60898–2, Edition 1.1, 2003–07, Edition 1:2000 consolidated with amendment 1:2003, *Electrical Accessories—Circuit-Breakers for Overcurrent Protection for Household and Similar Installations—Part 2: Circuit-Breakers for a.c. and d.c. Operation*, Geneva, Switzerland: International Electrotechnical Commission.

IEC (2005), International Standard 60479–1, Edition 4.0, 2005–07, *Effects of Current on Human Beings and Livestock, Part 1: General Aspects*, Geneva, Switzerland: International Electrotechnical Commission.

IEC (2009a), International Standard 60364-5-52, Edition 3.0, 2009–10, *Low-Voltage Electrical Installations- Part 5–52: Selection and Erection of Electrical Equipment—Wiring Systems*, Geneva, Switzerland: International Electrotechnical Commission.

IEC (2009b), International Standard 60038, Edition 7.0 2009–06, *IEC Standard Voltages*, Geneva, Switzerland: International Electrotechnical Commission.

IEC (2009c), International Standard 60947-4-1, Edition 3.0 2009–09, *Low-Voltage Switchgear and Controlgear—Part 4–1: Contactors and Motor-Starters—Electromechanical Contactors and Motor-Starters*, Geneva, Switzerland: International Electrotechnical Commission.

IEC (2011a), International Standard 60865–1: 2011, *Short-Circuit Currents—Calculation of Effects. Part 1: Definitions and Calculation Methods*, Geneva, Switzerland: International Electrotechnical Commission.

IEC (2011b), International Standard 60947–1, Edition 5.1 2011–03, *Low-Voltage Switchgear and Controlgear—Part 1: General Rules*, Geneva, Switzerland: International Electrotechnical Commission.

IEC (2013a), International Standard 60947–2, Edition 4.2 2013–01, *Low-Voltage Switchgear and Controlgear—Part 2: Circuit-Breakers*, Geneva, Switzerland: International Electrotechnical Commission.

IEC (2013b), International Standard 60934, Edition 3.2 2013–01, *Circuit-Breakers for Equipment (CBE)*, Geneva, Switzerland: International Electrotechnical Commission.

IEEE, Working Group on Electrostatic and Electromagnetic Effects (1978), "Electric and Magnetic Field Coupling from High Voltage AC Power Transmission—Classification of Effects on People," *IEEE Trans. Pwr. App. and Systems*, Vol. PAS-97, No. 6, Nov/Dec 1978, pp. 2243–2252.

IEEE (1993), IEEE Standard 141–1993, *IEEE Recommended Practice for Electric Power Distribution for Industrial Plants, (Red Book)*, New York, NY: Institute of Electrical and Electronic Engineers.

IEEE (1994), IEEE Standard 835–1994, *IEEE Standard Power Cable Ampacity Tables*, New York, NY: Institute of Electrical and Electronic Engineers.

IEEE (1997), C37.112-1996, *IEEE Standard Inverse-Time Characteristic Equations for Overcurrent Relays*, New York, NY: Institute of Electrical and Electronic Engineers.

IEEE (1998), IEEE Standard 902–1998, *IEEE Guide for Maintenance, Operation, and Safety of Industrial and Commercial Power Systems, (Yellow Book)*, New York, NY: Institute of Electrical and Electronic Engineers.

IEEE (1999a), IEEE Standard C37.09-1999, *IEEE Standard Test Procedure for AC High-Voltage Circuit Breakers Rated on a Symmetrical Current Basis*, New York, NY: Institute of Electrical and Electronic Engineers.

IEEE (1999b), IEEE Standard C37.04-1999, *IEEE Standard Rating Structure for AC High-Voltage Circuit Breakers*, New York, NY: Institute of Electrical and Electronic Engineers.

IEEE (2000), IEEE Standard 80–2000, *IEEE Guide for Safety in AC Substation Grounding*, New York, NY: Institute of Electrical and Electronic Engineers.

IEEE (2002a), IEEE Standard 837–2002, *IEEE Standard for Qualifying Permanent Connections Used in Substation Grounding*, New York, NY: Institute of Electrical and Electronic Engineers.

IEEE (2002b), IEEE Standard 1584–2002, *IEEE Guide for Performing Arc-Flash Hazard Calculations*, New York, NY: Institute of Electrical and Electronic Engineers.

IEEE (2002c), IEEE Standard C37.24-2002, *IEEE Standard for Low-Voltage DC Power Circuit Breakers Used in Enclosures*, New York, NY: Institute of Electrical and Electronic Engineers.

IEEE (2004), IEEE Standard 1584a-2004, *IEEE Guide for Performing Arc-Flash Hazard Calculations–Amendment 1*, New York, NY: Institute of Electrical and Electronic Engineers.

IEEE (2007a), IEEE Standard 142–2007, *IEEE Recommended Practice for Grounding of Industrial and Commercial Power Systems. (Green Book)*, New York, NY: Institute of Electrical and Electronic Engineers.

IEEE (2007b), IEEE Standard C37.110-2007, *IEEE Guide For The Application of Current Transformers Used for Protective Relaying Purposes*, New York, NY: Institute of Electrical and Electronic Engineers.

IEEE (2007c), IEEE Standard 525–2007, *IEEE Guide for the Design and Installation of Cable Systems in Substations*, New York, NY: Institute of Electrical and Electronic Engineers.

IEEE (2008), Standard 605–2008, *IEEE Guide for Design of Substation Rigid-Bus Structures.* New York, NY: Institute of Electrical and Electronic Engineers.

IEEE (2009a), IEEE Standard C37.06-2009, *AC High-Voltage Circuit Breakers Rated on a Symmetrical Current Basis—Preferred Ratings and Related Required Capabilities*, New York, NY: Institute of Electrical and Electronic Engineers.

IEEE (2009b) IEEE Standard C37.16-2009, *IEEE Standard for Preferred Ratings, Related Requirements, and Application Recommendations for Low-Voltage AC (635 V and below) and DC (3200 V and below) Power Circuit Breakers*, New York, NY: Institute of Electrical and Electronic Engineers.

IEEE (2011), IEEE Standard C37.122-2011, *IEEE Standard for Gas-Insulated Substations*, New York, NY: Institute of Electrical and Electronic Engineers.

IEEE (2012a), IEEE Standard C2-2012, *2012 National Electrical Safety Code (NESC)*, New York, NY: Institute of Electrical and Electronic Engineers.

IEEE (2012b), IEEE Standard. C62.11-2012, *IEEE Standard for Metal-Oxide Surge Arresters for AC Power Circuits (>1 kV)*, New York, NY: Institute of Electrical and Electronic Engineers.

IPC (undated), *Ground Fault Protection Ungrounded Systems to High Resistance Grounding Conversion Guide*, Mississauga, Ontario, Canada: IPC. (brochure).

Kersting, W.H. (2003), "The Computation of Neutral and Dirt Currents and Power Losses," *2003 IEEE PES Transmission and Distribution Conference and Exhibition, Vol. 3*, 7–12 Sept. 2003, pp. 978–983.

Kersting, W.H. (2007), *Distribution System Modeling and Analysis*, Boca Raton, FL: CRC Press

Kersting, W.H. (2009), "Center-Tapped Transformer and 120-/240-V Secondary Models," *IEEE Transactions on Industry Applications, Vol. 45*, No. 2, March/April 2009, pp. 575–581.

Kruse, G.C. and Pearce, H.T. (2000), "The Finite Element Simulation of Bundle Pinch of a Transmission Line Conductor Bundle," *IEEE Transactions on Power Delivery, Vol. 15*, No. 1, January 2000, 216–221.

Lee, R.H. (1971), "Electrical Safety in Industrial Plants," *IEEE Spectrum, Vol. 8*, No. 6, June 1971, pp. 51–55.

Lee, R.H. (1982), "The Other Electrical Hazard: Electrical Arc Blast Burns," *IEEE Transactions on Industry Applications,* Vol. *18*, No. 3, May/June 1982, pp. 246–251.

Lilien, J.-L.; Hansenne, E.; Papailiou, K.O.; and Kempf, J. (2000), "Spacer Compression for a Triple Conductor Bundle." *IEEE Transactions on Power Delivery*, Vol. *15*, No. 1, pp. 236–241.

Lilien, J.-L. and Papailiou, K.O. (2000), "Calculation of Spacer Compression for Bundle Lines Under Short-Circuit," *IEEE Transactions on Power Delivery,* Vol. *15*, No. 2, April 2000, pp. 839–845

Linders, J.R., *et al.* (1995), "Relay Performance Considerations with Low-Ratio CT's and High-Fault Currents," *IEEE Transactions on Industry Applications*, Vol. *31*, No. 2, March/April 1995, pp. 392–404.

Liu, H.; Mitolo, M.; and Qiu, J. (2013), "Ground Fault-Loop Impedance Calculations in Low-Voltage Single-Phase Systems," *IEEE Transactions on Industry Applications,* Vol. *50*, No. 2, March/April 2014, pp. 1331–1337.

Mason, C.R. (1956), *The Art and Science of Protective Relaying*, New York: John Wiley and Sons, Inc.

Miri, A.M. and Stein, N. (2002), "Calculated short-circuit behaviour and effects of a duplex conductor bus variation of the subconductor spacing." *In 8th Int. Conf. Optimiz. Elect. Electron.Equip.*, Brasov, Romania.

Miroshnik, R. (2003), "Force Safety Device for Substation With Flexible Buses." *IEEE Transactions on Power Delivery*, Vol. *18*, No. 4, pp. 1236–1240.

Mitolo, M. (2009a), "Is It Possible to Calculate Safety?" *IEEE Industry Applications Magazine,* Vol. *15*, No. 3, May/June 2009, pp. 31–35.

Mitolo, M. (2009b), *Electrical Safety of Low-voltage Systems*, New York, NY: McGraw-Hill.

Mitolo, M. (2010), "Of Electrical Distribution Systems With Multiple Grounded Neutrals", *IEEE Transactions on Industry Applications,* Vol. *46*, No. 4, July/August 2010, pp. 1541–1546.

Mitolo, M.; Sutherland, P.E.; and Natarajan, R. (2010), "Effects of High Fault Current on Ground Grid Design," *IEEE Transactions on Industry Applications,* Vol. *46*, No. 3, May/June 2010, pp. 1118–1124.

Mitolo, M.; Tartaglia, M.; and Panetta, S. (2010), "Of International Terminology and Wiring Methods Used in the Matter of Bonding and Earthing of Low-Voltage Power Systems," *IEEE Transactions on Industry Applications,* Vol. *46*, No. 3, May/June 2010, pp. 1089–1095.

Neher, J.H.; McGrath, M.H. (1957), "The Calculation of the Temperature Rise and Load Capability of Cable Systems," *AIEE Transactions*, Vol. *76*, pt. III, October 1957, pp. 752–772.

NFPA (2012), *Standard for Electrical Safety in the Workplace*, NFPA 70E-2012, Quincy, MA: National Fire Protection Association.

NFPA (2014), *National Electrical Code*, NFPA 70–2014, Quincy, MA: National Fire Protection Association.

Ozbek, O.; Birgul, O.; Eyuboglu, B.M.; and Ider, Y.Z. (2001), "Imaging Electrical Current Density Using 0.15 Tesla Magnetic Resonance Imaging System," *Proceedings (CD-ROM) of the IEEE/EMBS 23rd Annual Conference*, Istanbul - Turkey, 2001

Parsons, A.C.; Leuschner, W.B.; and Jiang, K.X. (2008), "Simplified Arc-Flash Hazard Analysis Using Energy Boundary Curves," *IEEE Transactions on Industry Applications*, Vol. *44*, No. 6, Nov/Dec 2008, pp. 1879–1885.

Paul, C.R. (2008), *Analysis of Multiconductor Transmission Lines,* Second Edition, Hoboken, NJ: John Wiley & Sons, Inc.

Paul, C.R. (2010), *Inductance Loop and Partial*, Hoboken, NJ: John Wiley & Sons, Inc.

Petit, A.; St-Jean, G.; and Fecteau, G. (1989), "Empirical Model of a Current-Limiting Fuse using EMTP," *IEEE Transactions on Power Delivery*, Vol. *4*, No. 1, January 1989, pp. 335–341.

Plonsey, R. and Barr, R.C. (2007), *Bioelectricity A Quantitative Approach*, New York, NY: Springer.

Power Systems Relaying Committee (1976), *Transient Response of Current Transformers*, IEEE Publication 76 CH 1130–4 PWR, New York, NY: Institute of Electrical and Electronic Engineers.

Pursula, A.; Nenonen, J.; Somersalo, E.; Ilmoniemi, R.J.; and Katila, T. (2000), "Bioelectromagnetic calculations in anisotropic volume conductors" *Biomag2000*, August 13–17, 2000, Helsinki, Finland.

Reilly, J. P. (1998), *Applied Bioelectricity: From Electrical Stimulation to Electropathology*, New York, NY: Springer-Verlag.

Roberts, J.; Fischer, N.; Fleming, B; and Taylor, A. (2003), "Obtaining a reliable polarizing source for ground directional elements in multisource, isolated-neutral distribution systems," *30th Annual Western Protective Relay Conference*, Spokane, WA, October 21–23, 2003.

Ruehli, A.H. (1972), "Inductance Calculations in a Complex Integrated Circuit Environment," *IBM Journal of Research Development*, Vol. *16*, No. 5, September 1972, pp. 470–481.

Rüger, W. (1989), "Mechanical Short-Circuit Effects of Single-Core Cables," *IEEE Transactions on Power Delivery*, Vol. *4*, No. 1, pp. 68–74.

Schweitzer (2003), *SEL-351-0, −1, −2, −3, −4 Directional Overcurrent Relay Reclosing Relay Fault Locator Instruction Manual*, 20030908, Pullman, WA: Schweitzer Engineering Laboratories, Inc.

Seveik, D.R. and DoCarmo, H.J. (2000), "Reliant Energy HL&P Investigation Into Protective Relaying CT Remanence," *Conference for Protective Relay Engineers*, Texas A&M University, College Station, Texas, April 11—13, 2000.

Shah, K. R.; Detjen, E. R.; and Epley, D. H. (1988), "Impact of Electric Utility and Customer Modifications on Existing Medium Voltage Circuit Breakers," *1988 IEEE Industry Applications Society Annual Meeting.*

Skipa, O.; Sachse, F.B.; and Dössel, O. (2000), "Linearization Approach for Impedance Reconstruction in Human Body from Surface Potentials Measurements." *Biomedizinische Technik*, Vols *45–1*, September 2000, pp. 410–411.

Southwire® (2005), *Power Cable Manual*, Fourth Edition, Carollton, GA: Southwire.

Southwire® (2010), *Data Sheet, 36–2 CT1-09ET 1c, 5/8kV, 115mil EPR (133/100%), TS, PVC, CT Rated,* Carrollton, GA: Southwire.

Standifer, M. (2004), *The Mark Standifer Story: Lessons Learned from an Arc Flash Tragedy*, Lexington, SC: ERI Safety Videos. http://www.youtube.com/watch?v=CzQX9IYLejI.

Stein, N.; Miri, A.M.; and Meyer, W. (2000), "400 kV Substation Stranded Conductor Buses—Tests and Calculations of Short-Circuit Constraints and Behaviour," *7th International Conference on Optimization of Electrical and Electronic Equipment OPTIM 2000*, Brasov (Romania), 11–12 May 2000. Proceedings pp. 251–258.

Storms, A.D. (1992), "Medium Voltage Circuit Breaker Retrofit Technology: The Current Control for Costly Outages," *1992 Doble Conference.*

Sutherland, P.E. (1997), "Surge Protection of a Large Medium Voltage Motor - A Case Study," *1997 IEEE Industrial and Commercial Power Systems Technical Conference,* Philadelphia, PA, May 11–16, 1997, pp. 36–40.

Sutherland, P.E. (1999), "DC Short-Circuit Analysis for Systems with Static Sources," *IEEE Transactions on Industry Applications*, Vol. *35*, No. 1, January/February 1999, pp. 144–151.

Sutherland, P.E. (2007), "On the Definition of Power in an Electrical System," *IEEE Transactions on Power Delivery*, Vol. *22*, No. 2, April 2007, pp. 1100–1107.

Sutherland, P.E. (2009), "Arc Flash and Coordination Study Conflict in an Older Industrial Plant," *IEEE Transactions on Industry Applications*, Vol. *45*, No. 2, March/April 2009, pp. 569–574.

Sutherland, P.E. (2011), "Stray Current Analysis," *2011 IEEE Electrical Safety Workshop*, Toronto, ON, January 24–28, 2011.

Sutherland, P.E. (2012), "Analysis of integral snubber circuit design for transformers in urban high rise office building," *2012 IEEE Industrial and Commercial Power Systems Technical Conference,* Louisville, KY, May 20–24, 2012.

Sutherland, P.E.; Dorr, D.; Gomatom, K. (2009), "Response to Electrical Stimuli," *IEEE Industry Applications Magazine*, Vol. *15*, No. 3, May/June 2009, pp.22–30.

Sverak, J.G. (1981), "Sizing Ground Conductors Against Fusing," *IEEE Transactions on Power Apparatus and Systems*, Vol. *PAS-100*, No. 1, January 1981, pp. 51–59.

Seveik, D.R. and DoCarmo, H.J. (2000), "Reliant Energy HL&P Investigation Into Protective Relaying CT Remanence", *Conference for Protective Relay Engineers*, Texas A&M University, College Station, Texas, April 11—13, 2000.

Swindler, D.L. (1989), "Medium Voltage Application of SF6 and Vacuum Circuit Breakers," 1989 *Tappi Engineering Conference.*

Swindler, D. L. (1993), "Applications of SF6 and Vacuum Medium Voltage Circuit Breakers," 1993 *IEEE PCIC Conference*, Denver, CO.

de Tourreil, C.; Pargamin, L.; Thévenet, G.; and Prat, S. (2000), ""Brittle Fracture" of Composite Insulators: Why and How they Occur," *IEEE PES Summer Meeting*, Vol. *4*, 16–20 July, pp. 2569–2574.

Trinh, N. G. (1992), "Risk of Burn-Through—A Quantitative Assessment of Gas-Insulated Switchgear to Withstand Internal Arcs." *IEEE Transactions on Power Delivery*, Vol. *7*, No. 1, pp. 225–236.

UL (2002), *Leakage Current for Appliances*, UL 101, Fifth Edition, April 29, 2002. Northbrook, IL: Underwriters Laboratories.

UL (2005), *UL Standard for Safety for Supplementary Protectors for Use in Electrical Equipment,* UL 1077, Sixth Edition, July 14, 2005, Northbrook, IL: Underwriters Laboratories.

UL (2009), *UL Standard for Safety for Molded-Case Circuit Breakers, Molded-Case Switches and Circuit-Breaker Enclosures* UL489 Eleventh Edition, September 1, 2009 Northbrook, IL: Underwriters Laboratories.

UL (2011), *UL Standard for Safety for Transfer Switch Equipment,* UL 1008 Sixth Edition, April 15, 2011, Northbrook, IL: Underwriters Laboratories.

USDA (1991), Agricultural Research Service, *Effects of Electrical Voltage/Current on Farm Animals: How to Detect and Remedy Problems,* Agricultural Handbook Number 696, Washington, DC: United States Department of Agriculture.

Valdes, M.; Cline, C.; Hansen, S.; and Papallo, T. (2010), "Selectivity Analysis in Low Voltage Power Distribution Systems with Fuses and Circuit Breakers," *IEEE Transactions on Industry Applications*, Vol. *46*, No. 2, pp. 593–602.

Valdes, M.E.; Crabtree, A.J.; and Papallo, T. (2010), "Method for Determining Selective Capability of Current-Limiting Overcurrent Devices Using Peak Let-Through Current—What Traditional Time–Current Curves Will Not Tell You," *IEEE Transactions on Industry Applications*, Vol. *46*, No. 2, pp. 603–611.

Valdes, M.E. ; Hansen, S.; and Sutherland, P., (2012) "Optimized Instantaneous Protection Settings: Improving Selectivity and Arc-Flash Protection," *IEEE Industry Applications Magazine*, Vol. *18*, No. 3, pp. 66–73.

Villas, J.E.T. and Portela, C. M. (2003), "Calculation of Electric Field and Potential Distributions into Soil and Air Media for a Ground Electrode of a HVDC System." *IEEE Transactions on Power Delivery*, Vol. *18*, No. 3, July, 2003, pp. 867–873.

Ward, D.J. (2003), "Overhead Distribution Conductor Motion Due to Short-Circuit Forces." *IEEE Transactions on Power Delivery*, Vol. *18*, No. 4, pp. 1534–1538.

INDEX

Principles of Electrical Safety, First Edition. Peter E. Sutherland.
© 2015 John Wiley & Sons, Inc. Published 2015 by John Wiley & Sons, Inc.

IEEE Press Series on Power Engineering

Series Editor: M. E. El-Hawary, Dalhousie University, Halifax, Nova Scotia, Canada

The mission of IEEE Press Series on Power Engineering is to publish leading-edge books that cover the broad spectrum of current and forward-looking technologies in this fast-moving area. The series attracts highly acclaimed authors from industry/academia to provide accessible coverage of current and emerging topics in power engineering and allied fields. Our target audience includes the power engineering professional who is interested in enhancing their knowledge and perspective in their areas of interest.